普通高等教育教材

分析化学实验

（英汉双语版）

娄振宁　许　旭　刘雪岩　主编

化学工业出版社

·北京·

内容简介

《分析化学实验》(英汉双语版)教材是基于新时期工科创新人才培养需要,根据分析化学实验教学大纲及多年的双语教学实践编写而成的。

全书共分 5 章,包括分析化学实验基础知识、常用样品制备及分离技术、定量分析基础实验(设置 21 个实验项目)、综合设计性实验(按化学类、环境类、生物类、制药类、食品类设置 31 个实验项目)、探索创新实验(设置 8 个实验项目),共计编入实验项目 60 个。其中,综合设计性实验旨在培养学生解决实际复杂问题的能力,满足不同专业学生的个性需求;探索创新实验重在激发学生创新意识、培养创造能力、提升科学素养。另外,本书以二维码的形式增加了丰富的数字资源,提供更多学习形式,增加学习趣味性。

本书可作为高等院校化学、化工、材料、环境、生物、食品、制药等相关专业的教材使用,也可供从事化学实验工作的科技人员参考。

图书在版编目(CIP)数据

分析化学实验:英、汉/娄振宁,许旭,刘雪岩主编. —北京:化学工业出版社,2024.5
ISBN 978-7-122-45645-8

Ⅰ.①分… Ⅱ.①娄…②许…③刘… Ⅲ.①分析化学-化学实验-教材-英、汉 Ⅳ.①O652.1

中国国家版本馆 CIP 数据核字(2024)第 096839 号

责任编辑:刘心怡 蔡洪伟　　文字编辑:曹 敏
责任校对:边 涛　　　　　　装帧设计:关 飞

出版发行:化学工业出版社
　　　　　(北京市东城区青年湖南街 13 号　邮政编码 100011)
印　　装:北京云浩印刷有限责任公司
787mm×1092mm　1/16　印张 23　字数 601 千字
2025 年 2 月北京第 1 版第 1 次印刷

购书咨询:010-64518888　　　　　售后服务:010-64518899
网　　址:http://www.cip.com.cn
凡购买本书,如有缺损质量问题,本社销售中心负责调换。

定　　价:59.80 元　　　　　　　　版权所有　违者必究

本书编写人员名单

主　编　娄振宁　许　旭　刘雪岩

副主编（按姓氏笔画排序）

　　　　王月娇　毛全兴　冯小庚
　　　　张渝阳　高　婧　康平利

参　编（按姓氏笔画排序）

　　　　王知微　邢志强　杨丽君
　　　　张　谦　陈　霞　矣　杰
　　　　林　觅　姜晓庆

主　审　张　蕾　王建国

前言

分析化学实验是化学相关专业的必修基础课，是分析化学课程的重要支撑。学生通过分析化学实验课程的学习，可熟练掌握基本的分析测试方法和测试技术，进一步培养理论联系实际、分析解决实际问题的能力，以及树立实事求是和严谨的科研作风。

双语教学是社会信息化和经济全球化的要求，也是高校教育教学改革与发展的必然趋势。为进一步推进双一流大学和一流课程建设进程，辽宁大学分析化学实验教学团队于2019年开展分析化学实验双语教学，本书是在多年双语教学实践的基础上编写而成的。在自编分析化学实验教材的基础上，我们根据分析化学实验课程教学大纲，调研目前已出版的分析化学实验双语教材，保留原有分析化学基础操作、基础实验部分，按化学类、环境类、生物类、制药类、食品类专业设置了综合设计性实验，并融入了课程思政、科技前沿、国家标准以及大学生化学实验竞赛和最新的样品分离技术等相关内容，使学生在掌握实验技能的同时，领悟实验教学中涉及的前沿科技和研究方法，通过实验培养学生更加宽广的科学视野和融会贯通的能力。本教材内容丰富，涉及面广，用书院校可结合各自专业针对性地选取相关实验开展教学。本书所选实验项目力求紧密联系社会和生活，联系科技前沿和各个交叉学科。同时双语教学能够让学生不断强化专业词汇，掌握大量的专业术语与英文表达，进一步践行"三位一体"的教育理念。

全书共分5章，第1章为分析化学实验基础知识；第2章为常用样品制备及分离技术；第3章为定量分析基础实验；第4章为综合设计性实验（按化学类、环境类、生物类、制药类、食品类设置）；第5章为探索创新实验（大学生化学竞赛及创新性实验），共计60个实验。使用本教材时，可根据学时数和实验的简繁情况，结合各院校、各专业的需求一次安排一个或两个内容相关的实验。

本书是辽宁大学分析化学研究所全体教师多年教学实践的结晶。全书由娄振宁、许旭、刘雪岩主编，王月娇、毛全兴、冯小庚、张渝阳、高婧、康平利任副主编，王知微、邢志强、杨丽君、张谦、陈霞、矣杰、林觅、姜晓庆参编，张蕾教授和王建国教授审阅了全书。本书在编写过程中得到了辽宁大学化学院的大力支持和诸多同行专家的指导和帮助，在此表示由衷的感谢！

由于编者水平有限，书中难免有不足之处，敬请读者批评和指正。

编者
2024 年 10 月

目录

Chapter 1 Basic Knowledge of Analytical Chemistry Experiments ········ 1

1.1　Tasks and Requirements of Analytical Chemistry Experiment ················ 1
1.2　Laboratory Safety Knowledge ················ 2
1.3　Basic Knowledge of Quantitative Analysis Experiments ················ 2
1.4　Instruments and Basic Operations of Quantitative Analysis ················ 6
1.5　Recording, Processing and Reporting of Experimental Data ················ 15

Chapter 2 Sample Preparation and Separation Techniques ············ 20

2.1　Preparation of Samples ················ 20
2.2　Separation Techniques of Samples ················ 20

Chapter 3 Basic Experiments for Quantitative Analysis ··············· 29

Experiment 3.1　Analytical Balance Weighing Exercise ················ 29
Experiment 3.2　Basic Operation for Titration Analysis ················ 30
Experiment 3.3　Calibration of Volumetric Instruments ················ 33
Experiment 3.4　Preparation and Standardization of HCl and NaOH Solution ········ 35
Experiment 3.5　Determination of Total Acidity in Edible Vinegar ················ 39
Experiment 3.6　Determination of Nitrogen Content in Ammonium Salts—Formaldehyde Method ················ 41
Experiment 3.7　Determination of Mixed Alkali—Double Indicator Method ············ 44
Experiment 3.8　Preparation and Standardization of EDTA Solution ················ 47
Experiment 3.9　Determination of Total Hardness of Tap Water ················ 50
Experiment 3.10　Determination of Aluminum Content in Industrial Aluminum Sulfate ················ 53
Experiment 3.11　Determination of Zn^{2+} and Bi^{3+} in Mixed Solution by Continuous Titration ················ 55
Experiment 3.12　Preparation and Standardization of Potassium Permanganate Solution ················ 58
Experiment 3.13　Determination of H_2O_2 Content in Hydrogen Peroxide Solution ················ 61
Experiment 3.14　Determination of Iron in Iron Ore by Potassium Dichromate Method ················ 63
Experiment 3.15　Preparation and Standardization of Iodine and Sodium Thiosulfate ················ 66

Experiment 3.16　Determination of Glucose in Glucose Oral Solution ……… 72
Experiment 3.17　Determination of Copper Content in Copper Alloys ……… 75
Experiment 3.18　Determination of Chlorine in Tap Water by Mohr Method ……… 77
Experiment 3.19　Determination of Nickel in Alloy Steel by Dimethylglyoxime Gravimetric Method ……… 79
Experiment 3.20　Determination of Trace Iron with Phenanthroline by UV-Vis Spectroscopy ……… 82
Experiment 3.21　Potentiometric Titration of Phosphoric Acid ……… 85

Chapter 4　Comprehensive Design Experiments ……… 90

4.1　Chemistry ……… 90

Experiment 4.1　Determination of Benzoic Acid Content ……… 90
Experiment 4.2　Determination of Phenol Content ……… 92
Experiment 4.3　Determination of Purity of Soluble Barium Salt by Gravimetric Method ……… 95
Experiment 4.4　Determination of Methanol Content in Industrial Alcohol ……… 98
Experiment 4.5　Determination of Acetic Acid Content by Potentiometric Titration ……… 101
Experiment 4.6　Determination of Iron, Aluminum, Calcium and Magnesium in Cement Clinker ……… 104

4.2　Environment ……… 109

Experiment 4.7　Determination of Total Alkalinity of Water ……… 109
Experiment 4.8　Determination of Chemical Oxygen Demand (COD) in River Water ……… 111
Experiment 4.9　Determination of Sulfate Content in Urban Sewage ……… 114
Experiment 4.10　Determination of Formaldehyde in Waterborne Coatings by Spectrophotometry ……… 116
Experiment 4.11　Determination of Glyphosate Content in Roundup ……… 120

4.3　Biology ……… 121

Experiment 4.12　Determination of Total Nitrogen Content for Compound Fertilizers ……… 121
Experiment 4.13　Determination of Soil Organic Matter Content ……… 124
Experiment 4.14　Determination of Chlorine Content for Compound Fertilizers ……… 127
Experiment 4.15　Determination of Potassium Content for Compound Fertilizers ……… 130
Experiment 4.16　Determination of Available Phosphorus Content for Compound Fertilizers ……… 132
Experiment 4.17　Determination of Available Phosphorus in Soil ……… 134
Experiment 4.18　Macroporous Resin Adsorption and Spectrophotometric Determination of Flavonoids from Gingko Leaves ……… 137

| Experiment 4.19 | Determination of Mixed Amino Acids Content | 140 |
| Experiment 4.20 | Determination of Reducing Sugar in Tobacco Leaf | 142 |

4.4　Pharmacy　144

Experiment 4.21	Determination of NH_4^+ in Chinese Medicine White Sal-Ammoniac	144
Experiment 4.22	Determination of Calcium in Calcium Supplement by $KMnO_4$ Method	147
Experiment 4.23	Determination of Al^{3+} and Mg^{2+} in Compound Aluminium Hydroxide Tablets	150
Experiment 4.24	Determination of the Content of Berberine Hydrochloride	154
Experiment 4.25	Determination of Acetaminophen in Ankahuangmin Capsules	156

4.5　Food　158

Experiment 4.26	Determination of Citric Acid Content in Commercially Available Citrus	158
Experiment 4.27	Determination of Calcium Content in Calcium Tablets/Milk Powder/Spinach	161
Experiment 4.28	Determination of NaCl Content in Soy Sauce	163
Experiment 4.29	Determination of Anthocyanin Content in Blueberries	165
Experiment 4.30	Determination of Melamine in Milk Products	167
Experiment 4.31	Determination of Vitamin B12 in Health Foods	169

Chapter 5　Innovation Experiments　172

5.1　Chemistry Experiment Competition for College Students　172

Experiment 5.1	Preparation of Rare Earth Europium (Eu) Complex and Determination of Coordination RatFio	172
Experiment 5.2	Preparation of Iron Oxide Nanoparticles and Its Application in the Determination of Melamine in Dairy Products	175
Experiment 5.3	Study on Synthesis of Complex [Ni(Me$_3$en) (acac)] BPh$_4$ and Its Solvent/Thermochromic Behavior	177
Experiment 5.4	Determination of Copper Content in Polynuclear Copper (Ⅰ) Complex	180

5.2　Research Training Experiments　182

Experiment 5.5	Gemini Ionic Liquid Modified Graphene Oxide Membranes for ReO_4^-/TcO_4^- Adsorption	182
Experiment 5.6	Colorimetric Determination of Uric Acid with Gold Nanoparticles	183
Experiment 5.7	A Green Chemistry Technique for Dissolution and Recovery of Gold from Electronic Wastes	184
Experiment 5.8	Extraction, Content Analysis and Cosmetic Application of Natural Pigment Betacyanin	185

Appendix ... 189

Appendix 1　Concentrations and Densities of Common Acid-Base Solutions 189
Appendix 2　Common Primary Standard Substances and Their Drying Conditions and Applications 189
Appendix 3　Preparation of Common Buffer 190
Appendix 4　Common Indicators 193
Appendix 5　Dissociation Constants of Weak Acids and Bases 194
Appendix 6　Stability Constants of Coordination Compounds Formed by Metal Ions and EDTA(18-25℃，$I=0.1$ mol/L) 198
Appendix 7　Standard Atomic Weights(1995，IUPAC) 198

第1章　分析化学实验基础知识 ... 200

1.1　分析化学实验的任务和要求 200
1.2　实验室安全知识 200
1.3　定量分析实验基础知识 201
1.4　定量分析实验仪器及基本操作 203
1.5　实验数据的记录、处理和实验报告 210

第2章　常用样品制备及分离技术 ... 214

2.1　样品的制备 214
2.2　样品的分离 214

第3章　定量分析基础实验 ... 220

实验3.1　分析天平的称量练习 220
实验3.2　滴定分析的基本操作练习——酸碱滴定 221
实验3.3　容量仪器的校准 223
实验3.4　酸碱溶液的配制与标定 225
实验3.5　食用醋中总酸度的测定 228
实验3.6　铵盐中氮含量的分析——甲醛法 230
实验3.7　混合碱的分析——双指示剂法 232
实验3.8　EDTA溶液的配制与标定 234
实验3.9　自来水硬度的测定 236
实验3.10　工业硫酸铝中铝含量的分析 238
实验3.11　锌铋混合溶液中Zn^{2+}、Bi^{3+}含量的连续测定 240
实验3.12　高锰酸钾标准溶液的配制与标定 243
实验3.13　双氧水中过氧化氢含量的测定 245
实验3.14　铁矿石中铁含量的测定 247
实验3.15　碘和硫代硫酸钠溶液的配制与标定 249
实验3.16　葡萄糖口服液中葡萄糖含量的测定 254

实验 3.17　铜合金中铜含量的测定 ·· 256
　实验 3.18　莫尔法测定自来水中的氯 ·· 258
　实验 3.19　丁二酮肟重量法测定合金钢中的镍 ·· 260
　实验 3.20　邻二氮菲分光光度法测定微量铁 ·· 262
　实验 3.21　磷酸的电位滴定 ··· 265

第 4 章　综合设计性实验 ·· 269

4.1　化学类 ·· 269
　实验 4.1　苯甲酸含量的测定 ··· 269
　实验 4.2　苯酚含量的测定 ·· 271
　实验 4.3　重量法测定可溶性钡盐的纯度 ·· 273
　实验 4.4　工业酒精中甲醇含量的测定 ··· 275
　实验 4.5　电位滴定法测定醋酸含量 ··· 278
　实验 4.6　水泥熟料中铁、铝、钙、镁含量的测定 ······································ 280

4.2　环境类 ·· 284
　实验 4.7　水质总碱度测定 ·· 284
　实验 4.8　河水中化学需氧量（COD）的测定 ·· 286
　实验 4.9　城市污水中硫酸盐含量的测定 ·· 288
　实验 4.10　分光光度法测定水性涂料中的甲醛 ·· 290
　实验 4.11　农达中草甘膦含量的测定 ·· 293

4.3　生物类 ·· 294
　实验 4.12　复合肥料中总氮含量的测定 ··· 294
　实验 4.13　土壤中有机质含量的测定 ·· 297
　实验 4.14　复合肥料中氯离子含量的测定 ··· 299
　实验 4.15　复合肥料中钾含量的测定 ·· 301
　实验 4.16　复合肥料中有效磷含量的测定 ··· 302
　实验 4.17　土壤中有效磷的测定 ··· 304
　实验 4.18　银杏叶总黄酮含量的测定 ·· 306
　实验 4.19　混合氨基酸的测定 ·· 309
　实验 4.20　烟叶样品中还原糖的测定 ·· 310

4.4　制药类 ·· 312
　实验 4.21　中药白硇砂中 NH_4^+ 含量的测定 ·· 312
　实验 4.22　高锰酸钾法测定补钙剂中钙含量 ·· 314
　实验 4.23　复方氢氧化铝片中 Al^{3+} 和 Mg^{2+} 含量的测定 ···························· 317
　实验 4.24　盐酸黄连素含量的测定 ··· 320
　实验 4.25　氨咖黄敏胶囊中对乙酰氨基酚含量的测定 ······························· 321

4.5　食品类 ·· 323
　实验 4.26　市售柑橘中柠檬酸含量的测定 ··· 323
　实验 4.27　钙片/奶粉/菠菜中钙含量的测定 ··· 325
　实验 4.28　酱油中 NaCl 含量的测定 ··· 327

  实验 4.29 蓝莓中花青素含量的测定 ………………………………………………… 329
  实验 4.30 奶制品中三聚氰胺的测定 ………………………………………………… 331
  实验 4.31 保健食品中维生素 B12 的测定 …………………………………………… 332

第 5 章 探索创新实验 ………………………………………………………………… 335

5.1 大学生化学竞赛（分析化学部分） ………………………………………………… 335
  实验 5.1 稀土铕配合物的制备及配位比的测定 …………………………………… 335
  实验 5.2 氧化铁纳米颗粒的制备及用于奶制品中三聚氰胺的测定 ……………… 337
  实验 5.3 配合物[Ni(Me$_3$en)(acac)]BPh$_4$ 的合成及其溶剂/热致变色行为研究 …… 339
  实验 5.4 多核铜（Ⅰ）配合物中铜含量的测定 ……………………………………… 341

5.2 创新性实验 ……………………………………………………………………………… 342
  实验 5.5 Gemini 离子液体改性氧化石墨烯薄膜在 ReO_4^-/TcO_4^- 吸附中的应用 …… 342
  实验 5.6 金纳米粒子比色法检测尿酸 ………………………………………………… 343
  实验 5.7 电子废弃物中金元素的绿色溶解与提取 ………………………………… 344
  实验 5.8 天然色素甜菜红的提取、含量分析及其美妆应用 ……………………… 344

附录 ……………………………………………………………………………………………… 349

  附录 1 常用酸碱的密度、含量和浓度 ……………………………………………… 349
  附录 2 常用基准物质的干燥条件和应用 …………………………………………… 349
  附录 3 常见缓冲溶液的配制 …………………………………………………………… 350
  附录 4 常用酸碱指示剂及配制方法 ………………………………………………… 352
  附录 5 弱酸、弱碱的解离常数 …………………………………………………………… 352
  附录 6 金属离子与 EDTA 生成配位化合物的稳定常数（18～25℃，
      $I=0.1$mol/L） …………………………………………………………………… 356
  附录 7 原子表（1995，IUPAC） ………………………………………………………… 357

参考文献 ………………………………………………………………………………………… 358

Chapter 1
Basic Knowledge of Analytical Chemistry Experiments

1.1 Tasks and Requirements of Analytical Chemistry Experiment

  Analytical chemistry is an important branch of chemistry, and it is a basic compulsory course for majors in chemistry, medicine, food, and environment. Analytical chemistry mainly studies the analytical methods and principles of substances chemical information such as the composition, structure, and content, with highly comprehensive, practical, and applicable. Analytical chemistry experiment is an important part of analytical chemistry course teaching. Through the study of this course, students can deepen their knowledge and understanding of the theoretical knowledge of analytical chemistry; master the basic operations and skills of analytical chemistry experiments correctly and proficiently; establish the concepts of "quantity" "error" and "significant number" and master the methods of experimental data recording and processing; develop the ability to analyze and solve practical problems independently through exploratory experiments; cultivate a rigorous scientific style and a scientific attitude of seeking truth from facts.

  In order to accomplish the above tasks, the following needs to be done:

  (1) Carefully preview before the experiment. Combined with experimental teaching materials and theoretical learning, clarify the purpose of the experiment, understand the principle of the experiment, be familiar with the content and method of the experiment, and be clear about the experimental procedures, operations, and precautions. Write a pre-report and list the data recording form.

  (2) Enter the laboratory on time and listen carefully. During the experiment, ensure the strictly standardized operation, observe the experimental phenomenon carefully and record the experimental data truthfully. Students should have a special experiment record book and use it with the preview report. It is never allowed to record data on a single page or small piece of paper. The text recording should be neat and clear. Do not tamper with data or make false statements.

  (3) Ensure the cleanliness of the experimental table and good experimental order during the experiment. Drugs and instruments should be returned to their original places in time af-

ter use to keep the laboratory clean and tidy.

1.2 Laboratory Safety Knowledge

In chemical laboratory, many easily broken glassware, water, electricity, gas and corrosive, flammable, explosive chemical reagents are often used. Therefore, to ensure the personal safety of experimenter and the normal conduct of the experiment, the following laboratory safety practice must be strictly obeyed.

(1) When entering the laboratory, you must wear a lab coat.

(2) Eating, drinking and smoking are prohibited in the laboratory. All chemicals are strictly prohibited from entering the mouth, and special equipment such as medicine spoons should be used to take chemical reagents that cannot be taken directly by hand. Wash hands promptly after the experiment is over.

(3) Pay attention to electrical safety, do not touch the power supply with wet hands. Cut off the power supply in time after the experiment.

(4) Experiments with toxic and irritating gases must be carried out in a fume hood. When sniffing the gas, use your hands to fan a small amount of the gas towards yourself, do not lean directly on the container to sniff.

(5) When using concentrated acid, alkali, and other strong corrosive reagents, do not splash on the skin and clothes, and wear safety goggles to prevent splashing into the eyes.

(6) Keep away from fire and heat sources when using flammable organic solvents (such as ethanol, benzene, acetone, ether, etc.). Organic reagents with low boiling points cannot be heated directly on an open flame, and can be heated in a water bath.

(7) When using toxic reagents (such as cyanide, arsenide, mercury salt, lead salt, etc.), strictly forbid to contact the wound or enter the mouth. The waste liquid cannot be poured into the sewer at will but should be poured into the designated recycling bottles for unified recycling treatment.

(8) It is strictly forbidden to mix various chemical reagents arbitrarily to avoid accidents.

(9) In the experiment, if a scald or cut occurs, it should be dealt with in time, and serious cases should be sent to the hospital for treatment immediately.

(10) In the event of a fire in the laboratory, immediately cut off the power supply and gas source, and take targeted fire-fighting measures according to the cause of the fire.

1.3 Basic Knowledge of Quantitative Analysis Experiments

1.3.1 Water for analytical chemistry experiments

1.3.1.1 Specification of water for analytical chemistry experiments

In laboratory, pure water of different specifications is selected according to the analysis tasks and requirements. The national standard of water for analytical laboratory in China stipulates the technical specifications, preparation and testing method of laboratory water. Table 1.1 shows the levels and main indicators of pure water used in laboratory.

Table 1.1 Level and main indicators of analysis laboratory water (GB/T 6682—2008)

Name	I	II	III
pH range(25℃)	—	—	5.0-7.5
Conductivity(25℃)/(mS/m)	≤0.01	≤0.10	≤0.50
Oxidizable material(calculated by O)/(mg/L)	—	≤0.08	≤0.40
Absorbance(254 nm, optical length of 1 cm)	≤0.001	≤0.01	—
Evaporation residue(105℃±2℃)/(mg/L)	—	≤1.0	≤2.0
Soluble silicon(calculated by SiO_2)/(mg/L)	≤0.01	≤0.02	—

Note: 1. Since it is difficult to determine the true pH of first-grade and second-grade water at their purity, the pH range of second-grade water is not specified.

2. As it is difficult to measure oxidizable material and evaporative residues at the purity of first-grade water, the limits are not specified. Other conditions and preparation methods can be used to ensure the quality of first-grade water

1.3.1.2 Preparation of pure water

Analytical chemistry experiments have high requirements for pure water, which must be purified for both dissolution and dilution. According to the specific experimental requirements, water purity requirements are also different. The commonly used methods of preparing pure water include distillation, ion exchange, electrodialysis.

(1) Distillation Distillation separates water and impurities according to their different boiling points. Distillation is simple in operation and low in cost. It can only remove non-volatile impurities in water, but cannot remove water-soluble gases and some low-boiling point volatile substances. Therefore, single-distilled water can only be used for qualitative analysis or general industrial analysis.

(2) Ion exchange Ion exchange resins are widely used in chemical laboratories to separate impurity ions in water. This method is called ion exchange method, and the pure water obtained is called deionized water. The advantages of this method are low cost, large amount of water prepared and strong ability to remove impurities. However, non-electrolyte impurities such as organic matter cannot be removed, and a trace of resin will be dissolved in water.

(3) Electrodialysis Electrodialysis method is developed on the basis of ion exchange method. It is a method of separating impurity ions from water by using anion and cation exchange membranes for selective permeation of anions and cations in water under the action of an external electric field. This method cannot remove non-ionic impurities and is only suitable for the experiments with low requirements.

1.3.1.3 Inspection of pure water

The inspection methods of pure water generally include physical method and chemical method. According to the requirements of the analytical experiment, the inspection of pure water is usually carried out by the following items.

(1) pH As CO_2 in the air is soluble in water, the pH of pure water is generally around 6.0.

(2) Conductivity At 25 ℃, the resistivity of pure water is $(1\text{-}10) \times 10^6$ Ω·cm, and that of ultrapure water is greater than 10×10^6 Ω·cm.

(3) Ca^{2+} and Mg^{2+} Take an appropriate amount of water sample, add $NH_3 \cdot H_2O$-

NH_4Cl buffer solution, and then add 1 drop of 0.2% eriochrome black T indicator, which does not appear red.

(4) Cl^- ion Take an appropriate amount of water sample, add HNO_3 to acidification, then add 2 drops of 1% $AgNO_3$ solution, shake well, no turbidity occurs.

1.3.2 Chemical reagents

1.3.2.1 Specifications of chemical reagents

The purity of reagent has a great influence on the accuracy of experimental results. Different experiments have different requirements for reagent purity. Therefore, it is necessary to understand the classification criteria of chemical reagents. Chemical reagents are classified into several grades according to the amount of impurities they contain. Its specifications and scope of use are shown in Table 1.2.

Table 1.2 Specifications of chemical reagents

Grade	Name	Sign	Label color	Scope of application
First grade	Guarantee reagent	G. R.	Green	Precise analytical experiments
Second grade	Analytical reagent	A. R.	Red	General analytical experiments
Third grade	Chemically pure	C. P.	Blue	General qualitative experiments
Fourth grade	Laboratory reagent	L. R.	Brown	General chemical preparation experiments
Biochemical reagent	Biological reagent	B. R.	Yellow	Biochemical experiments

In addition, there are some high purity reagents for special purposes, such as primary reagents, chromatographic reagents and spectrograde reagents. Among them, primary reagent, also known as standard reagent, is equal to or higher in purity than the first-grade reagent, and it can be used as reference substance in the titration analysis to prepare the standard solution directly. Spectral pure reagents are measured by the number and intensity of the interference lines that appear during spectral analysis; chromato graphically pure reagents are expressed as impurity-free peaks at maximum sensitivity of 10^{-10} g.

Reagents of different specifications should be selected according to the experimental requirements. In general, the reagents used in analytical experiments can be A. R. grade (analytical grade) and prepared with distilled or deionized water.

1.3.2.2 Storage of chemical reagents

Chemical reagents should be kept in a well-ventilated, clean and dry room, away from sources of ignition and protected from moisture, dust and other substances.

(1) The solid reagents are stored in wide-mouth bottle, and the liquid reagents are filled in the narrow-mouth bottle or the dropper bottle.

(2) Reagents that are easily decomposed by light (such as silver nitrate, hydrogen peroxide, potassium permanganate, etc.) should be stored in brown bottles and placed in the dark.

(3) Reagents that are easy to corrode the glass (such as hydrofluoric acid, fluoride salts, sodium hydroxide, etc.) should be kept in plastic bottles.

(4) Reagents with strong water absorption (such as anhydrous sodium carbonate, sodium hydroxide, etc.) should be tightly sealed.

(5) Reagents that can easily react with each other should be stored separately. Flammable and explosive reagents should be stored separately in a cool and ventilated place away from direct sunlight.

(6) Highly toxic reagents (such as cyanide, arsenic trioxide, etc.) must be kept by special personnel and strictly recorded when they are used.

(7) Each reagent bottle should be labeled with the name, concentration, purity of the reagent and the date of preparation, etc.

1.3.2.3 Use of chemical reagents

For the selection of chemical reagents, it is ought to reasonably choose the corresponding level of reagents on the premise of meeting the experimental requirements. The level of reagents should be low, so as not to exceed the level and cause waste, but also ensure the accuracy of analysis results. When taking reagents, note the following points:

(1) Any reagents cannot be taken directly by hand. Solid reagents should be taken with clean medicine spoons, and medicine spoons should not be mixed. When taking reagents, place the bottle caps upside down on the table and replace immediately after taking reagents to prevent contamination and deterioration.

(2) When the liquid reagent is taken from the drop bottle, the dropper should not touch the wall of the container used to avoid contamination of the original reagent. When taking liquid reagents from a narrow-mouth bottle, put the bottle stopper upside down on the table first, with the label facing the palm of the hand when pouring, so that the label will not be corroded by the small amount of liquid remaining on the mouth of the bottle.

(3) Reagents should be taken appropriately. Once the reagent is taken out, it shall not be poured back to avoid contamination of the reagent in the original reagent bottle.

1.3.3 Concentration of the solution and its preparation method

Concentration refers to the amount of solute contained in a certain amount of solution or solvent. The concentration of solution is often expressed in the following ways.

(1) Percentage concentration

① Mass-mass percentage concentration (m/M, %) refers to the number of grams of solute per 100 grams of solution, i.e., mass-mass percentage concentration(m/M, %) = solute mass(g)/[solute mass(g) + solvent mass(g)] × 100%. Commercially available acids and bases are usually expressed in this way. For example, 37% hydrochloric acid solution, i.e. 100 grams of hydrochloric acid solution contains 37 grams of pure HCl and 63 grams of water.

② Mass-volume percentage concentration (m/V, %) refers to the number of grams of solute per 100 mL of solution, i.e., mass-volume percentage concentration(m/V, %) = solute mass(g)/solution volume(mL) × 100%. For example, 1% silver nitrate solution means 1 gram of silver nitrate dissolved in an appropriate amount of water, and then diluted with water to 100 mL.

(2) Volume ratio concentration (V/V) refers to the concentration obtained when the liquid is diluted with water or mixed with other liquids. Volume ratio concentration (V/V) = liquid volume (solute volume, mL)/solvent volume (generally water, mL). For

example, 1∶2 hydrochloric acid is made by mixing 1 volume of concentrated hydrochloric acid and 2 volumes of water.

(3) Volume percentage concentration (V/V, %) refers to the volume (mL) of solute contained per 100mL of solution. For example, a 95% ethanol solution is a 100mL of solution containing 95mL of ethanol and 5mL of water.

(4) Amount of substance concentration (also known as molar concentration, mol/L) refers to the number of moles of solute in 1 liter of solution.

$$\text{Amount of substance concentration} = \frac{\text{moles of solute}}{\text{solution volume}} = \frac{\text{mass of solute/molar mass of solute}}{\text{solution volume}}$$

1.4 Instruments and Basic Operations of Quantitative Analysis

1.4.1 Analytical balances

1.4.1.1 How to use the analytical balance

(1) Take off the analytical balance cover and fold it well before weighing. Check spirit level of analytical balance. If the bubble of the spirit level is not in the center of the circle, adjust the leveling feet to make the bubble return to the center.

(2) Check whether the balance door is closed. Turn on the balance switch and press the "Tare" key, the balance displays 0.0000 g.

(3) Weigh according to the required method. Take and place the material from the side door during weighing and record it until the reading is stable. When reading, close the side door to avoid fluctuations of the reading caused by air flow.

(4) Turn off and clean the balance and cover it with the cover after weighing.

1.4.1.2 Weighing method

According to different weighing objects and requirements, corresponding weighing methods are adopted. There are three commonly used weighing methods:

(1) Direct weighing method　Direct weighing method is suitable for weighing clean and dry ware, blocks or rods of metal, etc.

Method: After pressing the "Tare" key to make it display "0.0000 g", place the weighed object on the weighing plate. After the balance reaches equilibrium, the reading is the mass of the weighed object.

(2) Fixed mass weighing method　Fixed mass weighing method is used to weigh a certain fixed mass of reagents. This method is suitable for weighing powdery or granular substances which are not easy to absorb water, stable in air and non-corrosive.

Method: put the dry vessel on the weighing plate of the balance, and press "Tare" to make the balance display zero after the balance is balanced. Gradually add the sample into the vessel to the desired mass.

(3) Decrement method　Decrement method is used to weigh samples within a certain range of mass. This method is suitable for weighing samples that are easily hygroscopic, oxidized

or reacted with CO_2.

Methods: Put an appropriate amount of samples into dry and clean weighing bottle, and close the bottle cap. Wrap the weighing bottle with a clean paper strip, as shown in Fig. 1.1(a), place the weighing bottle on the weighing plate of the analytical balance, and accurately weigh the total mass of the sample and the weighing bottle, denoted as m_1. Take out the weighing bottle and hold it above the container holding the sample. Wrap the bottle cap with a piece of paper, open the bottle cap, tilt the weighing bottle, tap the top of the bottle mouth with the cap to make the sample fall into the container slowly, as shown in Figure 1.1(b). When the poured sample is close to the required mass, tap the upper part of the bottle mouth gently with the bottle cap to make the sample stuck to the bottle mouth fall back into the bottle, and at the same time slowly lift the weighing bottle, and close the cap. Place the weighing bottle on the weighing plate of the balance and record the reading m_2. The difference between m_1 and m_2 is the mass of the sample. If the mass of the sample poured at one time exceeds the required range, the weighing must be discarded, and the sample cannot be put back into the weighing bottle.

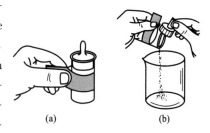

Fig. 1.1 Decrement method

Multiple samples can be continuously weighed according to the above method.

The mass of the first sample(g) = $m_1 - m_2$

The mass of the second sample(g) = $m_2 - m_3$

The mass of the third sample(g) = $m_3 - m_4$

1.4.2 Volumetric glassware

1.4.2.1 Volumetric flask and use

Volumetric flask is a type of volumetric vessel used to accurately measure the volume of the contained solution. It is a thin-necked, pear-shaped flat-bottomed glass bottle with ground glass or plastic stopper and a marking line on the neck. The volume of a volumetric flask is defined as: at the indicated temperature (generally 20℃), when the lower edge of the meniscus of the liquid is tangent to the marking line, the volume of the solution is equal to the volume marked on the bottle, in mL. Volumetric flask is mainly used to prepare standard solutions of accurate concentration or to dilute solutions to exact volumes.

(1) Leak detection of volumetric flasks　Volumetric flasks should be checked for leaks before use. Add water to the vicinity of the marking line, press the bottle stopper with the index finger of one hand, hold the neck above the marking line with the other fingers, and hold the bottom of the bottle with the fingertips of the other hand. Invert the volumetric flask for 2 minutes. If there is no water leakage, erect the bottle and turn the stopper 180°, then invert it for 2 minutes. If there is no water leakage, it can be used.

Leak detection of volumetric flasks

(2) Preparation of solution

① Dissolved sample: accurately weigh solid sample in a beaker, add a certain amount of water, stir to dissolve completely.

② Transfer solution: the solution is quantitatively transferred into a volumetric flask, as shown in Fig. 1.2(a). Hold the glass rod in one hand so that it stretches into the mouth of the flask, the lower end of rod is against the inner wall of the bottle neck below the marking line, and its rest part does not touch the bottle mouth; With the beaker in the other hand, the edge of the beaker mouth is close to the middle and lower part of the glass rod, and gently tilt the beaker so that the solution flows into the volumetric flask along the glass rod. After the solution runs out, lift the beaker up slowly along the glass rod, and erect the beaker and the glass rod at the same time, so that the droplets attached between the glass rod and the mouth of beaker flow back into the beaker, and then put the glass rod into the beaker. Rinse the glass rod and the inner wall of the beaker 3, 4 times with the water and transfer the washing liquid to the volumetric flask in accordance with the above method.

③ Constant volume: add water to 2/3 of the volume of the volumetric flask, pick up the volumetric flask and rotate it horizontally for several cycles to mix the solution initially. Continue to add water to 1 cm below the marking line, wait 1-2 minutes so that the solution attached to the inner wall of the bottleneck flows down. Add water with a washing bottle or dropper until the lower edge of the meniscus is tangent to the marking line [Fig. 1.2(b)], and cover the bottle stopper tightly.

④ Shake well: press the bottle stopper with the index finger of one hand, hold the bottom of the bottle with the fingertips of the other hand. Turn the volumetric flask upside down, so that the bubbles in the bottle rise to the top, shake it gently, and invert back again, as shown in Figure 1.2(c). Repeat the operation several times to mix the solution thoroughly.

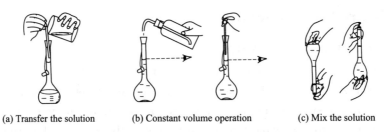

(a) Transfer the solution   (b) Constant volume operation   (c) Mix the solution

Fig. 1.2 Preparation of solution

(3) Dilution of the solution   Pipette a certain volume of solution into a volumetric flask, dilute and shake it well according to the above method.

Notes:

(1) Dissolution of solutes cannot be performed in volumetric flasks.

(2) Volumetric flasks can only be used to prepare solutions, not to store solutions. The prepared solution should be transferred to a dry and clean ground-mouth reagent bottle for storage.

(3) The hot solution should be cooled to room temperature before transferring, otherwise it will cause volume error.

(4) Volumetric flasks cannot be heated or dried in the oven.

(5) The solution to be protected from light should be prepared in a brown bottle.

(6) The volumetric flask should be washed with water immediately after use. If it is not used for a long time, the grinding mouth should be cleaned, dried and lined with paper pieces.

(7) The volumetric flask is used with the stopper that cannot be changed at will. Tie the stopper to the neck with a string or rubber band to prevent contamination or mixing.

1.4.2.2 Pipette and use

Pipette is a type of graduated vessel for accurately pipette a certain volume of liquid, including one-mark pipette and graduated pipette.

One-mark pipette is a slender, intermediate-expanded glass tube with a circular marking line is engraved on the upper part of the neck. The bulging part is marked with the volume at a specified temperature [Fig. 1.3(a)]. At the indicated temperature, the lower edge of the meniscus of the pipette solution is tangent to the marking line, and then the solution flows out freely, the volume of the outflow solution is the same as the volume marked on the pipettes.

Graduated pipette is a glass tube with a uniform scale [Fig. 1.3(b)], which can be used to accurately pipette any volumes of liquid within the labeled range. Its accuracy is slightly less than that of the one-mark pipette.

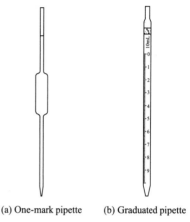

(a) One-mark pipette (b) Graduated pipette

Fig. 1.3 Pipette

(1) Use of one-mark pipettes

① Rinse. Wash with detergent, tap water and distilled water until there are no water droplets hung on the inner wall. Before pipetting the liquid, blot up the water inside and outside the tip of pipettes with filter paper, and rinse

Rinse

it three times with the pipetted solution to remove the residual water, and release the solution from the tip of the pipettes. The specific operation is to pour a little pipetted solution in a dry and clean beaker, use the ear-washing ball to suck the liquid to about 1/4 of the pipette, quickly remove the ear-washing ball and simultaneously block the tube mouth with the index finger of the right hand. Take out the pipette, hold the lower end of the tube with your left hand, slowly release the index finger of your right hand, and lower the upper nozzle while turning the pipette so that the solution touches the part of the marking line and the inner wall of the whole tube. Finally, the tube is upright so that the solution is released from the lower end of the tube and discarded.

② Pipette the solution. When pipetting the liquid, hold the top of the neck marking line by the right thumb and middle finger. Insert the pipette tip about 1-2 cm below the liquid surface. Too deep inserting will make too much solution adhere to the outer wall, affecting the accuracy of the solution volume; too shallow inserting will cause suction air due to the drop in the liquid level. Hold the ear-washing ball in the left hand, first press out the air inside the ball, then attach the tip of the ear-washing ball to the mouth of the pipette, loosen the ear-washing ball, and the solution is gradually sucked into the pipette. At this time, the tip of pipette should fall as the liquid level drops, as displayed in Fig. 1.4(a). When the liquid level in the pipettes rises to about 5 mm above the marking line, remove the ear-washing ball quickly and block the mouth of pipettes with the index finger of the right hand and lift the pipette away from the liquid surface. Tilt the container slightly so that the pipette tip is against the inner wall of the container, slightly loosen the index finger and gently twist the pipette with thumb and middle finger, so that the liquid level drops steadily until the lower edge of the meniscus of the solution is tangent to the marking line. Immediately press the mouth of pipettes with index finger to stop the solution from flowing out.

③ Release the solution. Wipe away the small amount of solution adhering to the bottom of the pipettes with filter paper, move the pipette to the container containing the solution, make the pipettes vertical, tilt the container about 30°, and the tip of the pipettes is close to the inner wall of the container. Release the index finger to let the solution flow freely along the pipettes wall, shown in Fig. 1.4(b). After the solution is released completely, wait for 15 s and take out the pipette. Do not blow out residual solution at the tip of the pipette if it is not marked "blow", as the volume of solution retained at the tip has been taken into account in producing the pipette.

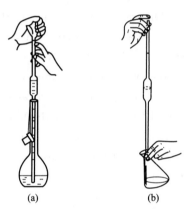

Fig. 1.4 Use of pipette

(2) Use of graduated pipettes The method of pipetting solution with the graduated pipette is generally the same as the above. It should be noted that each time the solution is sucked, the liquid level should be adjusted to the highest marking line, and then make the liquid level in the pipettes drop steadily until the required volume is reached. Press the mouth of pipettes with index finger tightly, and remove the pipette.

Note:

(1) Pipettes cannot be heated and dried in the oven, and cannot be used to transfer the superheated or supercooled solution.

(2) Pipette should be cleaned and placed on pipette rack after use.

(3) One-mark pipette and volumetric flask are generally used together, so the relative volume should be calibrated before use.

Use of pipettes

(4) Use the same pipette during the experiment to reduce experimental error.

1.4.2.3 Burette and use

Burette is engraved with precise scale and is a type of graduated vessel used to accurately measure the volume of the outflow solution during titration. The nominal volumes of burette used for constant analysis are 50 mL and 25 mL, as well as semi-microburettes or microburettes with nominal volumes of 10 mL, 5 mL, 2 mL and 1 mL. The burette with a nominal volume of 50 mL has aminimum scale of 0.1 mL and a reading accuracy of 0.01 mL.

Burette is divided into acid burette and basic burette according to the nature of the solution (Fig. 1.5). The lower end of the acid burette is equipped with a glass stopcock, which is used to hold acidic solutions and oxidizing solutions and cannot contain alkaline solutions, because alkaline solutions will make the glass stopcock bond with the groove, so that it is difficult to rotate. The lower end of the basic burette is connected with a rubber tube, in which there is a glass bead to control the outflow rate of the solution. Basic burette is mainly used to contain alkaline solutions. Oxidizing solutions or other solutions that can react with the rubber tube, such as $KMnO_4$, I_2, silver nitrate, etc., cannot be packed in the basic burette. There is also a general-purpose burette with a polytetrafluoroethylene (PTFE) stopcock at the lower end, and the appearance is the same as the acid burette, which is not limited by the acidity and alkalinity of the solution and can be used to hold various solutions.

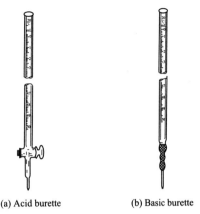

(a) Acid burette (b) Basic burette

Fig. 1.5 Burette

(1) Preparation of the burette before use Before using the acid burette, check whether the stopcock rotates flexibly, and then test for leaks. The leakage test method is as follows: close the stopcock, fill the burette with water, and fix the burette on the burette stand for 2 minutes. Observe whether there is water seepage from the mouth of burette and both ends of the stopcock. Turn the stopcock 180°, leave it for 2 min, and then observe whether there is water seepage. The stopcock can be used without water seepage. Otherwise, remove the stopcock, and spread vaseline and check the leak before use.

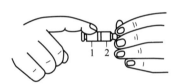

Fig. 1.6 Method of spreading vaseline

To spread vaseline, remove the stopcock and dry the stopcock and groove. Dip a small amount of vaseline with finger and apply a thin layer on both ends of the stopcock (1 and 2 in Fig. 1.6). Insert the stopcock into the stopcock groove. Then turn the stopcock in the same direction until the vaseline film between the stopcock and groove is uniform and transparent. After the above treatment, the stopcock should rotate flexibly. Once vaseline coated, secure the stopcocks with a rubber band.

For basic burette, appropriate size of glass beads and rubber tube should be selected. Check whether the burette is leaking, and whether the liquid droplets can be flexible controlled. If it does not meet the requirements, replace it.

For the burette with PTFE stopcock, check whether the stopcock rotates flexibly and leaks before use. Due to the elasticity of the PTFE stopcock, the tightness of the stopcock and groove can be adjusted by the nut at the end of the stopper, so these burettes do not need to be coated with vaseline.

(2) Loading of the solution

① Rinse. Before loading the solution, rinse the burette with 5-10 mL of this solution for 2-3 times to remove the residual water in the burette and ensure that the concentration of the titration solution remains unchanged. During operation, close the stopcock, hold the burette flat with both hands and rotate it slowly so that the solution fills the burette. Open the stopcock, allow a small amount of solution to flow out from the lower end of the burette and the rest of the solution is poured out from the mouth of the burette. The solution should

Rinse

be poured directly into the burette, and no other containers should be used to avoid changing the concentration of the solution or causing contamination.

② Exhaust. After filling the solution, make sure that the lower end of the burette is free of air bubbles, otherwise bubble will affect the accuracy of the measured volume during titration. For acid burettes, turn the stopcock quickly to allow the solution rush out and take away the air bubbles. For basic burettes, bend the rubber tube upward, tilt the glass tip upward, and squeeze the glass bead so that the solution can be ejected from the tip, and the air bubbles can be discharged (Fig. 1.7). Finally adjust the liquid level to the 0.00 mL or a little below and record the initial reading.

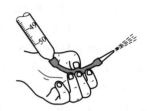

Fig. 1.7 Method of removing air bubble in basic burette

(3) Operation of the burette　The burette should be fixed vertically on the burette stand for titration. The titration is carried out preferably in an Erlenmeyer flask or, if necessary, in a beaker, iodine flask, etc.

① Operation of acid burette. When using an acid burette, control the stopcock of the burette with left hand, the thumb in front and the index and middle fingers behind, and gently turn the stopcock inward to allow the solution flow out drop by drop, as shown in Fig. 1.8(a). When turning the stopcock, be careful not to hold it in the palm of hand to prevent the stopcock from loosening or being pushed out, resulting in leakage.

② Operation of basic burette. When using the basic burette, the ring finger and little finger of the left hand clamp the outlet glass tube, the thumb is in front and the index finger is behind. Squeeze the upper part of the glass bead in the rubber tube slightly, so that a gap is formed between the rubber tube and the glass bead, and the solution can flow out [Fig. 1.8(b)]. Do not squeeze the glass bead forcefully, do not move the glass bead up and down. Do not squeeze the rubber tube under the glass ball to let air in and form air bubbles. At the end of the titration, release the thumb and index finger first and then the ring finger and little finger.

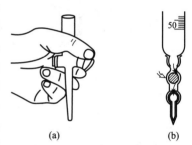

Fig. 1.8 Titration operation of burette

Titration operation

③ Titration operation method. During titration, hold the Erlenmeyer flask with the thumb, index finger and middle finger of right hand, and extend the tip of burette 1-2 cm below the mouth of flask. Shake the Erlenmeyer flask in a direction to make it move in a circular motion while adding the solution dropwise, so that the solution in the flask is evenly mixed, and the reaction is quick and complete [Fig. 1.9(a)]. The left hand should not leave the stopcock to allow the solution drop on its own during titration, and the Erlenmeyer flask cannot leave the burette tip. Pay attention to the color change of the solution in the Erlenmeyer flask. When approaching the end point, the titration speed should be slowed down. Drip a drop, shake a few times, and add half a drop each time until the solution has an obvious color change and meanwhile, the end point is reached. The method of adding a half drop solution is as follows: slightly turn the stopcock or squeeze the glass bead gently, so that the solution is suspended at the tip of the burette to form half drop. Touch the inner wall of the Erlenmeyer flask with the tip of burette to let the half drop drip down, and then wash the flask with a small amount of pure water to wash down the solution attached to the flask wall.

When titration is carried out in a beaker, the tip of the burette is extended into the beaker for 1-2 cm, and does not lean against the inner wall. Add the solution dropwise with the left hand, and continuously stir the solution in the beaker with the glass rod in the right hand. Note that the glass rod should make a circular motion, but can not touch the wall and the bottom of the beaker, as shown in Fig. 1.9(b).

When titrating in a stoppered Erlenmeyer flask such as an iodine flask, the stopper should be sandwiched between the middle finger and ring finger of the right hand, and should not be placed elsewhere to avoid contamination, as shown in Fig. 1.9(c).

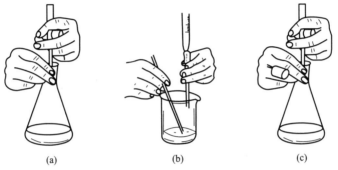

Fig. 1.9 Titration operation method

(4) Reading of the burette Inaccurate burette reading is usually one of the main sources of error in titration analysis. Therefore, the reading of burette should follow the following rules:

① After loading the solution or titration, let stand for 1-2 min so that the solution attached on the inner wall flows down before reading. Before each reading, check whether there are liquid drops hanging on the wall of the tube, whether there are bubbles at the lower part of the burette and whether there are droplets hanging on the tip of the burette.

② When reading, take the burette from the burette stand and pinch the ungraduated upper end of the burette with thumb and index finger of the right hand to keep the burette hanging down naturally before reading.

③ Due to the aqueous solution infiltrating the glass, the liquid level in the burette takes on a meniscus shape under the action of adhesion and cohesion. For colorless and light-colored solutions, the line of sight should be at the same level as the lowest point of the meniscus during reading, as shown in Fig. 1.10. For dark solutions, such as $KMnO_4$ and I_2 solutions, since the meniscus is not clear enough, the line of sight should be at the same level as the upper edge of the liquid level.

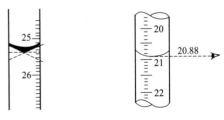

Fig. 1.10 Reading of the burette

④ Before each titration, the liquid level should be adjusted to 0.00 mL or a little lower, so that titration can be fixed within a certain volume range to reduce volume error.

⑤ The reading should be reached to the second decimal point, that is, accurate to ±0.01 mL.

1.4.3 pH meter

A pH meter, also known as acidity meter, is a precision instrument used to measure the pH of a solution. The pH meter consists of two parts: an electrode and potentiometer. The electrode is further divided into indicator electrode, reference electrode and combination electrode.

1.4.3.1 Basic electrodes

(1) Indicator electrode　The glass electrode is an indicator electrode for measuring pH, and the structure is shown in the Fig. 1.11. The glass ball at the lower end of the electrode is a pH-sensitive membrane blown from special glass, which can selectively respond to the activity of hydrogen ions. The electrode is equipped with a specific internal reference solution in which an Ag-AgCl internal reference electrode is inserted. The potential of a glass electrode changes with the pH of the solution.

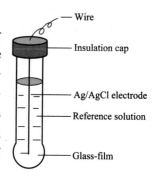

Fig. 1.11　Glass electrode structure

(2) Reference electrode　A saturated calomel electrode is usually used as a reference electrode, and the structure is shown in Fig. 1.12. The saturated calomel electrode consists of mercury, Hg_2Cl_2 and saturated KCl solution. The potential of saturated calomel electrode is stable at a certain temperature and does not change with the pH of solution. When the indicator electrode, reference electrode and tested solution constitute a working cell, the potentiometer can measure its electromotive force under the condition of zero current, and the pH of the solution can be determined.

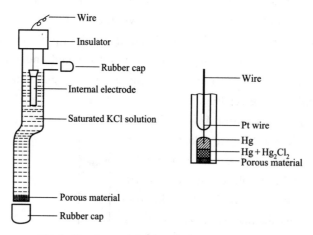

Fig. 1.12　Saturated calomel electrode structure

(3) Combination electrode　At present, combination electrode is widely used, which

is composed of indicator electrode and reference electrode, and the structure is shown in Fig. 1. 13. The glass ball membrane of the combination electrode is effectively protected, is not easily damaged, and is easy to use.

1. 4. 3. 2 Method of use

(1) Preparation of pH electrode Unplug the electrode protective bottle at the lower end of the electrode, pull off the rubber sleeve on the upper end of the electrode to expose the small hole at the upper end. Rinse the electrode with distilled water and blot the residual water with filter paper.

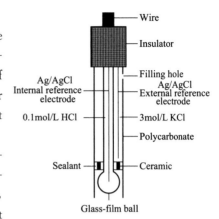

Fig. 1. 13 Combination electrode structure

(2) Calibration The instrument should be calibrated before use. The electrode slope is usually calibrated by the two-point calibration method. Generally, the buffer solution with pH=6.86 is used for the first time, and a buffer solution with the pH close to that of the tested solution is chosen for the second time. If the tested solution is acidic, the solution with pH=4.00 should be selected; when the tested solution is alkaline, the buffer solution with pH=9.18 is selected.

(3) Measurement of pH value The calibrated instrument can be used to measure the solution. Clean the electrode head with distilled water before each measurement and gently blot the residual water on the electrode with filter paper or rinse it with the tested solution. In order to ensure the accuracy during measurement, the electrode head ball should be completely immersed in the solution, the electrode should be 1-2 cm away from the container, and the solution maintains a uniform flow and no bubbles. The data can be read after the reading is stable.

(4) Measurement of potential (mV) value Adjust the display of the instrument to "mV" level, and other steps are the same as the above. It can be read after the reading is stable.

1. 5 Recording, Processing and Reporting of Experimental Data

The task of quantitative analysis is to determine the content of a certain component in the sample, so it is very important for the accuracy of the experimental results to record and process the original data in a standard manner and report the analysis results reasonably.

1. 5. 1 Recording of experimental data

(1) Data should be recorded in a special experiment notebook or preview report. It is never allowed to write data on a single sheet or small piece of paper.

(2) All kinds of measurement data and related phenomena during the experiment should be recorded in a timely, accurate and clear manner. Do not tamper with data or falsify

data.

(3) When recording the data in the experiment, pay attention to the number of significant digits. For example, when weighing with a platform scale, it is required to record to 0.01 g; when weighing with an analytical balance, record to 0.0001 g; the volume of burettes and pipettes should be recorded to 0.01 mL.

1.5.2 Processing of experimental data

Quantitative analysis experiments usually require to be measured 3-5 times in parallel, with the average value to represent the measurement results, and the relative standard deviation (or standard deviation) to evaluate the precision of the analysis results. In order to deal with the experimental data simply, clearly and correctly, it is usually expressed and processed in the form of tables.

For example, the data recording and processing table for calibrating the concentration of NaOH solution using potassium hydrogen phthalate (KHP) reference substance is as Table 1.3:

Table 1.3 Standardization of NaOH solution

No.	1	2	3
$m_{KHC_8H_4O_4}/g$			
$V_{NaOH}(initial)/mL$			
$V_{NaOH}(final)/mL$			
$V_{NaOH}(consumed)/mL$			
$c_{NaOH}/(mol/L)$			
$\bar{c}_{NaOH}/(mol/L)$			
$|d_i|$			
$\bar{d}_r/\%$			

1.5.3 Experiment report

The experiment report is the summary of the experiment, which is the process of making students rise from intuitive perceptual understanding to rational understanding, and is an effective way to cultivate students' ability to summarize, analyze problems and write. After the experiment, students should organize and summarize according to the experimental records and complete the writing of the experimental report. The experimental report should include the experimental topic, experimental purpose, experimental principle, experimental procedures, experimental data recording and processing, experimental results and thinking and discussion. The following is the template of experimental report format for reference.

Experiment report format template

Date: _____ Place: _____

Name: _____ No.: _____

Title: Determination of total acidity in edible vinegar

 Objectives

(1) Proficient in the use of burettes, volumetric flasks, and pipettes and titration operations.

(2) Master the preparation and standardization method of sodium hydroxide standard solution.

(3) Understand the reaction principle of titrating weak acid with strong base and the selection of indicator.

(4) Learn how to measure the total acidity in vinegar.

Principles

The main ingredient in vinegar is acetic acid (organic weak acid, $K_a = 1.8 \times 10^{-5}$). In addition, it also contains a small amount of other weak acids, such as lactic acid, etc., and the reaction product with NaOH is NaAc:

$$HAc + NaOH \longrightarrow NaAc + H_2O$$

At the stoichiometric point, pH ≈ 8.7, and the titration jump is in the alkaline range (for example, 0.1 mol/L NaOH with 0.1 mol/L HAc titration jumps in the pH range: 7.74-9.70). If an indicator that changes color in the acidic range such as methyl orange is used, it will cause a large titration error. (The solution is weakly alkaline at the stoichiometric point of the reaction, and when the indicator changes color in the acid range, the solution is also weakly acidic, and the titration is incomplete). Therefore, the indicator phenolphthalein (8.0-9.6) that changes color in the alkaline range should be selected here. (The selection of the indicator is mainly based on the titration jump range. The discoloration range of the indicator should be fully or partially included in the titration jump range, and the end point error should be less than 0.1%)

Therefore, using phenolphthalein as indicator, the content of HAc is determined with NaOH standard solution. The total acidity in vinegar is expressed by the content of HAc.

Materials

Instruments: Erlenmeyer flask, graduated cylinder, analytical balance (one ten thousandth), alkaline burette, pipette, volumetric flask.

Reagent: Potassium hydrogen phthalate; NaOH (s); phenolphthalein indicator.

Procedures

(1) Preparation and standardization of 0.1 mol/L NaOH solution

Preparation: Weighing 1.00 g NaOH using platform balance → dilute to 250 mL with deionized water.

Calibration: Weigh 0.4-0.6 g potassium hydrogen phthalate (three parts) on the analytical balance → 250 mL Erlenmeyer flask → add 20-30 mL distilled water, 2-3 drops of phenolphthalein →titrate with NaOH solution → the solution changes from colorless to light pink color and does not fade for 30 s →write down V_{NaOH} (initial).

(2) Determination of total acidity in edible vinegar

Dilution: Pipette accurately 25.00mL of edible vinegar → dilute to required volume and shake well.

Titration: Pipette 25.00mL of test solution in three portions → add 2-3 drops of phenolphthalein→ titrate with NaOH solution→ the solution changes from colorless to light pink color and does not fade for 30 s →write down V_{NaOH} (final).

Experimental data recording and processing

(1) Standardization of 0.1mol/L NaOH solution (see Table 1.4)

Table 1.4 Standardization of 0.1mol/L NaOH solution

No.	1	2	3
$m_{KHC_8H_4O_4}$/g			
V_{NaOH}(initial)/mL			
V_{NaOH}(final)/mL			
V_{NaOH}(consumed)/mL			
c_{NaOH}/(mol/L)			
$\bar{c}_{NaOH}$/(mol/L)			
$|d_i|$			
$\bar{d}_r$/%			

(2) Determination of total acidity in edible vinegar (Table 1.5)

Table 1.5 Determination of total acidity in edible vinegar

No.	1	2	3
$V_{vinegar}$/mL			
V_{NaOH}(initial)/mL			
V_{NaOH}(final)/mL			
V_{NaOH}(consumed)/mL			
c_{HAc}/(g/L)			
$\bar{c}_{HAc}$/(g/L)			
$|d_i|$			
$\bar{d}_r$/%			

 Attention

(1) The concentration of acetic acid in vinegar is relatively large, so it must be diluted before titration.

(2) When measuring the content of acetic acid, the distilled water used cannot contain carbon dioxide. Otherwise it will dissolve in water to generate carbonic acid, which will be

titrated at the same time.

Questions

(1) Why phenolphthalein is used as an indicator in acetic acid titrating?

(2) What is the determination principle of this method?

(3) After the phenolphthalein indicator turns the solvent red, it becomes colorless after being placed in the air for a period of time. What is the reason?

(4) How many significant number should be retained for the concentration of the standard solution?

Chapter 2
Sample Preparation and Separation Techniques

2.1 Preparation of Samples

Sample preparation generally refers to the collection and disintegration of the sample.

(1) The collection of samples refers to choosing a small amount of samples from a large number of materials as the original samples. The chosen samples should be highly representative, and the composition of the chosen samples can represent the average composition of all the materials. Because the solid samples (such as soil, sediments, metals) are inhomogeneity, the chemical composition of liquid samples is prone to change and so on, the accuracy and representativeness need to be ensure.

(2) Methods for sample decomposition are commonly as follow: dissolution method, melting method, ashing method and digestion method. In the sample decomposing, the measured components should be protected. Because of the separation of interfering components, the determination needs to be done simply and quickly.

This part of the content is explained in detail in the analytical chemistry textbook, so this book will not be repeated.

2.2 Separation Techniques of Samples

2.2.1 Selection of separation methods

In a separation system, different phases must usually be designed and substances are transferred between the different phases in order to spatially separate the different substances, while most separation processes choose two-phase systems. Finding a suitable two-phase system increases the difference in the potential energy of action of the various separated components between the two phases so that they are selectively partitioned into the different phases.

There are many types of separation methods and many ways to classify them, and they can be divided into several distinctive types from different perspectives. The following is a brief description of several separation methods that are commonly used in analytical chemistry experiments.

2.2.2 Precipitation

Precipitation is a classic separation method. Its basic principle is that two or more substances have a certain gap in the distribution ratio of solid and liquid phases. To achieve the purpose of separation, multiple precipitations are carried out to take advantage of this gap.

Some measures to improve the effect of precipitation and separation: (1) use organic precipitants (salting, chelation, adsorption); (2) use complex masking to improve the effect of precipitation and separation; (3) use selective redox; (4) use co-precipitation carrier to separate or enrich trace components; (5) homogeneous precipitation.

After precipitation separation, it is often necessary to cooperate with centrifugation or filtration for separation.

2.2.3 Distillation

The combined operation of heating a liquid to boiling, tuning the liquid into a gas, and then condensing the vapor into a liquid is called distillation. Distillation is one of the important methods for separating and purifying liquid organic mixtures, while reduced-pressure distillation is the process of distillation using a vacuum pump to pump the distillation system to a certain vacuum level. Distillation can separate volatile and non-volatile substances, as well as separate liquid mixtures with different electric charges. It is mainly used for the following (Fig. 2.1).

(1) Separation of liquid mixtures with a significant difference in boiling point (more than 30 ℃ difference).

(2) Constant method to determine the boiling point and determine the purity of the liquid.

(3) Removal of non-volatile substances contained in liquids.

(4) Recovery of solvent or evaporate part of the solvent in the need to concentrate the liquid.

2.2.4 Extraction

Broadly speaking, extraction separation method is a method to selectively transfer the target compounds in the sample phase to another phase or selectively retain them in the original phase (That is, transfer of non-target compounds), thus separating the target compounds and the original complex matrix from each other. Extractive separation methods are broadly classified as liquid phase extraction, solid phase extraction and supercritical fluid extraction.

Solvent extraction is a method to separate target and matrix substances from each other by the difference in partition coefficients of different substances between two mutually insoluble phases (aqueous and organic phases). Solvent extraction can be used for the separation of both organic and inorganic substances. Solvent extraction is still a fairly common separation technique in laboratories and plants, and its advantages are mainly in terms of simple instrumentation, easy operation, relatively high separation selectivity and wide range of applications. During the extraction process, the sample is transferred in a solution state, and its dissolution process is directly related to the intermolecular forces, which are related to

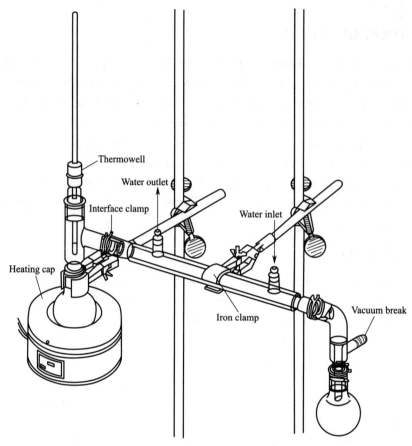

Fig. 2.1　Common distillation equipment

the molecular polarity. The general order: non-polar substances＜polar substances＜hydrogen-bonded substances＜ionic substances.

The solvent used to extract the desired substance is called an extractant. The purpose of adding extractants is to make it easy for those hydrophilic solutes to enter the organic phase by reacting with the extractant to form various types of hydrophobic compounds. Especially in the solvent extraction of metal ions, the variety of extractants is quite rich. There are the following basic requirements for extractants: (1) at least one extractive functional group; (2) sufficient hydrophobicity; (3) good selectivity; (4) high extraction capacity. The basic instruments for the extraction operation are the separating funnel, Soxhlet extractor (Fig. 2.2), etc.

Solvent microextraction, also known as liquid phase microextraction (LPME), is a new solvent extraction technology developed in 1996. It combines the

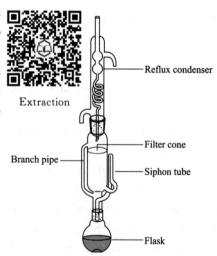

Fig. 2.2　Soxhlet extractor

advantages of liquid-liquid extraction and solid-phase microextraction, requires only a very small amount of organic solvent, simple device, easy operation and low cost. LPME technology is suitable for extracting trace target object with low solubility in aqueous solutions, and acidic or basic functional groups. LPME technology can also be easily connected to subsequent analytical instruments for on-line sample preparation. For example, in the subsequent gas chromatography analysis of LPME, a microinjection needle can be used to directly inject the sample, which overcomes the shortcomings of slow desorption speed of solid-phase microextraction (SPME), coating degradation and large memory effect. The main disadvantage of LPME compared to SPME that does not use solvents is the presence of solvent peaks in chromatographic analysis. Sometimes it will mask the chromatographic peaks of the target component. An example of using dispersion-liquid microextraction is to separate and enrich pyrethroid pesticide residues in water samples. Take 5.0 mL water sample filtered by vacuum filtration of 0.45 μm microporous membrane in a 10 mL tip-bottom centrifugal tube with stopper, add 10 μL of the extraction solvent chlorobenzene and 1.0 mL of dispersant acetone, and gently shake for 1 min to form an emulsion system of water/acetone/chlorobenzene. The chlorobenzene was uniformly dispersed in the aqueous phase, placed at room temperature for 2 min, and centrifuged at 5000 r/min for 5 min. Then the extraction solvent chlorobenzene dispersed in the aqueous phase was deposited to the bottom of the test tube, and 1 μL of the extraction solvent was directly injected with a micro-injector for gas chromatography.

When the extractant is fixed on other support carriers, the solvent loss and phase separation difficulties in solvent extraction are solved. However, the problems of poor stability of immobilized solvents and insufficient solvent resistance of support materials have led to the use of microencapsulation technology in solvent extraction, resulting in solvent microencapsulation technology. That is, the solvent used for extraction is coated in the cavity of the microcapsule during the microcapsule formation process. There are many preparation methods for solvent microcapsules, which generally include two steps: solvent dispersion and solvent coating. Due to the advantages of extracting resin, solvent microcapsules avoid the emulsification and phase separation problems of traditional solvent extraction, and have obvious advantages in extractant coating amount and prevention of extractant loss.

2.2.5 Recrystallization

Recrystallization is the process of heating and dissolving crystals in a solvent, and then reprecipitating after cooling, which is one of the most common methods for purifying solid organic compounds. The solubility of solid substances generally increases with increasing temperature. The purified crystal is dissolved in the hot solvent to make a saturated solution. And filter it while it is still hot, leaving the impurities with small solubility on the filter paper to be removed. During cooling, due to the decrease in solubility, it becomes a supersaturated solution to form crystals. After vacuum filtration, the crystals stay on the filter paper, and the impurities with high solubility stay in the solvent and separate from the crystals.

The choice of solvent is important to the recrystallization operation. Suitable solvents must meet the following conditions:

(1) No reaction with recrystallized substance;
(2) High solubility at high temperature, low solubility at low temperature;
(3) Impurity solubility is either very large or very small;
(4) Easily separated from recrystallized substances.

In addition, non-toxic and non-flammable should also be considered.

The general process of recrystallization operation is as follows.

(1) Dissolve Put the solid crude product containing impurities into a conical flask, add a small amount of solvent, install a reflux condenser, heat it to boiling, drop wise add the solvent until all the solids are dissolved, and continue to add about 20% solvent to compensate for the volatilization of the solvent and prevent crystal precipitation during thermal filtration.

(2) Decolorization If the crude product contains colored impurities, the above solution can be slightly cooled and added with about 5% activated carbon, and then heated to boil again, so that the colored impurities can be adsorbed on the activated carbon.

Do not add activated carbon into the boiling solution to prevent bumping. Do not add too much activated carbon, because it will absorb impurities and products at the same time, resulting in product loss.

(3) Thermal filtration The solution is filtered while hot to remove insoluble impurities. The key point of thermal filtration is that the system should be "hot" and the operation should be "fast". Therefore, sufficient preparation should be done before thermal filtration to make the operation steps detail and compact, andminimize the product loss.

(4) Cooling crystallization The thermally filtered solution is cooled to precipitate crystals, which are separated from some impurities in the solution.

The crystal obtained by rapid cooling is smaller, and the crystal obtained by natural cooling is larger. If the crystal is not easy to precipitate, the glass rod can be used to rub the inner wall of the container to initiate crystallization or add crystal seeds.

(5) Suction filtration and washing The crystals are separated from the solution (commonly referred to as the mother liquor) and washed with a suitable solvent.

(6) Dry.

2.2.6 Chromatography

Chromatography is a physicochemical separation and analysis method. It is based on slight differences in substances' physicochemical properties such as solubility, vapor pressure, adsorption capacity, stereoscopic structure or ion exchange, which make the partition coefficient in the mobile and stationary phases different. However, when the two phases are in relative motion, the components are separated from each other by successive distributions in the two phases.

Thin layer chromatography (TLC) is to evenly spread the appropriate stationary phase on the plane carrier, spot the sample, and then expand it with the appropriate solvent to achieve the purpose of separation, identification and quantification. TLC has the advantages of simple equipment, convenient operation, rapid separation, high sensitivity and resolution.

TLC can be used to separate mixtures, identify and refine compounds, track the progress of chemical reactions, and serve as a guide for exploring column chromatographic conditions. It has a high separation speed, high separation efficiency, and requires less samples. It is especially suitable for compounds with low volatility, easy to change at high temperature and unsuitable for gas chromatography analysis.

(1) Thin-layer plates are usually prepared with wet plates, either tiling or coating. Tiling method is to use both sides of the glass plate as the border, and then pour the blended adsorbent on the glass plate and scrape the surface, remove the border. The coating method is the method of laying a layer with a coating device. The adhesives commonly used in wet plates are calcined gypsum (10%-15%) and aqueous solution of sodium carboxymethyl cellulose (CMCNa, 0.2%-1%), optionally one or both of which may be added. The thin sheets can be used directly after drying naturally at room temperature. If the adsorption force is too weak, it can be used after activation at 105-120 ℃.

(2) Solvents commonly used as TLC developing agents include various types of organic solvents: ① Electron acceptor solvents benzene, toluene, ethyl acetate, acetone, etc. ②Proton donor solvents isopropanol, n-butanol, methanol and absolute ethanol, etc. ③Strong proton donor solvents such as chloroform, glacial acetic acid, formic acid and water. ④Proton acceptor solvents triethylamine, ether, etc. ⑤Dipole solvent dichloromethane. ⑥Inert solvents (non-polar solvents) cyclohexane, n-hexane, etc. In the selection of mixed solvents, the best selectivity can be obtained by changing one of the components on the premise of keeping the total polarity of the solvent unchanged.

(3) The optimum concentration of sample is that the sample can be evenly distributed on the surface of the adsorbent without precipitating, which is usually 5%-10%. Generally, the ribbon sample is used, and the sample band should be as narrow as possible to obtain a better separation effect. In addition, multiple spreads can be used to improve the separation effect of PTLC. That is, at the end of one spread of PTLC, the plate is put into the container for development after drying. This operation can be repeated several times depending on the R_f value of the ribbon.

(4) Samples with color development and collection of colored compounds can be directly observed spots, and substances that produce fluorescence can be observed under ultraviolet light. The commonly used thin-layer chromatography color developer preparation and color development methods are as follows:

① Sulfuric acid: spray the commonly used sulfuric acid ethanol (1:1) solution and let stand at 110 ℃ for 15 min, different organic compounds show different colors.

② 0.05% potassium permanganate solution: reducible compounds appear yellow on a light red background.

③ Acid potassium dichromate reagent: 5% potassium dichromate concentrated sulphuric acid solution, baked thin layer at 150℃ if necessary.

④ 5% phosphomolybdate ethanol solution was sprayed and baked at 120 ℃. The reducing compound was blue, and then smoked with ammonia, the background became colorless.

After determining the location of the band, connect the two sides with a pencil or needle tip to delineate the band distribution on the thin layer, and then scrape the band off the plate with a scraper or a tube scraper connected to a vacuum collector. Finally, the compound is eluted from the adsorbent with a solvent of the lowest polarity possible (Usually about 5 mL of solvent is used for 1g of adsorbent).

Column chromatography is a method of separating compounds by a column.

The adsorbent is packed in the chromatographic column, the liquid is added from the top of the column, and the components are adsorbed by the adsorbent, then add solvent (eluent) for elution. In this process, the component with strong adsorption force moves slowly, and the component with weak adsorption force moves fast, thus achieving separation.

Conventional column chromatography is a separation method in which the mobile phase flows through the stationary phase driven by gravity. For separation, the sample can be dissolved in a small amount of the initial eluting solvent and then added to the top of the stationary phase.

When the solubility of the sample to be separated in the eluent is not good, solid loading method can be used. That is, the sample is dissolved in a certain solvent, and 2-5 times the amount of stationary phase (or diatomaceous earth) is added. Then, the mixture is dried by rotary evaporation instrument or natural drying at low temperature, and the obtained powder is added to the upper part of the column. Before elution, the sample can be covered with sand or glass beads to prevent breakage of the sample interface. Generally, conventional column chromatography is used for the preparation of crude extracts or for the separation of mixtures with very different R_f values. Gradient elution can improve the resolution of conventional column chromatography. The order of polarity of commonly used solvents is: petroleum ether < carbon disulfide < carbon tetrachloride < trichloroethylene < benzene < methylene chloride < chloroform < ether < ethyl acetate < methyl acetate < acetone < n-propanol < methanol < water.

2.2.7　Ion exchange

Ion exchange separation is the exchange reaction between ion exchangers and ions to be separated in solution, which is a unit operation in solid-liquid phases.

Characteristics of ion exchange process: (1) High selectivity: the main removal of ionized material, and equal amount of material exchange; (2) High removal efficiency; (3) Strong applicability, and wide range of application coverage (Preparation of inorganic, organic and high purity materials); (4) Exchange agent can be used repeatedly; (5) Simple operation and easy separation.

The ion exchange reaction is generally reversible and the direction of the reaction depends mainly on the relative concentration of various ions in the resin phase and the solution phase. That is, the ion concentration difference acts as a driving force to promote the exchange reaction. When the reaction proceeds to a certain degree, the ion exchange equilibrium state can be reached. At this point, the chemical potential of each substance in the resin phase is equal to the chemical potential in the solution phase.

There are many factors that affect the exchange rate, such as: (1) particle size of resin: for small and uniform particle size resin exchange rate is high; (2) crosslinking degree of resin: the higher the degree of crosslinking, the worse the swelling property of the resin;

this affects the ion diffusion inside the resin; (3) temperature: increasing the temperature facilitates exchange; (4) solution concentration: increasing the concentration is beneficial to increasing the membrane diffusion rate; (5) stirring intensity: the diffusion rate can be increased by increasing the stirring speed appropriately; (6) properties of ions to be exchanged: valence state and radius.

2.2.8 Adsorption

Adsorption is a process in which a component in a liquid or gas can be selectively adsorbed and enriched on the surface of the adsorbent. The solid material with a certain adsorption capacity is called the adsorbent, and the adsorbed substance is named the adsorbate.

The main factors affecting the adsorption effect are as follow.

(1) Properties of the adsorbent Generally speaking, the adsorbent with the larger specific surface area and the higher porosity, has the larger adsorption capacity. Besides, the smaller the particle size, the faster the adsorption speed. The appropriate pore size is conducive to the diffusion of the adsorbate into the void. When adsorbing substances with large relative molecular mass, an adsorbent with a large pore size should be selected. For adsorbing substances with a small relative molecular mass, an adsorbent with a large specific surface area and a small pore size should be selected. Polar compounds should be selected for polar adsorption. For non-polar compounds, non-polar adsorbents should be selected.

(2) Properties of adsorbate Substances that can reduce surface tension are easily adsorbed by the surface, that is, the smaller the surface tension of the solid, the more the liquid is adsorbed. If the solute is absorbed by the more soluble solvent, the amount of adsorption is small.

(3) Temperature Adsorption is generally exothermic, so as long as the adsorption equilibrium is reached, increasing the temperature will reduce the amount of adsorption. However, at low temperature, some adsorption processes often fail to reach equilibrium in a short time, and increasing the temperature will speed up the adsorption rate and increase the amount of adsorption.

(4) pH value of the solution The pH value of the solution will affect the dissociation of the adsorbent or adsorbate, which in turn affects the amount of adsorption. Generally, the adsorption capacity is the largest near the isoelectric point. The optimum pH for adsorption of various solutes needs to be determined experimentally. For example, organic acids dissolve in bases, and amines dissolve in acids. Therefore, organic acids are more easily adsorbed by non-polar adsorbents under acidic conditions and amines under alkaline conditions.

(5) Adsorbate concentration and adsorbent dosage When the adsorption reaches equilibrium, the adsorbate concentration is called the equilibrium concentration. The general rule is that the larger the equilibrium concentration of adsorbate, the larger the adsorption capacity. For example, when using activated carbon for decolorization and depyrogenation, in order to avoid the adsorption of active ingredients, the feed liquid is often diluted appropriately.

2.2.9 Membrane separation

Membrane separation has the functions of separation, concentration, purification and refinement. According to its development, it has microporous filtration, dialysis, electrodialysis, reverse osmosis, ultrafiltration, gas separation and nanofiltration.

Compared with traditional separation technology, it has the following characteristics.

(1) High separation efficiency In the field of particle size separation, theminimum limit of gravity based separation technology is micron, while the particle size that can be separated by membrane separation is nanometer. Compared with the diffusion process, the ratio of the relative volatilities of substances in the distillation process is mostly less than 10, the difficult-to-separate mixture is sometimes just above 1, while the separation coefficient of membrane separation is much larger.

(2) Low energy consumption in the separation process Most membrane separation processes have no phase change, and the latent heat of phase change is large. In addition, many membrane separation processes are carried out near room temperature, and the consumption of heating or cooling the separated materials is very small.

(3) The working temperature of most membrane separation processes is around room temperature, which is especially suitable for the treatment of temperature-sensitive substances.

However, membrane separation technology also has some deficiencies, such as poor strength, short life, easy to be contaminated, and affecting the separation efficiency.

2.2.10 Electrochemical separation

Electrochemical separation is a method of chemical separation based on the electrical properties of atoms or molecules and the charged properties and behaviors of ions. In addition to the electrolytic separation method, some new electrochemical separation and analysis technologies have been created in recent years, such as electrophoresis separation method, chemically modified electrode separation method, and medium exchange voltammetry, which are high selectivity and high sensitivity separation and analysis methods. They play an important role in enriching and separating trace substances and eliminating the interference of similar substances. Compared with other chemical separation methods, the electrochemical separation method has the following characteristics: the chemical operation is simple, and it can separate multiple samples at the same time; in addition to the need for a certain amount of electric energy, the consumption of chemical reagents is small, and the radioactive contamination is also small; in addition to self generated deposition and electrodialysis, the separation speed of other electrochemical methods is relatively fast. Especially with the development of high-voltage electrophoresis in recent years, even the complex samples can be separated quickly and effectively.

For example, electrophoresis refers to the phenomenon that charged particles migrate in the direction opposite to their electrical properties under the action of an electric field. Electrophoretic technology makes use of the differences in the charged properties of various molecules of the sample to be separated, and in the size and shape of the molecules themselves, to make the charged molecules have different migration speeds in the electric field, so as to separate the sample.

Chapter 3

Basic Experiments for Quantitative Analysis

Experiment 3.1 Analytical Balance Weighing Exercise

 Objectives

(1) Understand the structure and weighing principle of the analytical balance.

(2) Master the weighing procedures and usage rules of the analytical balance.

(3) Master the operation of weighing samples using the subtraction method.

 Principles

See section 1.4.1.

 Materials

Instruments: Analytical balance (1/10000), platform scale (1/100), weighing bottle.

Reagents: quartz sand.

 Procedures

(1) Fixed weighing method (each weighs 0.5000 g of quartz sand) Take a clean watch glass and put it in the electronic balance. After the accurate mass is displayed, press the TARE (clear key), and then use the medicine spoon to slowly add the sample to the watch glass until 0.5000 g is displayed on the balance screen.

(2) Subtraction weighing method Weigh 3 parts of 0.3-0.4 g quartz sand and in all operations filter paper strips must be used to isolate fingers and glass instruments.

① Take three clean small beakers and accurately weigh their mass on the analytical balance, respectively.

② Take a clean weighing bottle, add about 1 g of quartz sand, and accurately weigh the mass on an analytical balance.

③ Take out the weighing bottle, tap it lightly with the cap, transfer 0.3-0.4 g of quartz sand into a small beaker, accurately weigh the weight of the weighing bottle and the remaining quartz sand, and calculate the reduced mass of the weighing bottle.

④ Accurately weigh the mass of the small beaker plus the sample, and calculate the added mass of the small beaker.

⑤ Transfer 0.3-0.4 g of the sample to the other two small beakers in the same way and then accurately weigh and calculate the corresponding mass.

Weighing method of electronic balance

Data recording and processing

Weighing exercise of quartz sand using the subtraction method (Table 3.1).

Table 3.1 Weighing of quartz sand

No.	1	2	3
Mass of empty beaker/g			
Mass of weighing bottle and sample/g			
Mass of the weighing bottle after pouring out the sample/g			
Mass of the beaker after pouring the sample/g			
Reduced mass of the weighing bottle/g			
Increased mass of the beaker/g			

Questions

(1) Describe the weighing method of electronic balance.

(2) The higher the sensitivity of the analytical balance, will the accuracy of weighing be higher?

(3) Can a small medicine spoon be used for sampling during the subtractive weighing proces?

(4) During the process of weighing out the sample by subtraction, if the sample in the weighing bottle absorbs moisture, what errors will be caused to the weighing? If the sample is poured into the crucible and then absorbs moisture, will it affect the weighing?

(5) Accurately weighing 1g $K_2Cr_2O_7$ sample with one ten thousandth analytical balance, how many significant figures can be recorded?

Experiment 3.2 Basic Operation for Titration Analysis

Objectives

(1) Master the washing and usage methods of commonly used instruments for titration analysis.

(2) Master the preparation method of acid-base standard solution, and compare the titration of acid-base solution with each other.

(3) Familiar with the use of methyl orange and phenolphthalein indicators and the correct judgment of the end point, and have a preliminary grasp of the selection method of acid-base indicators.

Principles

Titration is a common method to determine the concentration of a solution. Add the standard solution (with known concentration) from the burette to the solution to be tested (or vice versa) until the reaction reaches the end point, which is called titration.

The end point of the titration reaction is determined by the color change of the indicator. When strong acid titrates strong base, methyl orange is generally used as an indicator. On the contrary, phenolphthalein is generally used as an indicator.

Using the neutralization titration of acid-base neutralization reaction, it is easy to measure the molarity of acid base and calculate it according to the chemical equation. For example: $NaOH + HCl \longrightarrow NaCl + H_2O$.

Materials

Instruments: analytical balance, measuring cylinder, beaker, reagent bottle, acid burette, basic burette, Erlenmeyer flask, pipette.

Reagents: NaOH (s), hydrochloric acid (A.R. grade), phenolphthalein indicator (0.2% ethanol solution), methyl orange indicator (0.2%).

Procedures

(1) Preparation of 0.1 mol/L NaOH solution Use a balance to quickly weigh 1 g of NaOH solid in a 100 mL small beaker, add about 50 mL of CO_2-free deionized water to dissolve, then transfer to the reagent bottle, dilute to 500 mL with deionized water, shake well, and stopper with a rubber stopper. Affix the label and write the reagent's name and concentration (leave a space to fill in the exact concentration).

(2) Preparation of 0.1 mol/L HCl solution Use a clean graduated cylinder to measure about 9.0 mL of concentrated HCl (it should be calculated in the preview), pour it into a 500 mL reagent bottle, dilute to 500 mL with deionized water, cover it with a glass stopper, and shake well. Label it and set it aside.

(3) Practice on mutual titration of acid-base solutions

① Preparation of acid and basic burettes. Prepare an acid burette and a basic burette. Rinse acid and basic burettes 2 to 3 times with 5-10 mL HCl and NaOH solutions, respectively. Then fill the HCl and NaOH solutions separately, remove the bubbles, adjust the liquid level to a zero mark or a little lower position, let it stand for 1 minute, and record the initial reading.

② Titrate HCl with NaOH solution using phenolphthalein as an indicator. Discharge 25 mL HCl from the acid burette into an Erlenmeyer flask, add 1 to 2 drops of phenolphthalein, titrate with NaOH solution under constant shaking, and pay attention to control the titration speed. When the red around the drop point of the NaOH fades slowly, it indicates that the end point is approaching. Wash the inner wall of an Erlenmeyer flask with a washing bottle, and control the NaOH solution to drip out dropwise or half drop by half drop. When the solution is reddish and does not fade for half aminute, it is the end point. Write down the reading. Put 1-2 mL HCl from the acid burette, and then titrate with NaOH solution to the end point. Repeatedly practice titration, end point judgment and reading several times in this way.

③ Titrate NaOH with HCl solution using methyl orange as an indicator. Discharge 25 mL NaOH from the basic burette in an Erlenmeyer flask, add 1 to 2 drops of methyl orange, titrate with HCl solution under constant shaking until the solution turns from yellow to orange as the end point. Then put 1-2 mL NaOH into the basic burette, continue to titrate with HCl solution to the end point, and repeat the titration, end point judgment and reading several times in this way.

(4) Determination of the volume ratio of HCl and NaOH solutions V_{HCl}/V_{NaOH} Discharge 20 mL of HCl from the acid burette at a flow rate of 10 mL/min in an Erlenmeyer flask, add 1 to 2 drops of phenolphthalein, titrate with NaOH solution until the solution is reddish, and the end point is that the color does not fade in half aminute. Read and accurately record the volume of HCl and NaOH, and measure three times in parallel. To calculate V_{HCl}/V_{NaOH}, the relative average deviation is required to be no more than 0.3%. The volume ratio can also be determined by using methyl orange as an indicator, titrating NaOH with HCl solution, and measuring three times in parallel. If time permits, both mutual titrations can be carried out. Compare the results obtained and discuss them.

 Data Recording and Processing

Determination of the volume ratio of HCl and NaOH solution (Table 3.2)

Table 3.2 Determination of the volume ratio of HCl and NaOH solution

No.	1	2	3
V_{HCl}(initial)/mL			
V_{HCl}(final)/mL			
V_{HCl}(consumed)/mL			
V_{NaOH}(initial)/mL			
V_{NaOH}(final)/mL			
V_{NaOH}(consumed)/mL			
V_{HCl}/V_{NaOH}			
average of V_{HCl}/V_{NaOH}			
$|d_i|$			
$\bar{d}_r/\%$			

 Questions

(1) In the experiment, when phenolphthalein is used as an indicator, why is it required to titrate with NaOH solution until the solution is reddish, and the end point is that the color does not fade in half aminute?

(2) Are the following operations accurate?

① The waste liquid for each wash is poured out from the upper mouth of the pipette.

② In order to speed up the outflow of the solution, use the washing ear ball to blow out the solution in the pipette.

Preparation of NaOH solution

③ When sucking the solution, extend the end of the pipette too much into the solution; when transferring the solution, let it flow down in the air.

④ Rinse the beaker with tap water only.

⑤ Piston leaks during the titration process.

⑥ The bubble at the bottom of the burette is not eliminated.

⑦ During the titration, add a small amount of distilled water to the beaker.

⑧ There are droplets on the wall of the burette.

(3) Can NaOH and HCl directly prepare solutions of accurate concentration? Why?

(4) How to choose the indicator reasonably?

Experiment 3.3 Calibration of Volumetric Instruments

 Objectives

(1) Understand the meaning and method of calibration of volumetric instruments.

(2) Preliminarily master the calibration of pipette and relative calibration between volumetric flask and pipette.

(3) Master the use of burettes, volumetric flasks, and pipettes.

(4) Get familiar with the weighing operation of the analytical balance.

 Principles

Burettes, pipettes, and volumetric flasks are commonly used volumetric glass ware in analytical experiments, and they all have scales and nominal capacities. All volumetric wares are allowed to have a certain capacity error. In analytical tests that require high accuracy, it is absolutely necessary to calibrate a set of volumetric wares used by yourself.

The calibration methods include weighing method and relative calibration method. The principle of the weighing method is: weigh the mass m of the pure water poured in and out of the calibrated volumetric wares with an analytical balance, and then calculate the actual capacity of the calibrated volumetric wares according to the density ρ of the pure water. Due to the thermal expansion and contraction of glass, the volume of volumetric wares is also different at different temperatures. Therefore, the standard temperature of a volumetric glass ware is 20 ℃. The scale and capacity marked on the various volumetric wares are called the nominal scale and capacity at a standard temperature of 20 ℃. However, in the actual calibration work, the mass of water in the container is weighed at room temperature and in the air. Therefore, the following three aspects must be considered:

(1) Air buoyancy changes mass.

(2) The density of water changes with temperature.

(3) The volume of the glass container itself changes with temperature.

In spite of the above influence, the water mass in a glass container with a capacity of 1 L (20 ℃) at different temperatures can be obtained (by the data in Table 3.3). It is very convenient to calculate the calibration value of the volumetric wares in this way.

Table 3.3 Density of pure water at different temperatures (ρ_w)

$t/℃$	$\rho_w/(g/mL)$	$t/℃$	$\rho_w/(g/mL)$	$t/℃$	$\rho_w/(g/mL)$
8	0.9886	15	0.9979	22	0.9968
9	0.9985	16	0.9978	23	0.9966
10	0.9984	17	0.9976	24	0.9963
11	0.9983	18	0.9975	25	0.9961
12	0.9982	19	0.9973	26	0.9959
13	0.9981	20	0.9972	27	0.9956
14	0.9980	21	0.9970	28	0.9954

For example, the mass of pure water released by a 25 mL pipette at 25 ℃ is 24.921 g and the density is 0.99617 g/mL, and then calculate the actual volume of the pipette at 20 ℃. $V_{20} = 24.921/0.99617 = 25.02(\text{mL})$. Then the correction value of this pipette is 25.02 mL − 25.00 mL = +0.02 mL.

It needs to be pointed out that both improper calibration and use are the main causes of volume error, which may even exceed the allowable error or the error of the volumetric ware itself. Therefore, you must perform the operation correctly and carefully during the calibration tominimize the calibration error. When the calibration value is used, the allowable times should not be less than twice, and the deviation of the two calibration data should not exceed 1/4 of that the volumetric wares allowed, and the average value shall be taken as the calibration value.

Sometimes, it is only required to have a certain proportional relationship between the two kinds of containers, without knowing their respective accurate volumes. In this case, the relative volume calibration method can be used. For pipettes and volumetric flasks that are often used together, it is more important to adopt the relative calibration method. For example, use a 25 mL pipette to take distilled water into a clean and dry 100 mL volumetric flask. After the 4th repetition, observe whether the lower edge of the meniscus of the water at the bottleneck is just tangent to the upper edge of the marking line. If not, a new mark should be made as the marking line, and the standard new marking line will be used when the pipette and volumetric flask are used together.

 Materials

Equipment: analytical balance, burette (50 mL), volumetric flask (250 mL), pipette (25 mL), Erlenmeyer flask (50 mL), thermometer.

Reagent: distilled water.

 Procedures

(1) Absolute calibration of burettes (weighing method)

① Weigh a washed and dry Erlenmeyer flask with ground glass stopper on an analytical balance to obtain the mass of the empty bottle m_0, and record it to 0.001 g.

② Fill the washed burette with pure water, adjust it to the 0.00 mL scale, and release a certain volume (denoted as V_1) from the burette into the weighed Erlenmeyer flask and close the stopper, and then weigh the mass of "bottle + water" m_1. The difference ($m_1 − m_0$) between the two masses is the mass of the released water m_w. Use the same method to weigh m_w in the burette from 0 to 10.00 mL, 0 to 15.00 mL, 0 to 20.00 mL, 0 to 25.00 mL, 0 to 30.00 mL, etc., and divide each m_w by the density of water at the experimental water temperature. Finally, the actual volume V_2 of each part of the burette can be obtained.

Repeating the calibration once, the water mass difference between the two corresponding intervals should be less than 0.02 g. Obtain the average value, and calculate the calibration value $\Delta V = (V_2 − V_1)$. With V_1 as the abscissa and ΔV as the ordinate, draw the calibration curve of the burettes. Pipettes and volumetric flasks can also be calibrated by weighing.

(2) Relative calibration of pipettes and volumetric flasks   Use a clean 25 mL pipette to

transfer pure water into a clean and dry 250 mL volumetric flask, repeat the operation 10 times, and observe whether the lower edge of the meniscus of the liquid is just tangent to the upper edge of the marking line. If not, mark it on the bottle neck with adhesive tape. In future experiments, when the pipette and volumetric flask are used together, the new mark should be used.

 Data Recording and Processing

Absolute calibration of the burette (weighing method, see Table 3.4)

Table 3.4 Calibration of burette

Volume of water released by the burette, V_1/mL	Mass of empty bottle, m_0/g	Mass of bottle + water, m_1/g	Actual mass of water ($m_w = m_1 - m_0$)/g	Real capacity V_2/mL	Calibration value $\Delta V = V_2 - V_1$
0.00-10.00					
0.00-15.00					
0.00-20.00					
0.00-25.00					
0.00-30.00					

Notes

(1) Do not take the Erlenmeyer flask directly by hand, but use a paper strip (more than three layers) to take it.

(2) Do not get water on the ground part of the Erlenmeyer flask.

(3) When you measure the temperature of the experimental water, the thermometer needs to be inserted into the water before reading, and the thermometer ball is still immersed in the water when reading.

 Questions

(1) When you calibrate the burette, why should the mass of the Erlenmeyer flask and water only be weighed to 0.001 g?

(2) Why should the volumetric flask be air-dried when calibrated Should it be air-dried when you prepare the standard solution in the volumetric flask?

(3) When calibrating burettes in different sections, why do you start at 0.00 mL each time?

(4) What are the main factors affecting the calibration of volumetric wares?

(5) Why should volumetric wares be calibrated? How to apply calibration values in capacity analysis?

Experiment 3.4 Preparation and Standardization of HCl and NaOH Solution

 Objectives

(1) Master the preparation and standardization of acid and base standard solutions.

(2) Practice the correct use of pipettes.

(3) Further master the titration operations proficiently.

(4) Accurately judge the change of acid and base indicator color.

Principles

Standard solutions are solutions of known exact concentrations. There are usually two preparation methods: direct method and standardization method.

(1) Direct method   Accurately weigh a certain mass of substances, dissolve them, quantitatively transfer them to a volumetric flask, dilute to the scale, and shake well. The exact concentration of the standard solution can be calculated according to the mass of the weighed substance and the volume of the volumetric flask. The substance used to prepare the standard solution in this way must be the primary standard substance.

(2) Standardization method   The standard solution of most substances should not be prepared by the direct method, and the standardization method can be used, which needs to firstly prepare a solution with an approximate concentration, and then use the primary standard substance or a standard solution with known accurate concentration to standardize its exact concentration. HCl and NaOH standard solutions are most commonly used in acid-base titration. Because concentrated hydrochloric acid is volatile, and NaOH solid easily absorbs CO_2 and water vapor in the air, it can only be prepared by standardization method. Its concentration is generally between 0.01 and 1 mol/L, and a solution of 0.1 mol/L is usually prepared.

(3) Primary standard substances for standardization of alkali solution   Commonly used primary standard substances for standardization of alkali standard solutions are potassium hydrogen phthalate and oxalic acid, etc.

① Potassium hydrogen phthalate. It has the advantages of being easy to produce pure product and store, not easy to absorb water in the air, with a high molar mass. So it is a good primary standard substance. The standardization reaction is as follows:

$$\text{C}_6\text{H}_4(\text{COOH})(\text{COOK}) + \text{NaOH} \longrightarrow \text{C}_6\text{H}_4(\text{COONa})(\text{COOK}) + \text{H}_2\text{O} \quad (3.1)$$

At the stoichiometric point, the solution is weakly alkaline (pH=9.20). Therefore, phenolphthalein can be used as an indicator.

Potassium hydrogen phthalate is usually dried at 105-110 ℃ for 2 hours. If the drying temperature is too high, it will be dehydrated into phthalic anhydride.

② Oxalic acid ($H_2C_2O_4 \cdot 2H_2O$). It will not be weathered and lose water when the relative humidity is 5%-95%, so it can be stored in a ground glass bottle. The solid state of oxalic acid is relatively stable, but the stability of the solution state is poor. And the air can slowly oxidize the oxalic acid solution, while light and Mn^{2+} can catalyze its oxidation. Therefore, the oxalic acid solution should be stored in a dark place.

The standardization reaction is as follows:

$$2\text{NaOH} + \text{H}_2\text{C}_2\text{O}_4 \Longrightarrow \text{Na}_2\text{C}_2\text{O}_4 + 2\text{H}_2\text{O} \quad (3.2)$$

The reaction product is $Na_2C_2O_4$, which is alkaline in aqueous solution and phenolphthalein can be used as an indicator.

(4) Primary standard substances for standardization of acid solution   The commonly

used primary standard substances for acid standardization are anhydrous sodium carbonate and borax. Its concentration can also be standardized with NaOH standard solutions of known exact concentrations.

① Anhydrous sodium carbonate. It is easy to absorb moisture in the air, so first put it at 270-300 ℃ to dry for 1 h, and then keep in the dryer before use. The standardization reaction is as follows:

$$Na_2CO_3 + 2HCl =\!\!=\!\!= 2NaCl + H_2O + CO_2 \uparrow \tag{3.3}$$

At the stoichiometric point, the pH of H_2CO_3 saturated solution is 3.9. HCl should be dripped until the solution turns orange as the end point with methyl orange as an indicator. In order to make the saturated part of H_2CO_3 decompose and escape continuously, the solution should be vigorously shaken or heated near the end point.

② Borax ($Na_2B_4O_7 \cdot 10H_2O$). It is easy to produce pure product, having low hygroscopicity and high molar mass. However, due to the presence of crystal water, when the relative humidity in the air is less than 39%, there will be obvious weathering and water loss. It is usually stored in a thermostat with a relative humidity of 60% (saturated sucrose solution below). The standardization reaction is as follows:

$$Na_2B_4O_7 + 2HCl + 5H_2O =\!\!=\!\!= 2NaCl + 4H_3BO_3 \tag{3.4}$$

The product is H_3BO_3, the pH of its aqueous solution is about 5.1, and methyl red can be used as an indicator.

③ Standardization by comparing with the NaOH standard solution of known exact concentration. The comparative standardization of 0.1 mol/L HCl and 0.1 mol/L NaOH solutions is the titration of strong acid and strong base. The pH is 7.00 at the stoichiometric point, and the titration jump range is relatively large (pH=4.30-9.70). Therefore, any indicator whose color change range falls entirely or partially within the jumping range, such as methyl orange, methyl red, phenolphthalein, and methyl red-bromocresol green mixed indicator, can be used to indicate the end point. In comparative titration, an acid solution can be used to titrate an alkaline solution, and vice versa. If the NaOH solution is titrated with HCl solution, methyl orange is used as the indicator.

 Materials

Equipment: analytical balance, graduated cylinder, burette, Erlenmeyer flask, beaker.

Reagents: hydrochloric acid (0.1 mol/L), NaOH solution (0.1 mol/L), phenolphthalein indicator (0.2% ethanol solution), methyl orange indicator (0.1%), potassium hydrogen phthalate (primary standard substance), anhydrous sodium carbonate (primary standard substance).

 Procedures

(1) Standardization of 0.1 mol/L HCl solution   Method 1: Anhydrous sodium carbonate as the primary standard substance.

Accurately weigh ( ) to ( ) g three samples of anhydrous Na_2CO_3 (molar mass 105.99 g/mol) by an analytical balance with a weighing bottle, and place them into 250 mL Erlenmeyer flasks respectively (number the Erlenmeyer flasks before weighing). Add 20-30 mL of water to the graduated cylinder to dissolve, add 1-2 drops of the methyl orange indicator, and titrate with HCl solution to be standardized until the color of the solution changes from yellow to orange, which is the end point. When approaching the end point, in order to remove the dissolved CO_2 in the solution, it should be shaken vigorously, or heated to drive out CO_2, and then titrate to the end point after cooling. Calculate the concentration c_{HCl} and the relative average deviation of the HCl solution according to the mass of the primary standard substance Na_2CO_3 and the volume of HCl consumed.

Method 2: Borax as the primary standard substance.

Accurately weigh ( ) to ( ) g three samples of $Na_2B_4O_7 \cdot 10H_2O$ (molar mass 381.37 g/mol), place them into 250 mL Erlenmeyer flasks, add 20-30 mL water to dissolve, add 1-2 drops of methyl orange indicator, and titrate with HCl solution to be standardized until the color of the solution changes from yellow to orange, which is the end point. Calculate the concentration c_{HCl} and relative average deviation according to the mass of borax and the volume of HCl consumed.

(2) Standardization of 0.1 mol/L NaOH solution   Method 1: Potassium hydrogen phthalate as the primary standard substance.

Accurately weigh ( ) to ( ) g three samples of $KHC_8H_4O_4$ (molar mass 204.23 g/mol), place them into 250 mL Erlenmeyer flasks respectively, add 20-30 mL water to dissolve, add 2-3 drops of the phenolphthalein indicator, and titrate with NaOH solution to be standardized until the color of the solution changes from colorless to reddish and does not fade for 30 s, which is the end point. Calculate the concentration c_{NaOH} and the relative average deviation according to the mass of primary standard substance and the volume of NaOH consumed.

Method 2: Oxalic acid as the primary standard substance.

Accurately weigh ( ) to ( ) g three samples of $H_2C_2O_4 \cdot 2H_2O$ (molar mass 126.07 g/mol), place them into 250 mL Erlenmeyer flasks respectively, add 20-30 mL water to dissolve, add 2-3 drops of phenolphthalein indicator, and titrate with NaOH solution to be standardized until the color of the solution changes from colorless to reddish and does not fade for 30 s, which is the end point. Calculate the concentration of c_{NaOH} and the relative average deviation according to the mass of primary standard substance and the volume of NaOH consumed.

 Data Recording and Processing

(1) Standardization of 0.1 mol/L HCl solution (Table 3.5)

Table 3.5 Standardization of 0.1 mol/L HCl solution

No.	1	2	3
m (primary standard substance)/g			
V_{HCl}(initial)/mL			
V_{HCl}(final)/mL			
V_{HCl}(consumed)/mL			
c_{HCl}/(mol/L)			
$\bar{c}_{HCl}$/(mol/L)			
$|d_i|$			
$\bar{d}_r$/%			

(2) Standardization of 0.1 mol/L NaOH solution (Table 3.6)

Table 3.6 Standardization of 0.1 mol/L NaOH solution

No.	1	2	3
m (primary standard substance)/g			
V_{NaOH} (initial)/mL			
V_{NaOH} (final)/mL			
V_{NaOH} (consumed)/mL			
c_{NaOH}/(mol/L)			
$\bar{c}_{\text{NaOH}}$/(mol/L)			
$|d_i|$			
$\bar{d}_r$/%			

Questions

(1) How do you calculate the mass range of the primary standard substance potassium hydrogen phthalate or Na_2CO_3? What is the effect if the weighing mass is too much or too little?

(2) Which volumetric glass ware should be used to add 20-30 mL of water to dissolve the primary standard substance, a graduated cylinder or a pipette? Why?

(3) If the primary standard substance is not dried, will the standardization result of the standard solution be higher or lower?

(4) When standardization the concentration of HCl solution with NaOH standard solution, phenolphthalein is used as an indicator. If the NaOH solution absorbs CO_2 due to improper storage, how will it affect the measurement result?

(5) Can accurate concentrations of HCl and NaOH solutions be obtained through direct preparation? Why?

Experiment 3.5   Determination of Total Acidity in Edible Vinegar

Objectives

(1) Be proficient in the use of burettes, volumetric flasks, and pipettes and titration operations.

(2) Master the preparation and standardization method of sodium hydroxide standard solution.

(3) Understand the reaction principle of titrating weak acid with strong base and the selection of indicator.

(4) Learn how to measure the total acidity in vinegar.

Principles

The main ingredient in vinegar is acetic acid (organic weak acid, $K_a = 1.8 \times 10^{-5}$). In addition, it also contains a small amount of other weak acids such as lactic acid, etc, and the reaction product with NaOH is NaAc:

$$HAc + NaOH \rightleftharpoons NaAc + H_2O \qquad (3.5)$$

At the stoichiometric point, pH $\approx$ 8.7, and the titration jump is in the alkaline range (for example, 0.1 mol/L NaOH titrating 0.1 mol/L HAc jumps in the pH range: 7.74-9.70). If an indicator that changes color in the acidic range such as methyl orange is used, it will cause a large titration error. (The solution is weakly alkaline at the stoichiometric point of the reaction, and when the indicator changes color in the acid range, the solution is also weakly acidic, and the titration is incomplete). Therefore, the indicator phenolphthalein (8.0-9.6) that changes color in the alkaline range should be selected here (The selection of the indicator is mainly based on the titration jump range. The discoloration range of the indicator should be fully or partially included in the titration jump range, and the end point error should be less than 0.1%).

Therefore, using phenolphthalein as the indicator, the content of HAc was determined with the NaOH standard solution. The total acidity in vinegar is expressed by the content of HAc.

 Materials

Instruments: erlenmeyer flask, graduated cylinder, analytical balance (one ten thousandth), alkaline burette, pipette, volumetric flask.

Reagents: potassium hydrogen phthalate, NaOH (s), phenolphthalein indicator.

 Procedures

(1) Preparation and standardization of 0.1 mol/L NaOH solution (See Exp. 3.4).

(2) Rinse the alkali burette with the standardized NaOH solution, and then pourinto the NaOH solution. Accurately pipette 10.00 mL of edible vinegar sample and place it in a 250 mL volumetric flask, dilute to the mark with distilled water, and shake well.

(3) Pipette 25.00 mL of the diluted solution into a 250 mL Erlenmeyer flask. Add 25 mL of freshly boiled and cooled distilled water, and add 2 drops of phenolphthalein indicator. Use the standardized NaOH standard solution to titrate until the solution turns reddish and does not fade for 30 seconds, which is the end point. Parallel experiments need to be done 3 times. Calculate the total acidity of edible vinegar according to the consumption of NaOH standard solution.

(4) Using methyl orange as the indicator, titrate with the above method. Calculate the results and compare the difference between the two indicator results.

Note:

(1) The concentration of acetic acid in vinegar is high and the color is darker, so it must be diluted before titration.

(2) When the acetic acid content is measured, distilled water containing carbon dioxide cannot be used, otherwise it will dissolve in water to generate carbonic acid, which will be titrated at the same time.

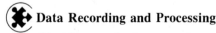 **Data Recording and Processing**

(1) The standardization of 0.1 mol/L NaOH solution (Table 3.7)

Table 3.7 Standardization of NaOH solution

No.	1	2	3
$m_{KHC_8H_4O_4}$/g			
V_{NaOH}(initial)/mL			
V_{NaOH}(final)/mL			
V_{NaOH}(consumed)/mL			
c_{NaOH}/(mol/L)			
$\bar{c}_{NaOH}$/(mol/L)			
$|d_i|$			
$\bar{d}_r$/%			

(2) Determination of total acidity in edible vinegar (Table 3.8)

Table 3.8 Determination of total acidity in edible vinegar

No.	1	2	3
$V_{vinegar}$/mL			
V_{NaOH}(initial)/mL			
V_{NaOH}(final)/mL			
V_{NaOH}(consumed)/mL			
c_{HAc}/(g/L)			
$\bar{c}_{HAc}$/(g/L)			
$|d_i|$			
$\bar{d}_r$/%			

Questions

(1) Why is phenolphthalein used as an indicator when titrating acetic acid?

(2) What is the measurement principle of this method?

(3) After the phenolphthalein indicator turns the solution reddish, it becomes colorless after being placed in the air for a period of time. What is the reason?

(4) How many significant digits should be retained for the concentration of the standard solution?

Experiment 3.6 Determination of Nitrogen Content in Ammonium Salts—Formaldehyde Method

Objectives

(1) Master the principle and method of determining nitrogen in ammonium salt by the formaldehyde method.

(2) Be proficient in titration operation and judgment of titration end point.

Principles

Ammonium salt is a common inorganic fertilizer, strong acid and weak base salt. Its content can be determined by acid-base titration. However, it is difficult to titrate directly with NaOH standard solution due to the weak acidity of NH_4^+ ($K_a = 5.6 \times 10^{-10}$). The formaldehyde method is widely used in production and laboratories to determine nitrogen content in ammonium salt.

The formaldehyde method is based on the reaction of formaldehyde with a certain amount of ammonium salt to generate a same amount of acid (H^+) and hexamethylene tetraammonium salt ($K_a = 7.1 \times 10^{-6}$), the reaction is as follows:

$$4NH_4^+ + 6HCHO \longrightarrow (CH_2)_6N_4H^+ + 6H_2O + 3H^+ \tag{3.6}$$

According to the equation, 4 mol NH_4^+ react with formaldehyde to quantitatively produce 3 mol of H^+ and 1 mol of protonated $(CH_2)_6N_4H^+$ ($K_a = 7.1 \times 10^{-6}$), that is, 1 mol of NH_4^+ is converted into 1 mol of strong acid, which reacts completely with 1 mol of NaOH.

At the stoichiometric point, the hexamethylenetetramine in the solution is a very weak base ($K_a = 1.5 \times 10^{-9}$), and the pH of the solution is about 8.7, so phenolphthalein is selected as the indicator.

The reaction of ammonium salt with formaldehyde proceeds slowly at room temperature. After adding formaldehyde, it needs to stand for a few minutes to make the reaction complete.

Formaldehyde often contains a small amount of formic acid, so it is necessary to use phenolphthalein as an indicator and neutralize it with NaOH solution before use, otherwise the measurement result will be high.

If the ammonium salt contains free acid, it needs to be neutralized and removed, that is, with methyl red as the indicator, titrate to orange with NaOH solution, and deduct it.

 Materials

Instruments: burette, volumetric flask, pipette, Erlenmeyer flask, weighing flask, graduated cylinder, analytical balance.

Reagents: 0.1 mol/L NaOH, 0.1% methyl red ethanol solution, 0.2% phenolphthalein ethanol solution, 20% formaldehyde solution: formaldehyde (40%) mixed with equal volume of water, $(NH_4)_2SO_4$ sample.

 **Procedures**

(1) Standardization of the NaOH solution (see Exp. 3.4, standardized with potassium hydrogen phthalate).

(2) Treatment of formaldehyde solution Formaldehyde often contains a small amount of formic acid due to the oxidation of formaldehyde by air, so it should be removed, otherwise a positive error will occur. The treatment method is as follows: take the supernatant of original formaldehyde (40%) in a beaker, dilute it to double volume with water, add 1-2 drops of 0.2% phenolphthalein indicator, and neutralize it with 0.1mol/L NaOH solution until the color of the formaldehyde solution turns light red.

(3) Determination of nitrogen content in the sample Accurately weigh (   ) to (   ) g $(NH_4)_2SO_4$ in a beaker, dissolve with an appropriate amount of distilled water, then quantitatively transfer it to a 250 mL volumetric flask, and finally dilute to the mark with distilled water, shake well. Pipette 25 mL of the test solution into an Erlenmeyer flask, add 1-2 drops of methyl red indicator, the solution turns red. Neutralize with 0.1 mol/L NaOH solution until red turns to orange, then add 8 mL of neutralized formaldehyde solution (1∶1), then add 1-2 drops of phenolphthalein indicator and shake well. After standing

for one minute, the solution is titrated with 0.1mol/L NaOH standard solution until the solution is light red for half aminute and does not fade, which is the end point. The results were recorded and the samples were measured 3 times in parallel. Calculate the nitrogen content in the sample based on the concentration of the NaOH standard solution and the volume consumed by titration.

Data Recording and Processing

(1) Standardization of 0.1 mol/L NaOH (Table 3.9)

Table 3.9 Standardization of 0.1 mol/L NaOH

No.	1	2	3
$m_{KHC_8H_4O_4}/g$			
V_{NaOH}(initial)/mL			
V_{NaOH}(final)/mL			
V_{NaOH}(consumed)/mL			
$c_{NaOH}/(mol/L)$			
$\bar{c}_{NaOH}/(mol/L)$			
$|d_i|$			
$\bar{d}_r/\%$			

(2) Determination of nitrogen content in ammonium salts

Table 3.10 Determination of nitrogen content in ammonium salts

No.	1	2	3
$V_{ammonium\ salts}/mL$			
V_{NaOH}(initial)/mL			
V_{NaOH}(final)/mL			
V_{NaOH}(consumed)/mL			
$X_N/\%$			
$\bar{X}_N/\%$			
$|d_i|$			
$\bar{d}_r/\%$			

Questions

(1) Why not use NaOH direct titration for determination of nitrogen in ammonium salts?

(2) Why is phenolphthalein used as an indicator for neutralizing formic acid in the formaldehyde reagent, methyl red as indicator for neutralizing free acid in the ammonium salt?

(3) Can formaldehyde method be used for the determination of nitrogen content in NH_4HCO_3?

(4) After $(NH_4)_2SO_4$ sample is dissolved in water, can NaOH solution be used to directly determine nitrogen content? Why?

(5) For the determination of nitrogen content in $CO(NH_2)_2$ urea, H_2SO_4 were first added for heating and digestion, and all of them became $(NH_4)_2SO_4$, according to the formaldehyde method of the same determination, try to write the calculation formula of nitrogen.

Experiment 3.7 Determination of Mixed Alkali—Double Indicator Method

Objectives

(1) Understand the application of acid-base titration.

(2) Master the principle of the double-indicator method for the determination of mixed alkali, the identification and calculation methods of the components.

Principles

Mixed bases are mixtures of Na_2CO_3 and NaOH or Na_2CO_3 and $NaHCO_3$. To determine the content of each component in the same sample, titrate with HCl standard solution, and select two different indicators to indicate the arrival of the first and second stoichiometric points respectively. According to the volume of HCl standard solution consumed when reaching the two stoichiometric points, the composition of the sample can be determined and the content of each component can be calculated.

Add phenolphthalein indicator to the mixed alkali sample. At this time, the solution is red. Titrated with HCl standard solution until the solution changes from red to light pink (refer to the control color bottle provided by the teacher), then the NaOH contained in the sample is completely neutralized and Na_2CO_3 is neutralized to $NaHCO_3$. If the solution contains $NaHCO_3$, it is not titrated. The volume of HCl standard solution consumed by the titration is V_1 (mL). The reaction is as follows:

$$NaOH + HCl = NaCl + H_2O \qquad (3.7)$$
$$Na_2CO_3 + HCl = NaCl + NaHCO_3 \qquad (3.8)$$

Then add methyl orange indicator and continue titration with HCl standard solution until the color of the sample changes from yellow to orange. At this time, the $NaHCO_3$ in the sample solution (either produced from the Na_2CO_3 by the first neutralization, or in the original sample) is neutralized to produce CO_2 and H_2O. At this time, the volume of HCl standard solution consumed (that is, consumed from the first metering point to the second metering point) is V_2 (mL). The reaction is as follows:

$$NaHCO_3 + HCl = NaCl + CO_2 \uparrow + H_2O \qquad (3.9)$$

When $V_1 > V_2$, the sample is a mixture of Na_2CO_3 and NaOH. The HCl needed to neutralize Na_2CO_3 is added in two batches. The two dosages should be equal, that is, the volume of HCl consumed by titrating Na_2CO_3 is $2V_2$. The volume of HCl consumed by neutralizing NaOH is $(V_1 - V_2)$, so the formula for calculating the content of NaOH and Na_2CO_3 should be as follows:

$$c(NaOH) = \frac{(V_1 - V_2)c_{HCl}M_{NaOH}}{1000m_s} \times 100\% \qquad (3.10)$$

$$c(Na_2CO_3) = \frac{V_2 c_{HCl} M_{Na_2CO_3}}{1000m_s} \times 100\% \qquad (3.11)$$

When $V_1 < V_2$, the sample is a mixture of Na_2CO_3 and $NaHCO_3$. In this situation, V_1 is the volume of HCl consumed in the first step of neutralizing Na_2CO_3, so the total volume of HCl consumed in the two-step neutralization of Na_2CO_3 is $2V_1$. The volume of HCl consumed by neutralizing $NaHCO_3$ is $(V_2 - V_1)$, and the formula for calculating the content of $NaHCO_3$ and Na_2CO_3 should be as follows:

$$c(NaHCO_3) = \frac{(V_2 - V_1) c_{HCl} M_{NaHCO_3}}{1000 m_s} \times 100\% \qquad (3.12)$$

$$c(Na_2CO_3) = \frac{V_1 c_{HCl} M_{Na_2CO_3}}{1000 m_s} \times 100\% \qquad (3.13)$$

Where, m_s is the mass of the mixed alkali sample (g).

Materials

Instruments: analytical balance, burette, Erlenmeyer flask, beaker, weighing flask.

Reagents: 0.1 mol/L HCl standard solution, 0.2% phenolphthalein ethanol solution, 0.2% methyl orange aqueous solution, NaOH and Na_2CO_3 or Na_2CO_3 and $NaHCO_3$ mixed samples, reference material Na_2CO_3, mixed alkali sample, the reference solution with pH=8.3: 0.05 mol/L $Na_2B_4O_7$ solution and 0.1 mol/L HCl standard solution, prepare a buffer solution in a ratio of 6 : 4, add phenolphthalein indicator, and place it in a ground flask. The solution is light red.

Procedures

(1) Prepare 0.1 mol/L HCl standard solution (Take Na_2CO_3 as primary standard substance).

(2) Determination of mixed alkali (Double indicator method)　Accurately weigh 2.0000 g sample into a 250 mL beaker, add water to dissolve it, quantitatively transfer it into a 250 mL volumetric flask, dilute it with water to the mark, and shake well.

Pipette 3 parts of the above solution into an Erlenmeyer flask in parallel, each 25.00 mL. Add 3 drops of phenolphthalein indicator and titrate with HCl solution until the solution turns light pink. When approaching the end point, take the reference color bottle as a reference, slowly add the HCl solution, and shake it sufficiently for each drop, until the color of the solution is the same as that of the reference solution. Record the volume V_1 (mL) of the consumed HCl standard solution.

Then add 2 drops of methyl orange indicator to the above solution, and continue titrating with HCl solution until the solution changes from yellow to orange. When approaching the end point, the test solution should be shaken vigorously to avoid the formation of CO_2 supersaturated solution and bring forward the end point. Record the consumed HCl volume V_2 (mL).

Measure in parallel for 3 times, judge the composition of the sample according to the consumption of HCl volume V_1 and V_2, calculate the percentage content (%) of Na_2CO_3 and $NaHCO_3$ or Na_2CO_3 and NaOH, and calculate the total alkali content of the sample expressed as Na_2O (g/L).

Data Recording and Processing

(1) Standardization of 0.1 mol/L HCl solution (Table 3.11)

Table 3.11 Standardization of 0.1 mol/L HCl solution

No.	1	2	3
m(primary standard substance)/g			
V_{HCl}(initial)/mL			
V_{HCl}(final)/mL			
V_{HCl}(consumed)/mL			
c_{HCl}/(mol/L)			
$\bar{c}_{HCl}$/(mol/L)			
$|d_i|$			
$\bar{d}_r$/%			

(2) Determination of mixed alkali samples (Table 3.12)

Table 3.12 Determination of mixed alkali samples

No.	1	2	3
m_{Mix}/g			
V_1(initial)/mL			
V_1(final)/mL			
V_1(consumed)/mL			
V_2(initial)/mL			
V_2(final)/mL			
V_2(consumed)/mL			
Judge the composition of mixed alkali	A	B	
w_A/%			
$\bar{w}_A$/%			
$\bar{d}_r$/%			
w_B/%			
$\bar{w}_B$/%			
$\bar{d}_r$/%			
w_{Na_2O}/%			
$\bar{w}_{Na_2O}$/%			
$\bar{d}_r$/%			

Questions

(1) In the same solution, use the double indicator method to measure the mixed alkali, try to judge the following five cases, what is the composition of the mixed alkali?
① $V_1=0$; ② $V_2=0$; ③ $V_1>V_2$; ④ $V_1<V_2$; ⑤ $V_1=V_2$

(2) Calculate pH at the two stoichiometric points.

(3) Before the first stoichiometric point in the titration of mixed bases, if the titration speed is too fast and the Erlenmeyer flask is not shaken enough, what will be the impact on

the determination? Why?

(4) Briefly describe the advantages and disadvantages of the double indicator method.

Experiment 3.8 Preparation and Standardization of EDTA Solution

 Objectives

(1) Master the principles and characteristics of coordination titration.

(2) Master the basic principles and methods of standardizing EDTA.

(3) Familiar with the use of metal indicators and end point judgment.

 Principles

Ethylenediamine tetraacetic acid (EDTA) is insoluble in water, and its standard solution is usually prepared by indirect method with its disodium salt ($Na_2H_2Y \cdot 2H_2O$, $M = 392.28$). The reference materials for standardizing EDTA solution include Zn, ZnO, $CaCO_3$, Cu, $MgSO_4 \cdot 7H_2O$, Hg, Ni, Pb, etc.

The EDTA solution used to determine the content of Pb^{2+} (or Zn^{2+}) and Bi^{3+} can be standardized with ZnO or metal Zn as the reference material. Xylenol orange was used as an indicator. In the solution of pH=5-6, the xylenol orange indicator (XO) itself was yellow, while the complex with Zn^{2+} was purple-red. EDTA can form a more stable complex with Zn^{2+}. When the EDTA solution is used to titrate to the near end point, EDTA will displace Zn^{2+} complexed with xylenol orange, and make xylenol orange free. Therefore, the solution changes from purple-red to yellow. The discoloration principle can be expressed as follows:

$$XO(yellow) + Zn^{2+} = Zn\text{-}XO(purple\text{-}red) \tag{3.14}$$

$$Zn\text{-}XO(purple\text{-}red) + EDTA = Zn\text{-}EDTA(colorless) + XO(yellow) \tag{3.15}$$

If the EDTA solution is used to determine the content of CaO and MgO in limestone or dolomite and to determine the hardness of water, it is best to use $CaCO_3$ as the reference material for standardization. Just like this, the reference substance and the sample to be tested contain the same composition, so that the titration conditions are consistent and system errors can be reduced. After $CaCO_3$ is dissolved in HCl to make a standard solution of Ca, acidity is adjusted to pH$\geqslant$12, calcium indicator is used, and EDTA is used to titrate the solution until the color changes from purple-red to pure blue.

 Materials

Instruments: analytical balance, burette, Erlenmeyer flask, volumetric flask, beaker, weighing flask.

Reagents: disodium EDTA ($Na_2H_2Y \cdot 2H_2O$), metal Zn (A.R.), XO (0.2% aqueous solution), 1:1 hydrochloric acid, 1:1 ammonia water, primary standard substance $CaCO_3$, chrome black T indicator (1% ethanol solution); Ammonia buffer solution (pH=10): dissolve 20g NH_4Cl in a small amount of water, add 100 mL ammonia water, dilute to 1 L with water; K-B indicator: mix 0.2 g acid chrome blue K and 0.4 g naphthol green B in 100 mL water; 20% hexamethylenetetramine (pH=5.5): dissolve 20 g of reagent in water, add 4 mL of concentrated HCl, and dilute to 100 mL.

 Procedures

(1) Preparation of 0.02 mol/L EDTA standard solution   2 g ethylenediamine tetraacetic acid disodium salt was weighed in a 250 mL beaker, 10 mL water was added, heated and dissolved, diluted to 250 mL, and stored in a reagent bottle or polyethylene plastic bottle.

(2) Preparation of standard zinc solution   Accurately weigh (　) to (　) g of pure zinc in a 250 mL beaker and cover with a watch glass. Add 5-10 mL of 1∶1 HCl from the mouth of the beaker, put it aside until Zn is completely dissolved, quantitatively transfer it to a 250 mL volumetric flask, dilute with water to the mark, and shake well. Calculate its exact concentration $c_{Zn^{2+}}$.

(3) Preparation of $CaCO_3$ standard solution   Accurately weigh the $CaCO_3$ (　) to (　) g dried at 120 ℃ in a small beaker. Wet with a small amount of water, cover with a watch glass, drop 10 mL of 1∶1 HCl from the mouth of the beaker, after the $CaCO_3$ is completely dissolved, quantitatively transfer it to a 250 mL volumetric flask, dilute to the mark, and shake well. Calculate its exact concentration $c_{Ca^{2+}}$.

(4) Standardization of EDTA solution concentration

① Standardize EDTA solution with zinc standard solution

a. Use chrome black T indicator. Pipette 25.00 mL of zinc standard solution into a 250 mL Erlenmeyer flask, and add 1∶1 ammonia water dropwise until white precipitate begins to appear. Add 10mL ammonia buffer solution (pH=10), add 20mL water, and add 2-3 drops of chrome black T indicator. Titrate with EDTA standard solution until the solution changes from purple-red to pure blue, that is, the end point. Measure 3 copies in parallel, and the difference in titration volume does not exceed 0.04 mL each time. Calculate the concentration of c_{EDTA} based on the volume of EDTA standard solution consumed.

b. Xylenol orange was used as indicator. Pipette 25.00 mL of zinc standard solution into a 250 mL Erlenmeyer flask, add 2 to 3 drops of 0.5% xylenol orange indicator, then add 20% hexamethyltetramine until the solution is a stable magenta color, and add an additional 5 mL. The EDTA standard solution was titrated until the solution changed from purple-red to bright yellow, and that is the end point. Measure 3 copies in parallel, and the difference in titration volume does not exceed 0.04 mL each time. Calculate the concentration of EDTA based on the volume of EDTA standard solution consumed.

② EDTA solution was calibrated with $CaCO_3$ standard solution. Pipette 25.00 mL of $CaCO_3$ standard solution into a 250 mL Erlenmeyer flask, add 10 mL ammonia buffer solution (pH=10) and 3 drops of K-B indicator, and titrate with EDTA standard solution until the solution turns from purple-red to blue. Measure 3 copies in paraller. The difference in titration volume does not exceed 0.04 mL each time. Calculate the concentration of c_{EDTA} based on the volume of EDTA standard solution consumed.

Data Recording and Processing

(1) Standardization of 0.02 mol/L EDTA solution ($CaCO_3$ as primary standard substance, see Table 3.13)

Table 3.13　Standardization of 0.02 mol/L EDTA solution (CaCO₃)

No.	1	2	3
m_{CaCO_3}/g			
V_{EDTA}(initial)/mL			
V_{EDTA}(final)/mL			
V_{EDTA}(consumed)/mL			
c_{EDTA}/(mol/L)			
$\bar{c}_{EDTA}$/(mol/L)			
$|d_i|$			
$\bar{d}_r$/%			

(2) Standardization of 0.02 mol/L EDTA solution (Zn as primary standard substance, see Table 3.14)

Table 3.14　Standardization of 0.02 mol/L EDTA solution (Zn)

No.	1	2	3
m_{Zn}/g			
V_{EDTA}(initial)/mL			
V_{EDTA}(final)/mL			
V_{EDTA}(consumed)/mL			
c_{EDTA}/(mol/L)			
$\bar{c}_{EDTA}$/(mol/L)			
$|d_i|$			
$\bar{d}_r$/%			

Questions

(1) What are the differences between the preparation methods of EDTA standard solution and Zn standard solution?

(2) What should be paid attention to when preparing Zn standard solution?

(3) Using Zn as reference and xylenol orange as indicator to calibrate the concentration of EDTA solution, what range should the acidity of the solution be controlled in? How to control it? What if the solution is more acidic?

(4) When CaCO₃ is used as the reference and the calcium indicator is used to indicate the end point of EDTA calibration, how much acidity of the solution should be controlled? Why is that? How to control it?

(5) Why should a buffer solution be added to coordination titration?

Experiment 3.9　Determination of Total Hardness of Tap Water

Objectives

(1) Master the principle and method for the determination of water hardness by the EDTA method.

(2) Learn the expression method of water hardness and the calculation method of common hardness.

Principles

The hardness of water is mainly the presence of calcium salts and magnesium salts in water, and other metal ions such as iron, aluminum, manganese, zinc also contribute to hardness, but the general contents of them are very small, and can be ignored in the measurement of the total hardness of industrial water.

There are two types of hardness: temporary hardness and permanent hardness. Calcium salts present in the form of bicarbonate, form carbonate precipitates and lose their hardness when heated. The hardness formed by these salts is called temporary hardness, and the reaction is as follows:

$$Ca(HCO_3)_2 \xrightarrow{\Delta} CaCO_3 \downarrow + CO_2 \uparrow + H_2O \tag{3.16}$$

$$Mg(HCO_3)_2 \xrightarrow{\Delta} MgCO_3 (\text{incompletely precipitate}) + CO_2 \uparrow + H_2O \tag{3.17}$$
$$\phantom{Mg(HCO_3)_2 \xrightarrow{\Delta}} \Big\downarrow {+H_2O}$$
$$\phantom{Mg(HCO_3)_2 \xrightarrow{\Delta}} Mg(OH)_2 \downarrow + CO_2 \uparrow$$

The hardness of sulfates, chlorides, nitrates, and other forms of calcium and magnesium in water is called permanent hardness, and they do not precipitate during heating (but at the operating temperature of the boiler, the solubility decreases and they precipitate, which is called boiler scale).

The sum of temporary hardness and permanent hardness is called "total hardness", the hardness formed by magnesium ions is called "magnesium hardness", and the hardness formed by calcium ions is called "calcium hardness".

The total hardness of water is often measured by complexometric titration, in the buffer solution of NH_3-NH_4Cl at pH=10, with chrome black T (EBT) or K-B as the indicator. Titrate with EDTA standard solution until the solution changes from purple to pure blue, which reachs the end point.

Masking of interfering ions: If there are trace impurities such as Fe^{3+} and Al^{3+} in the water sample, they can be masked by triethanolamine, and heavy metal ions such as Cu^{2+}, Pb^{2+}, Zn^{2+} can be masked by Na_2S or KCN.

Measure the hardness of calcium and magnesium respectively: The pH can be controlled between 12 and 13 (at this time, magnesium hydroxide is precipitated), and calcium indicator is used for determination. Magnesium hardness can be calculated by subtracting calcium hardness from total hardness.

There are many ways to express the hardness of water. Degree (°) is commonly used in China to represent it. 1° indicates that 100000 parts of water contain one part of CaO, which means that 1L of water contains 10mg of CaO, i.e. 1°＝10mg/L.

Materials

Equipments: Analytical balances, burettes, pipettes, Erlenmeyer flasks, weighing flasks, beakers.

Reagents: 0.02 mol/L EDTA solution; hydrochloric acid (1:1) solution; triethanolamine (1:1) solution; NH_3-NH_4Cl buffer solution (pH＝10); chrome black T indicator (0.5%); $CaCO_3$ (primary standard substance); K-B indicator.

Procedures

(1) Preparation and standardization of 0.02 mol/L EDTA standard solution

① Preparation: Weigh 2 g of EDTA in a beaker on a balance, dissolve it by heating with a small amount of water, transfer it to a 250 mL polyethylene plastic bottle after cooling and dilute to 250 mL with deionized water.

② Standardization: Accurately weigh 0.25 g of primary standard $CaCO_3$, place it in a 100 mL beaker, wet it with a small amount of water, cover the beaker with a watch glass, and slowly add 1:1 HCl 5 mL dropwise into the beaker. After it is completely dissolved, add 50 mL of deionized water, boil for a few minutes to remove CO_2, then rinse the watch glass and the inner wall of the beaker with a small amount of water after cooling. Quantitatively transfer it into a 250 mL volumetric flask, dilute to the mark with water, and shake well. Pipette 25.00 mL of Ca^{2+} standard solution into a 250 mL Erlenmeyer flask (add 1 drop of methyl red, neutralize it with ammonia water until the solution turns from red to yellow, if the ammonia buffer solution has enough buffer capacity, this step can be omitted). Add 20 mL water and 5 mL of Mg^{2+}-EDTA solution, and then add 10 mL of ammonia buffer solution, 3 drops of chrome black T indicator, immediately titrate with the EDTA solution to be standardized until the solution changes from purple red (wine red) to pure blue (purple blue), which is the end point. Parallel standardize 3 times to calculate the exact concentration of EDTA solution.

(2) Determination of total hardness of tap water Pipette 100.00 mL of water sample into a 250 mL Erlenmeyer flask, add 1-2 drops of 1:1 HCl and slightly boil for a few minutes to remove CO_2. After cooling, add 3 mL of 1:1 triethanolamine (if the water sample contains heavy metal ions, add 1 mL 2% Na_2S solution for masking), 10 mL ammonia buffer solution, 2-3 drops of K-B indicator. Titrate with EDTA standard solution until the solution changes from purple to pure blue, which is the end point. Note that it should be slowly dripped and shaken as it approaches the end point. Three parallel measurements are made to calculate the total hardness of water, and the analysis results are expressed in (°).

Notes:

(1) Chrome black T has high color sensitivity with Mg^{2+}, but low color sensitivity with Ca^{2+}. When the Ca^{2+} content in the water sample is high and Mg^{2+} is very low, an insensitive end point is obtained, and K-B mixed indicator can be used to replace it.

(2) When the iron content in the water sample exceeds 10 mg/mL., it is difficult to

mask with triethanolamine. It is necessary to dilute the water sample with distilled water until Fe^{3+} does not exceed 10 mg/mL.

Data Recording and Processing

(1) Standardization of 0.02 mol/L EDTA solution ($CaCO_3$) (see Table 3.15)

Table 3.15 Standardization of 0.02 mol/L EDTA solution ($CaCO_3$)

No.	1	2	3
m_{CaCO_3}/g			
V_{EDTA}(initial)/mL			
V_{EDTA}(final)/mL			
V_{EDTA}(consumed)/mL			
$c_{EDTA}/(mol/L)$			
$\bar{c}_{EDTA}/(mol/L)$			
$\lvert d_i \rvert$			
$\bar{d}_r/\%$			

(2) Determination of total hardness of water (Table 3.16)

Table 3.16 Determination of total hardness of water

No.	1	2	3
$V_{\text{water sample}}/mL$			
V_{EDTA}(initial)/mL			
V_{EDTA}(final)/mL			
V_{EDTA}(consumed)/mL			
$H/(°)$			
$\bar{H}/(°)$			
$\lvert d_i \rvert$			
$\bar{d}_r/\%$			

Questions

(1) When $CaCO_3$ solution and EDTA solution are prepared, what kind of balance should be used? Why?

(2) How does the chrome black T indicator indicate the end point of the titration?

(3) When the primary standard $CaCO_3$ is dissolved in HCl solution, what should be paid attention to in the operation?

(4) Why should buffer solution be added in complexometric titration?

(5) When the hardness of water is measured by EDTA method, which ions interfere? How to eliminate them?

(6) What are the differences between complexometric titration and acid-base titration? What should be paid attention to in the operation?

Experiment 3.10 Determination of Aluminum Content in Industrial Aluminum Sulfate

 Objectives

(1) Understand the principle of the back titration and the displacement titration method.

(2) Master the principle and method of determination of aluminum in industrial aluminium sulfate.

 Principles

As an inorganic raw material, industrial aluminium sulfate is widely used in many fields such as food, pharmacy, papermaking, water treatment, etc. Al^{3+} is easy to form a series of polynuclear hydroxyl complexes, which are complexd with EDTA slowly, and so back titration method is usually used to determine aluminum. Firstly, add a quantitative and excess EDTA standard solution, and boil it for a few minutes at pH≈3.5, so that the Al^{3+} and EDTA are fully coordinated. At pH=5-6, with xylenol orange as an indicator, excess EDTA was back titrated with Zn^{2+} salt standard solution to obtain the content of aluminum.

However, the selectivity of back titration for the determination of aluminum is poor, and all ions that can form stable complexes with EDTA affect the detection. For aluminum in complex samples such as alloys, silicates, cement and slag, displacement titration is often used to improve selectivity. After back-titrating the excess EDTA with Zn^{2+}, add excess NH_4F and heat it to boiling. Perform the displacement reaction between AlY^- and F^-, release H_2Y^{2-} (EDTA) that is equal to the amount of Al^{3+}, and the aluminum content is obtained by titrating the released EDTA with a standard solution of Zn^{2+} salt.

Aluminum content is determined by displacement titration. If the sample contains Ti^{4+}, Zr^{4+}, Sn^{4+} ions, the same displacement reaction as Al^{3+} will occur and interfere with the determination of Al^{3+}. At this time, the method of masking should be adopted to mask the above-mentioned interfering ions. For example, mask Ti^{4+} with mandelic acid etc. The impurities contained in aluminum alloys mainly include Si, Mg, Cu, Mn, Fe, Zn, and some also contain Ti, Ni, Ca, etc., which are usually dissolved in HNO_3-HCl mixed acid, or decomposed in silver crucible or plastic beaker with $NaOH$-H_2O_2 and then acidified with HNO_3.

 Materials

Equipments: Analytical balance, burette, pipette, Erlenmeyer flask, weighing flask, beaker.

Reagents: 20% NaOH solution, 1∶1 HCl solution, 1∶3 HCl solution, 0.02 mol/L EDTA solution, 0.2% xylenol orange solution, 1∶1 ammonia water, 20% hexamethylenetetramine solution, 0.02 mol/L Zn^{2+} standard solution, 20% NH_4F solution (stored in a plastic bottle).

 Procedures

(1) Standardization of 0.02 mol/L EDTA solution (see experiment 3.8, standardize

with Zn^{2+} standard solution).

(2) Sample handling   Accurately weigh 0.10-0.11 g industrial aluminium sulfate into a 50 mL beaker, add 10 mL NaOH, dissolve it completely in a boiling water bath. After a little cooling, add 1∶1 HCl solution until flocculent precipitation occurs, then add 10mL 1∶1 HCl solution, quantitatively transfer the solution to a 250 mL volumetric flask, add water to the mark, and shake well.

(3) Determination of aluminum content in samples   Accurately pipette 25.00 mL of the above solution into a 250 mL Erlenmeyer flask, add 30.00 mL EDTA and two drops of xylenol orange. At this time, the solution is yellow. Add ammonia water until the solution turns purple, and then add 1∶3 HCl solution, making the solution yellow. After boiling for 3 minutes and cooling, add 20 mL of hexamethylenetetramine. At this time, the solution should be yellow. If the solution is red, add 1∶3 HCl solution dropwise to make it yellow. The Zn^{2+} standard solution is used for titration until the solution changes from yellow to purplish red, which is the endpoint.

Add 10mL NH_4F to the above solution, heat until it's slightly boiling, cool with running water, and then add 2 drops of xylenol orange, at this time, the solution is yellow. If it is red, add 1∶3 HCl solution dropwise to make it become yellow. Then titrate with the Zn^{2+} standard solution. When the solution changes from yellow to purple-red, it is the end point. Calculate the mass fraction of Al according to the volume consumed by this Zn^{2+} standard solution.

Data Recording and Processing

(1) Standardization of 0.02 mol/L EDTA solution (Table 3.17)

Table 3.17   Standardization of EDTA solution

No.	1	2	3
m_{Zn}/g			
V_{EDTA}(initial)/mL			
V_{EDTA}(final)/mL			
V_{EDTA}(consumed)/mL			
$c_{EDTA}/(mol/L)$			
$\bar{c}_{EDTA}/(mol/L)$			
$\lvert d_i \rvert$			
$\bar{d}_r/\%$			

(2) Determination of aluminum content in aluminum alloys (Table 3.18)

Table 3.18   Determination of aluminum content in aluminum alloys

No.	1	2	3
m_s/g			
V_{EDTA}/mL			

(continued)

No.	1	2	3
$V_{Zn^{2+}}$ (initial)/mL			
$V_{Zn^{2+}}$ (final)/mL			
$V_{Zn^{2+}}$ (consumed)/mL			
$w_{Al}/\%$			
$\overline{w}_{Al}/\%$			
$|d_i|$			
$\overline{d}_r/\%$			

Questions

(1) Why is back titration sufficient for the determination of Al^{3+} in simple samples, but displacement titration is required for the determination of Al^{3+} in complex samples?

(2) When back titration is used to determine Al^{3+} in a simple sample, must the concentration of excess EDTA solution added be accurate? Why?

(3) Should the EDTA solution used in this experiment be standardized?

(4) Why would you add excess EDTA? Is it possible to ignore the volume consumed when titrating with Zn^{2+} standard solution for the first time? Is it necessary to accurately titrate the solution from yellow to purple at this time? Why?

Experiment 3.11 Determination of Zn^{2+} and Bi^{3+} in Mixed Solution by Continuous Titration

Objectives

(1) Master principle and method for determining multi ions by controlling the acidity in continuous titration.

(2) Be familiar with the application of xylenol orange.

Principles

If two ions M and N can be determined via titration, there must be premises, such as $\lg(c_M K'_{MY}) \geqslant 5$, $\lg(c_N K'_{NY}) \geqslant 5$, $\Delta\lg(cK') \geqslant 5$ ($\Delta pM' = 0.2, E_t \leqslant 0.3\%$). The suitable pH range for determination of M can be derived from the following formulas.

The highest acidity: $\alpha_{Y(H)} \approx \alpha_{Y(N)} = 1 + K_{NY}[N]$ (3.18)

The lowest acidity: $[OH^-] = n\sqrt{\dfrac{K_{sp}}{c_M}}$ (3.19)

Both Zn^{2+} and Bi^{3+} can react with EDTA and form stable complexes with different formation constants ($\lg K_{ZnY} = 16.50$, $\lg K_{BiY} = 27.94$). The difference between their formation constants is higher than 5, i.e., $\Delta\lg(cK) > 5$, their contents can be determined respectively by controlling the acidity of solution.

Bi^{3+} react with xylenol orange (XO) and form purple-red complex in strong acid condition, e.g., pH=1. Under this condition, Zn^{2+} can't form stable complex with XO and thus hardly interfere with the analysis of Bi^{3+}. The conditional formation constant of Bi^{3+}-XO is 9.6 and Bi^{3+} can be determined accurately. After that, adjust the pH of solution by adding hexamethylenetetramine to 5-6, Zn^{2+} reacts with XO and forms purple-red complex. Titrate the solution with EDTA till it turns light yellow, which is the end point of Zn^{2+}.

Reactions at pH=1:

Before titration $Bi^{3+} + H_3In^{4-}$ (yellow) ⟶ BiH_3In^- (purple-red)

Before stoichiometric point $Bi^{3+} + H_2Y^{2-}$ ⟶ $BiY^- + 2H^+$

At the stoichiometric point $H_2Y^{2-} + BiH_3In^-$ (purple-red) ⟶ $BiY^- + H_3In^{4-}$ (yellow) $+ 2H^+$

Reactions at pH=5-6:

Before titration $Pb^{2+} + H_3In^{4-}$ (yellow) ⟶ PbH_3In^{2-} (purple-red)

Before stoichiometric point $Pb^{2+} + H_2Y^{2-}$ ⟶ $PbY^{2-} + 2H^+$

At the stoichiometric point $H_2Y^{2-} + PbH_3In^{2-}$ (purple-red) ⟶ $PbY^{2-} + H_3In^{4-}$ (yellow) $+ 2H^+$

 Materials

Instruments: balance, burette (50 mL), pipette, volumetric flask (250 mL), conical flask, weighing bottle, cylinder, beaker.

Reagents: ① 0.02 mol/L EDTA; ② standard ZnO; ③ xylenol orange solution (0.2%); ④ 1∶1 HCl; ⑤ hexamethylenetetramine (20%); ⑥ 1∶1 ammonium solution; ⑦ mixed solution of Zn^{2+} and Bi^{3+}.

 Procedures

(1) Preparation of EDTA (0.02 mol/L) $Na_2H_2Y \cdot 2H_2O$ (4 g) is dissolved in 100 mL of deionized water. If it dissolves slowly, heat the solution. Then, the solution is diluted to 500 mL and mixed homogenously before usage. Keep the solution in a glass bottle or polyethylene bottle for long-term storage.

(2) Calibration of EDTA (0.02 mol/L) ( )-( ) g of standard ZnO is transferred into a 250 mL beaker, and then the beaker was covered by a watch glass. After that, 5-10 mL of 1∶1 HCl was added in drop by drop from the gap between beaker and watch glass. Then transfer the solution into a volumetric flask (250 mL) if ZnO is dissolved completely, and the solution is diluted till the liquid level reaches the markline. After mixed homogenously, a standard Zn^{2+} solution is prepared and the accurate concentration of Zn^{2+} [c_{Zn}/(mol/L)] can be derived.

Then, 25.00 mL of the standard Zn^{2+} solution is transferred into a conical flask (250 mL), followed by the addition of 2 or 3 drops of XO solution (0.2%) and hexamethylenetetramine solution (20%) till the solution turns purple-red. Another 5 mL of hexamethylenetetramine solution is added. Subsequently, titration using EDTA is conducted till the solution becomes light yellow, which is the end point. The titration should be repeated 3 times, and the difference of volumes of EDTA should be less than 0.04 mL. The average concentration of EDTA can be derived and used in next steps.

(3) Determination of Zn^{2+} and Bi^{3+} in mixed solution Mixed solution of Zn^{2+} and Bi^{3+} (25.00 mL) is transferred into a conical flask (250 mL), followed by the addition of 1 or 2 drops of XO solution (0.2%). Then, titration by using EDTA is performed till the solution turns yellow from purple-red. Three replicates are conducted and the concentration of Bi^{3+} (g/L) can be derived from the volume of consumed EDTA.

After the determination of Bi^{3+}, 2 or 3 drops of XO solution are added in the above solution, followed by the addition of ammonia (1 : 1) drop by drop to make the solution orange. The pH value is about 5, which can be verified by using pH test paper. Then, hexamethylenetetramine solution (20%) is added in till the solution is purple-red and another 5 mL of hexamethylenetetramine is added in to make the pH value stable. Titration is then started till the solution becomes light yellow. Three replicates are conducted and the concentration of Bi^{3+} (g/L) can be derived from the volume of consumed EDTA.

Notes:

① $Bi(NO_3)_3$ doesn't precipitate and XO doesn't react with Zn^{2+} at pH = 1. XO doesn't react with Bi^{3+} if the acidity is too high and the solution is yellow.

② The reaction between Bi^{3+} and EDTA is very slow. Therefore, the titration speed should be controlled slowly and the mixed solution needs fierce vibration to be homogenous quickly.

Data Recording and Processing

(1) Standardization of EDTA (0.02 mol/L) (Table 3.19)

Table 3.19 Standardization of EDTA (0.02 mol/L)

Number	1	2	3
m_{Zn}/g			
V_{EDTA}(initial)/mL			
V_{EDTA}(final)/mL			
V_{EDTA}(consumed)/mL			
c_{EDTA}/(mol/L)			
$\bar{c}_{EDTA}$/(mol/L)			
$|d_i|$			
$\bar{d}_r$/%			

(2) Continuous titration of mixed solution (Table 3.20)

Table 3.20 Continuous titration of mixed solution

Number	1	2	3
Determination of Bi^{3+} at pH=1.0			
V_1(initial)/mL			
V_1(final)/mL			
V_1(consumed)/mL			
$w_{Bi^{3+}}$/%			
$\bar{w}_{Bi^{3+}}$/%			

(continued)

Number	1	2	3
$\lvert d_i \rvert$			
$\overline{d}_r/\%$			
Determination of Zn^{2+} at pH=5.0			
V_2(initial)/mL			
V_2(finial)/mL			
V_2(consumed)/mL			
$w_{Zn^{2+}}/\%$			
$\overline{w}_{Zn^{2+}}/\%$			
$\lvert d_i \rvert$			
$\overline{d}_r/\%$			

Questions

Continuous titration

(1) Is it practicable to determine Zn^{2+} first and determine Bi^{3+} subsequently?

(2) If there is Fe^{3+}, ascorbic acid is usually used as masking agent. Can triethanolamine be used as an alternative to ascorbic acid?

(3) Why the co-existed Zn^{2+} interferes hardly with the determination of Bi^{3+} under strong acid circumstance, e.g., pH≈1?

(4) Is it practicable to determine respectively the content of Bi^{3+} (controlling pH value around 1) and total content of Zn^{2+} and Bi^{3+} (controlling pH value ranges 5-6) in two identical sample solutions? Why?

Experiment 3.12　Preparation and Standardization of Potassium Permanganate Solution

Objectives

(1) Correctly prepare a relatively stable $KMnO_4$ standard solution.

(2) Master the principle and method of using $Na_2C_2O_4$ to calibrate $KMnO_4$ solution.

Principles

Potassium permanganate is a strong oxidant, and its oxidizing ability and reduction products have a great relationship with the acidity of the solution.

In strongly acidic solutions:

$$MnO_4^- + 5e + 8H^+ \Longrightarrow Mn^{2+} + 4H_2O \quad E^\ominus = 1.51V \quad (3.20)$$

In weakly acidic, neutral and weakly alkaline solutions:

$$MnO_4^- + 3e + 2H_2O \Longrightarrow MnO_2 + 4OH^- \quad E^\ominus = 0.59V \quad (3.21)$$

In strongly alkaline solutions:

$$MnO_4^- + e \Longrightarrow MnO_4^{2-} \quad E^\ominus = 0.564V \quad (3.22)$$

When applying the potassium permanganate method, different methods can be used according to the properties of the tested substances.

The purity of commercially available $KMnO_4$ reagent is generally about 99%-99.5%, which often contains a small amount of MnO_2 and other impurities. Distilled water often contains trace amounts of reduced organic substances, which can react with MnO_4^- to precipitate of $MnO(OH)_2$. These products and changes of external conditions such as light, heat, acid, and alkali will promote the further decomposition of $KMnO_4$. Therefore, $KMnO_4$ standard solution cannot be obtained by direct weighing method.

In order to obtain a relatively stable $KMnO_4$ solution, it must be prepared as described in the experimental procedure. There are many reference materials for standardizing $KMnO_4$ solution, such as $Na_2C_2O_4$, $H_2C_2O_4 \cdot 2H_2O$, As_2O_3, $(NH_4)_2Fe(SO_4)_2$ and pure iron wire. Among them, $Na_2C_2O_4$ is more commonly used.

In H_2SO_4 solution, the reaction of MnO_4^- and $C_2O_4^{2-}$ is as follows:

$$2MnO_4^- + 5C_2O_4^{2-} + 16H^+ = 2Mn^{2+} + 10CO_2 \uparrow + 8H_2O \qquad (3.23)$$

In order for this reaction to proceed quantitatively and rapidly, the following titration conditions should be noted:

(1) Temperature The reaction rate is slow at room temperature, and the solution is often heated to about 70-80 ℃ and titrated while hot. The temperature should not be lower than 60 ℃ when the titration is completed. However, the temperature should not be too high. If it is higher than 90 ℃, the $H_2C_2O_4$ will be partially decomposed, resulting in a high standardization result.

$$H_2C_2O_4 = CO_2 \uparrow + CO \uparrow + H_2O \qquad (3.24)$$

(2) Acidity If the acidity is too low, MnO_4^- will be partially reduced to MnO_2; if the acidity is too high, $H_2C_2O_4$ will be decomposed. Generally, the acidity at the beginning of the titration should be controlled at 0.5-1 mol/L. In order to prevent the reaction that induces the oxidation of Cl^-, it should be carried out in H_2SO_4 medium.

(3) Titration speed At the beginning of the titration, the reaction rate of MnO_4^- and $C_2O_4^{2-}$ is very slow, and the dripped $KMnO_4$ fades slowly. Therefore, the titration speed should not be too fast at the beginning of the titration, otherwise the dropwise $KMnO_4$ solution will not have time to react with $C_2O_4^{2-}$, and it will decompose in the hot acidic solution, resulting in a low calibration result.

$$4MnO_4^- + 12H^+ = 4Mn^{2+} + 5O_2 + 6H_2O \qquad (3.25)$$

(4) Catalyst At the beginning of the titration, a few drops of $KMnO_4$ solution faded slowly, but after the action of these drops of $KMnO_4$ and $Na_2C_2O_4$ was completed, the titration product Mn^{2+} was formed, and the reaction rate gradually accelerated. If a few drops of $MnSO_4$ solution are added before the titration, the reaction speed will be very fast at the beginning of the titration. It can be seen that Mn^{2+} acts as a catalyst in this reaction.

(5) Indicator Because $KMnO_4$ has its own color, a slight excess of MnO_4^- in the solution can show pink color. Generally it is not necessary to add another indicator, $KMnO_4$ can be its own indicator.

(6) Titration end point After titration with $KMnO_4$ solution to the end point, the pink color that appeared in the solution could not be maintained. This is because the reducing

gas and dust in the air can slowly reduce MnO_4^-, so the pink color of the solution gradually disappeared. Therefore, if the pink color in the solution during titration does not fade within 0.5 to 1 min, it can be considered that the titration end point has been reached.

After the calibrated $KMnO_4$ solution is placed for a period of time, if $MnO(OH)_2$ is found to be precipitated, it should be re-filtered and calibrated.

Materials

Instruments: microporous glass funnel, acid burette, Erlenmeyer flask, weighing bottle, measuring cylinder, beaker, etc.

Reagents: $KMnO_4$ (A.R.), 3 mol/L H_2SO_4 solution, $Na_2C_2O_4$ (dry at 105 ℃ for 2 h for use).

Procedures

(1) Preparation of $KMnO_4$ solution   Weigh a little more than the theoretically calculated amount of $KMnO_4$ solid about ( ) g, dissolve it in 500 mL of water, cover with a watch glass, heat it to boiling, and keep it in a slightly boiling state for about 1 h. Replenish the evaporated water at any time and cool it. After the solution was allowed to stand at room temperature for 2-3 days, it was filtered with a microporous glass funnel. The filtrate was stored in a brown reagent bottle and stored in a dark place for later use.

(2) Calibration of 0.02 mol/L $KMnO_4$ solution (Semi-micro titration)   Accurately weigh ( ) to ( ) g of the reference substance $Na_2C_2O_4$, dissolve the $Na_2C_2O_4$ and transfer the solution to a volumetric flask (25 mL), dilute it until the liquid level reaches scale line. Transfer 5.0 mL of the $Na_2C_2O_4$ solution to a 25 mL Erlenmeyer flask, add 4 mL of water and 3 mL of 3 mol/L H_2SO_4, and heat it in a water bath to about 70-80 ℃ (The solution begins to emit steam). Titrate with 0.02 mol/L $KMnO_4$ solution while it is still hot. After the first drop of $KMnO_4$ is added, the color fades very slowly, and it needs to be shaken vigorously. After Mn^{2+} is produced in the solution, the titration speed can be accelerated until the solution turns reddish. The reddish color lasts for 1 min and does not fade, which is the end point. Repeat the experiment 3 times. Record the volume of $KMnO_4$ solution consumed by the titration.

Calculate the mass concentration of the $KMnO_4$ solution and the relative mean deviation of the results.

Data Recording and Processing

Standardization of 0.02 mol/L $KMnO_4$ solution (Table 3.21)

Table 3.21   Standardization of 0.02 mol/L $KMnO_4$ solution

No.	1	2	3
$m_{Na_2C_2O_4}$/g			
V_{KMnO_4} (initial)/mL			
V_{KMnO_4} (final)/mL			
V_{KMnO_4} (consumed)/mL			

(continued)

No.	1	2	3
c_{KMnO_4}/(mol/L)			
$\bar{c}_{KMnO_4}$/(mol/L)			
$\|d_i\|$			
$\bar{d}_r$/%			

Questions

(1) What should be paid attention to when preparing $KMnO_4$ solution? Why?

(2) When preparing $KMnO_4$ solution, what are the residual products on the filter after filtration? What material should be used to clean it?

(3) The $KMnO_4$ solution should be filtered with a microporous glass funnel during the preparation of the $KMnO_4$ solution. Can it be filtered with quantitative filter paper? Why?

(4) When preparing $KMnO_4$ solution, the $KMnO_4$ solution must be boiled for 1-2 hours. What is the purpose?

(5) What are the reaction conditions for standardizing $KMnO_4$ solution with $Na_2C_2O_4$?

(6) Why is 3 mol/L H_2SO_4 used to control the acidity of the solution?

Experiment 3.13 Determination of H_2O_2 Content in Hydrogen Peroxide Solution

Objectives

(1) Master the standardization method of $KMnO_4$ solution.

(2) Master the principle and method of direct determination of H_2O_2 content by the $KMnO_4$ method.

Principles

Hydrogen peroxide is a widely used disinfectant in the medical and health industry. Its main component H_2O_2 has a peroxy bond —O—O— in its molecule. It is a strong oxidant in acidic solution, but H_2O_2 exhibits reducibility when it encounters $KMnO_4$, which is more oxidative. To determine the content of hydrogen peroxide, $KMnO_4$ is used to oxidize H_2O_2 in diluted sulfuric acid solution at room temperature, and the reaction formula is:

$$5H_2O_2 + 2MnO_4^- + 6H^+ = 2Mn^{2+} + 5O_2 \uparrow + 8H_2O \tag{3.26}$$

The reaction rate is very slow at the beginning, and the first drop of solution is not easy to fade, and the reaction rate accelerates after Mn^{2+} is formed (Mn^{2+} plays an autocatalytic role). When the stoichiometric point is reached, a slight excess of $KMnO_4$ makes the solution reddish and the titration ends.

Materials

Instruments: acid burette, volumetric flask, pipette, erlenmeyer flask, weighing flask, graduated cylinder, etc.

Reagents: $Na_2C_2O_4$ (reference substance), 3 mol/L H_2SO_4 solution, 0.02 mol/L

KMnO$_4$ solution, 3% H$_2$O$_2$ test solution (from diluted commercial H$_2$O$_2$ solution).

Procedures

(1) Prepare 250 mL of 0.02 mol/L KMnO$_4$ solution (Finish a week ahead of schedule, see Exp. 3.12).

(2) Standardization of KMnO$_4$ solution with Na$_2$C$_2$O$_4$ (see Exp. 3.12).

(3) Determination of H$_2$O$_2$ content in hydrogen peroxide (Semi-micro titration)   Accurately remove 1.00 mL of H$_2$O$_2$ test solution with a pipette and place it in a 25 mL volumetric flask, dilute it with water to the scale line, and shake well.

Pipette 3 parts of 5.00 mL of the above test solution, put them in 25 mL Erlenmeyer flask respectively, add 4 mL of water and 2 mL of 3 mol/L H$_2$SO$_4$ solution, and titrate with KMnO$_4$ standard solution until the solution turns reddish and does not fade within 30 s, which is the end point.

Calculate the H$_2$O$_2$ content (g/100 mL) in the sample and the relative mean deviation of result.

Data Recording and Processing

(1) Standardization of 0.02 mol/L KMnO$_4$ solution (Table 3.22)

Table 3.22 Standardization of KMnO$_4$ solution

No.	1	2	3
$m_{Na_2C_2O_4}$/g			
V_{KMnO_4} (initial)/mL			
V_{KMnO_4} (final)/mL			
V_{KMnO_4} (consumed)/mL			
c_{KMnO_4}/(mol/L)			
$\bar{c}_{KMnO_4}$/(mol/L)			
$|d_i|$			
$\bar{d}_r$/%			

(2) Determination of H$_2$O$_2$ content (Table 3.23)

Table 3.23 Determination of H$_2$O$_2$ content

No.	1	2	3
$V_{H_2O_2}$/mL			
V_{KMnO_4} (initial)/mL			
V_{KMnO_4} (final)/mL			
V_{KMnO_4} (consumed)/mL			
$\rho_{H_2O_2}$/(g/100 mL)			

(continued)

No.	1	2	3
$\bar{\rho}_{H_2O_2}$/(g/100 mL)			
$\mid d_i \mid$			
$\bar{d}_r$/%			

Questions

(1) For this experiment, what are the advantages and disadvantages of the $KMnO_4$ titration method?

(2) The determination of H_2O_2 content in hydrogen peroxide by the $KMnO_4$ method, why should it be carried out under acidic conditions? Can HNO_3, HCl or HAc be used to adjust the acidity of the solution? Why?

Determination of H_2O_2 content

(3) Describe the principles of the cerium method and iodometric determination of H_2O_2 content.

Experiment 3.14 Determination of Iron in Iron Ore by Potassium Dichromate Method

Objectives

(1) Learn the acid dissolution method of iron ore.

(2) Master the principle and operation of the mercury-free method for iron determination.

(3) Master the color-change principle of redox indicator.

(4) Understand the treatment of waste potassium dichromate.

Principles

Potassium dichromate is a commonly used oxidant, which can be reduced to Cr^{3+} by reducing agent in an acidic solution. The semi reaction formula is:

$$Cr_2O_7^{2-} + 14H^+ + 6e \rightleftharpoons 2Cr^{3+} + 7H_2O \quad (3.27)$$

The standard electrode potential ($E^{\ominus}$) of potassium dichromate is 1.33 V, and the conditional potential $E^{\ominus\prime}$ is smaller than $E^{\ominus}$ in acidic solution. For example, $E^{\ominus\prime}=1.08$ V in 3 mol/L HCl solution, $E^{\ominus\prime}=1.15$ V in 3 mol/L H_2SO_4 solution. The conditional potential increases along with the increase of solution acidity. That is to say, the oxidation capacity of $K_2Cr_2O_7$ can be regulated to a suitable degree to avoid side reactions and enhance selectivity by controlling the acidity of solution. Moreover, $K_2Cr_2O_7$ is easily purified (99.99%) and stable for direct usage. All these advantages facilitate the wide application of $K_2Cr_2O_7$ in redox titration analysis.

Potassium dichromate method is national standard method for determination of iron content in ore, alloy, silicate, and other products. A typical sample pretreatment procedure includes the reduction of Fe^{3+} to Fe^{2+} by $SnCl_2$ in hot concentrated HCl solution and elimination of excessive SnCl$_2$ by HgCl$_2$. Every test produces waste water containing about 480 mg of mercury, which needs dilution by 9.6-10.0 tons of pure water to satisfyies the discharge standard. Even if the waste water satisfies the discharge standard, the mercury will be deposited in environment and cause serious pollution or potential health hazards. Recently, new mercury-free determination method for iron has been developed, which includes the new potassium dichromate method, cerium sulfate method and EDTA method.

In this experiment, a mercury-free potassium dichromate method is used to determine the iron content in iron ore.

The sample is first dissolved by using mixed acid of sulphuric acid and phosphoric acid. Then, most of the Fe^{3+} is reduced by using SnCl$_2$, and the residual Fe^{3+} is reduced by using TiCl$_3$.

$$2Fe^{3+} + SnCl_4^{2-} + 2Cl^- \rightleftharpoons 2Fe^{2+} + SnCl_6^{2-} \tag{3.28}$$

$$Fe^{3+} + Ti^{3+} + H_2O \rightleftharpoons Fe^{2+} + TiO^{2+} + 2H^+ \tag{3.29}$$

Before adding TiCl$_3$ into the solution, sodium tungstate is added to indicate the excessive addition of TiCl$_3$. When Fe^{3+} is reduced to Fe^{2+} totally, excessive TiCl$_3$ will react with colorless tungsten (Ⅵ) and produce blue tungsten (Ⅴ). Then, the excessive TiCl$_3$ is eliminated by titration of K$_2$Cr$_2$O$_7$ till the blue tungsten disappears.

Afterwards, the Fe^{2+} can be analyzed by titration with K$_2$Cr$_2$O$_7$ as titrant and sodium diphenylamine sulfonate as indicator. When Fe^{2+} is oxidized to Fe^{3+} totally, violet complex produced by the reaction between sodium diphenylamine sulfonate and Fe^{3+} indicates the end point.

$$6Fe^{2+} + Cr_2O_7^{2-} + 14H^+ \rightleftharpoons 6Fe^{3+} + 2Cr^{3+} + 7H_2O \tag{3.30}$$

 Materials

Instruments: acid burette, volumetric flask, pipette, conical bottle, weighing bottle, cylinder, beaker.

Reagents: (1) solid K$_2$Cr$_2$O$_7$ (AR); (2) H$_2$SO$_4$-H$_3$PO$_4$ (1:1); (3) HCl (3mol/L); (4) concentrated HNO$_3$; (5) Na$_2$WO$_4$ (10%): 10 g of Na$_2$WO$_4$ is dissolved in deionized water, mixed with 2-5 mL commercial H$_3$PO$_4$, and diluted to 100 mL; (6) SnCl$_2$ (10%): 10 g of SnCl$_2$ · H$_2$O is dissolved in 40 mL of concentrated hot HCl, the solution is diluted to 100 mL with deionized water; (7) TiCl$_3$ (1.5%): 10 mL of original packed TiCl$_3$ solution is diluted with HCl solution (2 mol/L) to 100 mL. A few petroleum ether is added to isolate the TiCl$_3$ from air to avoid the oxidation of TiCl$_3$; (8) sodium diphenylamine sulfonate (0.2%); (9) iron ore sample to be tested.

Procedures

(1) Preparation of standard K$_2$Cr$_2$O$_7$ solution (0.01667 mol/L) ( )g of K$_2$Cr$_2$O$_7$ is dissolved in deionized water and transferred into a 250 mL volumetric flask, followed by dilution till the liquid level reaches the scale line.

(2) Pre-treatment of iron ore

① 0.2 g of iron ore (contains about 60% of iron) is accurately weighed and added in a conical bottle (250 mL).

② Deionized water (about 5 mL) is added in to disperse the iron powder. Then 10 mL of mixed acid of H_2SO_4-H_3PO_4 (1 : 1) is added in and the sample is heated in ventilation to decompose the sample till there is white smoke (SO_3)[1]. The solution should be limpid with white or pale slag at the bottom when the sample is totally decomposed [2].

③ 30 mL of HCl solution (3 mol/L) is then added in and the mixed solution is heated and stopped to be heated as soon as the solution is boiled.

④ $SnCl_3$ is added drop by drop when the solution is cooled down to about 60 ℃ to reduce Fe^{3+} until the solution turns from brown to light yellow [3]. Subsequently, 1 mL of Na_2WO_4 (10%) is added in before $TiCl_3$ is added in drop by drop till the solution turns blue (tungsten blue). Then, 60 mL of fresh deionized water (boiled to eliminate dissolved oxygen [4]) is added in and the solution is placed for 10-20 s.

(3) Determination of iron in iron ore Before the titration of Fe^{2+}, tungsten blue is eliminated by using the standard $K_2Cr_2O_7$ solution till the blue color fades. The volume of $K_2Cr_2O_7$ solution consumed in this step doesn't count. After that, the end-point indicator sodium diphenylamine sulfonate is added in and the solution is immediately titrated by using $K_2Cr_2O_7$ solution till the solution becomes violet. Three replicates should be determined. The replicate sample should be titrated as soon as possible after reduction treatment, instead of pretreating several samples at the same time and then titrating them one by one [5].

Notes:

[1] The decomposition temperature of H_2SO_4 (338 ℃) is much higher than that of HNO_3 (125 ℃). The white smoke indicates that the decomposition of H_2SO_4 begins, which proves that HNO_3 has been decomposed totally. Otherwise, the residual HNO_3 will hinder the reduction of Fe^{3+} in the following pretreatment procedure. It is noteworthy that the heating of sample solution should be stopped immediately when there is white smoke. If the sample solution is heated for too long, H_3PO_4 will form pyrophosphate precipitates and the sample solution can be partially trapped in the precipitates.

[2] The white or pale slag is usually $SiO_2 \cdot nH_2O$ or $H_3SiO_3 \cdot nH_2O$.

[3] Too much $SnCl_2$ will induce serious positive error. If too much $SnCl_2$ is added carelessly, the error can be eliminated by adding $KMnO_4$ (2%) till the solution is light yellow.

[4] Tungsten blue is an unstable oxide of tungsten, which will be oxidized and fade if there is residual oxygen in water.

[5] The Fe^{2+} is easily oxidized in phosphoric acid medium, which should be titrated immediately in 1 minute after tungsten blue fades. Otherwise, negative error will be induced.

Data Recording and Processing

(1) Preparation of standard $K_2Cr_2O_7$ solution (Table 3.24).

Table 3.24 Preparation of standard $K_2Cr_2O_7$ solution

$m_{K_2Cr_2O_7}$ (initial)/g	
$m_{K_2Cr_2O_7}$ (final)/g	
$m_{K_2Cr_2O_7}$/g	

(2) Determination of iron in iron ore sample (Table 3.25).

Table 3.25 Determination of iron in iron ore sample

Number	1	2	3
V_{initial}/mL			
V_{final}/mL			
V_{titrant}/mL			
$w_{\text{Fe}_2\text{O}_3}$/%			
$\overline{w}_{\text{Fe}_2\text{O}_3}$/%			
$|d_i|$			
$\overline{d}_r$/%			

Questions

(1) Why are mixed acid of sulfur-phosphorus and nitric acid solution added in the decomposition process of iron ore sample?

(2) Why should the sample be heated until there is white smoke? If there is nitric acid in the titration process, what will happen to the experimental results?

(3) How do you prepare $SnCl_2$ solution? Why should HCl be added in the preparation process? How do you prepare $SnCl_2$ solution for long-term storage?

(4) Why should $SnCl_2$ solution be added in when the sample solution is hot? And when can the addition of $SnCl_2$ be stopped?

(5) Why is it necessary to titrate one sample after reduction treatment immediately, instead of pretreating several samples at the same time and then titrating them one by one?

Experiment 3.15 Preparation and Standardization of Iodine and Sodium Thiosulfate

Objectives

(1) To master the preparation method and storage conditions of I_2 and $Na_2S_2O_3$ solutions.

(2) To master the principle and method of standardization of I_2 and $Na_2S_2O_3$ solutions.

Principles

(1) Iodometry Iodometry is a determination method based on the oxidability of I_2 and the reducibility of I^-. Since I_2 is volatile and difficult to dissolve in water, I_2 is usually dissolved in KI solution to increase the solubility, in which it exists in the form of I_3^- complex ion. The half-reaction is:

$$I_3^- + 2e \rightleftharpoons 3I^- \quad (3.31)$$

To simplify and emphasize the stoichiometric relationship, it is still abbreviated as I_2 generally. The standard potential is 0.545 V, indicating that I_2 is a weak oxidant and I^- is a moderate-strong reducing agent.

Direct iodometry (or iodometry) is the determine method that directly titrate $S_2O_3^{2-}$, SO_3^{2-}, As(III), Sn(II), vitamin C and other strong reducing agents with I_2 standard solution. In indirect iodometry (or titration iodometry), I_2 is first produced by reactions between I^- and MnO_4^-, $Cr_2O_7^{2-}$, H_2O_2, Cu^{2+}, Fe^{3+} and then titrated with $Na_2S_2O_3$ standard solution. Indirect iodometry has wider applications.

The I_3^-/I^- electric pair shows excellent reversibility, and its potential is not affected by acidity or other complexants within a large pH range (pH<9). Therefore in selection of the measurement conditions, it is only necessary to consider the properties of the tested substance.

The iodometry exhibits high sensitivity with starch as the indicator, which shows blue when the concentration of I_2 is 1×10^{-5} mol/L. When the solution turns blue (direct iodometry) or fades to colorless (indirect iodometry), the titration end point shows up.

The analysis error in iodometry mainly comes from the volatilization of I_2 and the oxidation of I^- by air. To prevent the volatilization of I_2, excessive KI can be added to form I_3^- complex ions, the solution temperature should be controlled, and the reaction is suggested to be carried out in an iodine volumetric flask with a stopper. After the reaction is complete, titrate immediately, and do not shake vigorously during titration. To prevent I^- from being oxidized by air, the reaction bottle in which I_2 precipitated should be placed in a dark place and impurities such as Cu^{2+} and NO_2^- that can catalyze the oxidation of I^- by air should be removed in advance. After taking the above measures, the measurement error of iodometry is greatly reduced, and the accuracy of the measurement results is guaranteed.

(2) Preparation of iodine standard solution Although pure I_2 can be prepared via sublimation method, the high volatility of I_2 makes it difficult to weigh it accurately. Generally, a solution with approximate concentration is prepared before calibration. First, a certain amount of I_2 is mixed with excess of KI, and the mixture is grinded with a small amount of water to dissolve all the I_2. Then the solution is diluted and stored in a brown bottle in dark. The contact between I_2 solution and organic substances such as rubber should be avoided, exposure to light and heat is also prevented, otherwise the concentration may change.

Iodine solutions are also prepared indirectly in laboratories. For example, KIO_3 and KI undergo comproportionation in acid solution and generate I_2, and excess I^- and I_2 generate I_3^- ions to increase the solubility of iodine. Since potassium iodate is the primary standard substance, the accurate concentration of iodine standard solution prepared by this method can be calculated without subsequent calibration.

$$KIO_3 + 5KI + 6HCl = 3I_2 + 3H_2O + 6KCl \qquad (3.32)$$
$$I_2 + I^- = I_3^- \qquad (3.33)$$

(3) Standardization of iodine standard solution The iodine standard solution directly prepared with iodine needs calibration.

As_2O_3 can be used as the primary substance to calibrate I_2. As_2O_3 is hardly dissolved in water, but it is easily dissolved in alkaline solution. If the pH value ranges from 8 to 9, I_2 reacts with $HAsO_2$ quickly and quantitatively. The reaction formula is:

$$HAsO_2 + I_2 + 2H_2O \Longrightarrow HAsO_4^{2-} + 2I^- + 4H^+ \tag{3.34}$$

Due to the high toxicity of As_2O_3, the application of this calibration method is limited. In fact, such I_2 solutions can also be calibrated with $Na_2S_2O_3$ standard solution.

$$2S_2O_3^{2-} + I_2 \Longrightarrow S_4O_6^{2-} + 2I^- \tag{3.35}$$

(4) Preparation and standardization of sodium thiosulfate solution   Generally, crystalline $Na_2S_2O_3 \cdot 5H_2O$ contains a small amount of impurities, such as S, Na_2SO_3, Na_2SO_4, Na_2CO_3 and NaCl, and it is also prone to weathering and deliquescence. Therefore, its standard solution cannot be directly prepared. The prepared $Na_2S_2O_3$ solution is also unstable for three main reasons:

① it is easily decomposed due to the presence of acid, even weak acid such as carbonic acid:

$$Na_2S_2O_3 + CO_2 + H_2O \Longrightarrow NaHCO_3 + NaHSO_3 + S\downarrow \tag{3.36}$$

② it is easily decomposed by microorganism:

$$Na_2S_2O_3 \Longrightarrow Na_2SO_3 + S\downarrow \tag{3.37}$$

③ it is easily oxidized by oxygen in air:

$$2Na_2S_2O_3 + O_2 \Longrightarrow 2Na_2SO_4 + 2S\downarrow \tag{3.38}$$

To avoid the consumption of $Na_2S_2O_3$, it is vital to follow these steps. First, newly boiled and cooled water should be used to get rid of CO_2, O_2 and microorganism. Second, small amount of Na_2CO_3 should be added to maintain a weak alkaline circumstance to inhibit the growth of bacterium. Third, the solution should be stored in brown bottle and placed in dark to avoid exposure to light. The solution should be recalibrated after storage for a period of time. If it becomes turbid, it indicates sulfur precipitation, and the solution should be discarded. Each step of preparation should be very meticulous, and the instruments used must be clean.

The primary standard substances used to calibrate $Na_2S_2O_3$ include $K_2Cr_2O_7$, KIO_3, $KBrO_3$ and copper. Indirect iodometry is usually adopted. Take the $K_2Cr_2O_7$ as an example, it reacts with I^- in acid solution:

$$Cr_2O_7^{2-} + 6I^- + 14H^+ \Longrightarrow 2Cr^{3+} + 3I_2 + 7H_2O \tag{3.39}$$

The produced I_2 is titrated using $Na_2S_2O_3$ with starch as indicator:

$$I_2 + 2S_2O_3^{2-} \Longrightarrow S_4O_6^{2-} + 2I^- \tag{3.40}$$

The reaction (3.39) is slow. To accelerate this reaction, excess KI is usually added and the acidity is improved. However, too high acidity facilitates the oxidation of I^-:

$$4I^- + 4H^+ + O_2 \Longrightarrow 2I_2 + 2H_2O \tag{3.41}$$

Therefore, it is vital to control the acidity of the solution. Usually, the acidity is controlled at 0.4 mol/L.

Acidity can also affect the reaction (3.41), which is one of the most important reaction of iodometry. The molar ratio of I_2 and $S_2O_3^{2-}$ is 1:2. However, the ratio will change to 1:1 if the acidity is too high, due to the following reactions:

$$S_2O_3^{2-} + 2H^+ \Longrightarrow H_2SO_3 + S\downarrow \tag{3.42}$$

$$I_2 + H_2SO_3 + H_2O \Longrightarrow SO_4^{2-} + 4H^+ + 2I^- \tag{3.43}$$

If the acidity is too low, I_2 will transform into OI^- and IO_3^-, which will turn $S_2O_3^{2-}$ to SO_4^{2-} partially:

$$4I_2 + S_2O_3^{2-} + 10OH^- \rightleftharpoons 2SO_4^{2-} + 8I^- + 5H_2O \tag{3.44}$$

Namely, the reaction ratio of I_2 and $S_2O_3^{2-}$ is 4 : 1 at this condition, which brings about significant errors.

To be concluded, the reaction circumstance of I_2 and $S_2O_3^{2-}$ is suggested to be neutral or weakly acidic. Before the first step of the titration with $Na_2S_2O_3$, the solution should be diluted to reduce the acidity because the acidity of the first step of the reaction is high. On the one hand, reducing the acidity can reduce the oxidation of I^- by air. On the other hand, it can weaken the color of Cr^{3+}, which is convenient to observe the end point. Starch should be added just before the end point, otherwise the starch adsorbs partial I_2, leading to false and indistinct end point. If the solution rapidly turns blue after titration to the end point, it indicates that the reaction between $Cr_2O_7^{2-}$ and I^- has not been quantitatively completed. In this case, the experiment should be redone.

Materials

Instruments: mortar, iodine flask.

Reagents: $Na_2S_2O_3 \cdot 5H_2O$, I_2, KIO_3 or $K_2Cr_2O_7$ (dried, solid), As_2O_3 (solid, dried at 105℃ for 2h), KI, Na_2CO_3, $NaHCO_3$, 6 mol/L HCl, 6 mol/L NaOH, starch solution (0.5 %), 7 mol/L $NH_3 \cdot H_2O$, 20% NH_4HF_2, 8 mol/L HAc, 10% NH_4SCN.

Procedures

(1) Preparation of I_2 solution (about 0.05 mol/L) Accurately weigh (　) g of I_2 and 5 g KI into mortar in fume hood. Add some water and grind the mixture until the I_2 has dissolved in water completely. Transfer the solution into a brown bottle and dilute into 250 mL. Before calibration, the solution is shaken thoroughly and placed in a dark place overnight.

(2) Preparation of $Na_2S_2O_3$ solution (about 0.1 mol/L) Accurately weigh (　) g $Na_2S_2O_3 \cdot 5H_2O$ into a beaker, add 200 mL newly boiled and cooled distilled water, stir the solution till all the $Na_2S_2O_3$ has dissolved, and add 0.1 g Na_2CO_3. Then the solution is diluted into 500 mL with newly boiled and cooled distilled water and stored in brown bottle for 3-5 days before calibration.

(3) Standardization of I_2 solution (about 0.05 mol/L)

① Standardization with As_2O_3. Accurately weigh(　)-(　)g As_2O_3 into small beaker (100 mL) and add into 10 mL of NaOH solution (6 mol/L). After the As_2O_3 is dissolved, 2 drops of phenolphthalein is added in and the solution is neutralized with 6 mol/L HCl solution till the solution turns colorless. Then, 2-3 g $NaHCO_3$ is added in and the solution is stirred till the $NaHCO_3$ is dissolved. Subsequently, the solution is transferred into 250 mL volumetric flask, diluted until the liquid level reaches the mark, and shaken thoroughly.

Transfer 25.00 mL of the above solution using a pipette into a 250 mL clean conical flask, then add 50 mL of water, 5g of $NaHCO_3$, and 2 mL of starch solution. Titrate the solution with I_2 standard solution until the solution turns blue and the blue does not fade within 30s. Record the volume of consumed I_2 solution, and calculate its concentration. Repeat the calibration three times.

② Standardization with $Na_2S_2O_3$. Transfer 25.00 mL of the standard $Na_2S_2O_3$ solution using a pipette into a 250 mL clean conical flask, then add 50 mL of water, 5g of $NaHCO_3$, and 2 mL of starch solution. Titrate the solution with I_2 standard solution until the solution turns blue and the blue does not fade within 30s. Record the volume of consumed I_2 solution, and calculate its concentration. Repeat the calibration three times.

(4) Standardization of $Na_2S_2O_3$ solution (0.1 mol/L)

① Standardization with KIO_3. Accurately weigh( )-( )g KIO_3 into a beaker (100 mL), and add water to dissolve the KIO_3. Then, transfer the solution into a 250 mL volumetric flask, dilute until the liquid level reaches the mark, and shake thoroughly.

Transfer 25.00 mL of the above solution using a pipette into a 250 mL clean conical flask, then add 10 mL of HCl solution (6 mol/L), 1g KI, and water till the total volume is about 120 mL. Titrate the solution with $Na_2S_2O_3$ standard solution immediately until the solution turns light yellow. Then, 2 mL of starch solution is added in and the solution turns blue. Continue the titration till the solution becomes colorless. Record the volume of consumed $Na_2S_2O_3$ solution, and calculate its concentration. Repeat the calibration three times.

② Standardization with $K_2Cr_2O_7$. Accurately weigh( )-( )g $K_2Cr_2O_7$ into a beaker (100 mL), and add 30 mL of water to dissolve the $K_2Cr_2O_7$. Then, transfer the solution into a 250 mL volumetric flask, dilute until the liquid level reaches the mark, and shake thoroughly.

Transfer 25.00 mL of the above solution using a pipette into a 250 mL clean conical flask, then add 10 mL of HCl solution (6 mol/L) and 1g of KI, immediately cap the glass stopper, shake the flask and place it in a dark place for 5-10 minutes. Then, rinse the bottle cap and inner wall with distilled water, and add 100 mL of distilled water. Titrate the solution with $Na_2S_2O_3$ standard solution until the solution turns light yellow. Then, 2 mL of starch solution is added in and the solution turns blue. Continue the titration till the solution becomes colorless. Record the volume of consumed $Na_2S_2O_3$ solution, and calculate its concentration. Repeat the standardization three times.

③ Standardization with copper. Accurately weigh( )-( )g copper in to a 250 mL conical flask, add 2 mL 6 mol/L HCl solution in the flask. After heating the mixture, slowly add 2-3 mL of 30% H_2O_2 (as little as possible, as long as the copper can be completely decomposed), heat it at low temperature until the sample is completely dissolved, and then heat it fiercely to decompose the excess H_2O_2, and add 20 mL of water after cooling. After that, add 1∶1 ammonia dropwise until precipitation appears. Then, add 8 mL of 1∶1 HAc, 10 mL of 20% NH_4HF_2, and 1 g of KI into the solution. Subsequently, titrate the solution with $Na_2S_2O_3$ until the solution turns pale yellow. Then add 3 mL of starch, and continue to titrate till the solution turns light blue. Ten milliliter of 10% NH_4SCN solution is then added and titration is continued until the solution becomes colorless. Record the volume of the consumed $Na_2S_2O_3$ and calculate the concentration. Repeat the standardization three times.

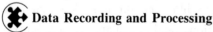
Data Recording and Processing

(1) Preparation of I_2 solution (0.05 mol/L) (Table 3.26)

Table 3.26 Preparation of I_2 solution (0.05 mol/L)

m_{I_2}/g	

(2) Standardization of I_2 solution with As_2O_3 (Table 3.27)

Table 3.27 Standardization of I_2 solution with As_2O_3

Number	1	2	3
$m_{As_2O_3}$ (initial)/g			
$m_{As_2O_3}$ (final)/g			
$m(As_2O_3)$/g			
$c_{As_2O_3}$/(mol/L)			
$V(As_2O_3)$/mL			
V_{I_2} (initial)/mL			
V_{I_2} (final)/mL			
V_{I_2} (consumed)/mL			
c_{I_2}/(mol/L)			
$\bar{c}_{I_2}$/(mol/L)			
$|d_i|$			
$\bar{d}_r$%			

(3) Standardization of I_2 solution with $Na_2S_2O_3$ (Table 3.28)

Table 3.28 Standardization of I_2 solution with $Na_2S_2O_3$

Number	1	2	3
$V_{Na_2S_2O_3}$/mL			
V_{I_2} (initial)/mL			
V_{I_2} (final)/mL			
V_{I_2} (consumed)/mL			
c_{I_2}/(mol/L)			
$\bar{c}_{I_2}$/(mol/L)			
$|d_i|$			
$\bar{d}_r$%			

(4) Preparation of $Na_2S_2O_3$ solution (about 0.1 mol/L) (Table 3.29)

Table 3.29 Preparation of $Na_2S_2O_3$ solution (about 0.1 mol/L)

$m_{Na_2S_2O_3}$/g	

(5) Standardization of Na$_2$S$_2$O$_3$ solution (about 0.1mol/L) (Table 3.30)

Table 3.30　Standardization of Na$_2$S$_2$O$_3$ solution (about 0.1 mol/L)

Number	1	2	3
$m_{Na_2S_2O_3}$ (initial)/g			
$m_{Na_2S_2O_3}$ (final)/g			
$m_{Na_2S_2O_3}$/g			
$c_{standard}$/(mol/L)			
$V_{standard}$/mL			
V(initial)/mL			
V(final)/mL			
V(titrant)/mL			
$c_{Na_2S_2O_3}$/(mol/L)			
$\bar{c}_{Na_2S_2O_3}$/(mol/L)			
$\lvert d_i \rvert$			
$\bar{d}_r\%$			

Questions

(1) Why can't Na$_2$S$_2$O$_3$ standard solution be directly prepared? How do you prepare a relatively stable Na$_2$S$_2$O$_3$ solution? Why are freshly boiled and cooled distilled water used?

(2) Why is solid NaHCO$_3$ added in calibration of I$_2$ with As$_2$O$_3$? Can NaHCO$_3$ be replaced by Na$_2$CO$_3$, why?

(3) In calibration of Na$_2$S$_2$O$_3$ solution with K$_2$Cr$_2$O$_7$, why should it be sealed and placed in a dark place for 5-10min before titration?

(4) Why is excess KI added in standardization of Na$_2$S$_2$O$_3$?

Experiment 3.16　Determination of Glucose in Glucose Oral Solution

Objectives

(1) Master principle and method for determining glucose via indirect iodometry.

(2) Master principle and operation of back titration.

Principles

Glucose (C$_6$H$_{12}$O$_6$) reacts quantitatively with I$_2$ and produces gluconic acid in alkali solutions due to its aldehyde group. The involved reactions are as follows:

$$I_2 + 2OH^- = IO^- + I^- + H_2O \qquad (3.45)$$

$$CH_2OH(CHOH)_4CHO + IO^- + OH^- = CH_2OH(CHOH)_4COO^- + I^- + H_2O \qquad (3.46)$$

Then, the disproportionation reaction of the residual IO^- occurs.
$$3IO^- = IO_3^- + 2I^- \tag{3.47}$$
After adding acid into the solution to make an acid environment, I_2 is produced,
$$IO_3^- + 5I^- + 6H^+ = 3I_2 \downarrow + 3H_2O \tag{3.48}$$
and the generated I_2 can be quantitatively analyzed by titration with the standard $Na_2S_2O_3$ solution.
$$I_2 + 2S_2O_3^{2-} = 2I^- + S_4O_6^{2-} \tag{3.49}$$
This method can be applied for the determination of glucose in oral liquid.

Materials

Instruments: burette (50 mL), volumetric flask (250 mL), pipette, iodimetric flask, weighing bottle, cylinder, beaker.

Reagents: $Na_2S_2O_3 \cdot 5H_2O$, KIO_3, KI, HCl solution (6 mol/L), NaOH solution (20%), starch solution (0.5%), commercial glucose oral liquid.

Procedures

(1) Preparation and standardization of $Na_2S_2O_3$ solution (0.1 mol/L) Please see Experiment 3.15.

(2) Determination of glucose Glucose sample solution is first prepared by transferring 25.00 mL commercial glucose oral liquid into volumetric flask (250 mL) and adding deionized water to the mark.

The standard KIO_3 solution is prepared by dissolving 1.2-1.6 g of KIO_3 in deionized water and fixing the volume in volumetric flask (250 mL). After mixed homogeneously, 25.00 mL of the standard KIO_3 solution is transferred into an iodimetric flask, which is followed by the addition of 10 mL of HCl solution (6 mol/L) and 1 g KI. Immediately, the iodimetric flask is capped tightly by using a matched glass plug before shaking the solution to make a uniform solution. After the glass plug and the inner wall of flask are washed by deionized water to dissolve I_2 as much as possible, add in 25.00 mL glucose sample solution, followed by adding NaOH (20%) drop by drop while shaking the solution till it turns yellow. Then the iodimetric flask is capped tightly and placed in dark environment for 10 min. After that, the glass plug and the inner wall of flask are washed again with deionized water. Subsequently, 4 mL HCl (6 mol/L) solution is added and the solution is diluted with deionized water till the total volume is about 120 mL. Then, the solution is titrated with standard $Na_2S_2O_3$ solution till the solution turns yellow. Two milliliters of starch solution is added in and the solution turns blue. Then, titration is continued until the solution becomes colorless, which is the end point. The calibration is repeated for three times. The volume of the consumed $Na_2S_2O_3$ should be recorded and the content of glucose (m/V, %) can be calculated from that. In addition, the relative RSD should also be given out in the report.

Data Recording and Processing

(1) Preparation of KIO_3 solution (Table 3.31)

Table 3.31　Preparation of KIO₃ solution

m_{KIO_3} (initial)/g	
m_{KIO_3} (final)/g	
$m(KIO_3)$/g	
c_{KIO_3} /(mol/L)	

(2) Standardization of $Na_2S_2O_3$ by using KIO_3 (Table 3.32)

Table 3.32　Standardization of $Na_2S_2O_3$ by KIO_3

Number	1	2	3
V_{KIO_3}/mL			
$V_{initial}$/mL			
V_{final}/mL			
$V_{titrant}$/mL			
$c_{Na_2S_2O_3}$/(mol/L)			
$\overline{c}_{Na_2S_2O_3}$/(mol/L)			
$\lvert d_i \rvert$			
$\overline{d}_r \%$			

(3) Determination of glucose (Table 3.33)

Table 3.33　Determination of glucose

Number	1	2	3
$V_{Glucose}$/mL			
$V_{initial}$/mL			
V_{final}/mL			
$V_{titrant}$/mL			
$m_{Glucose}$/g			
$w_{Glucose}/\%$			
$\overline{w}_{Glucose}/\%$			
$\lvert d_i \rvert$			
$\overline{d}_r \%$			

Questions

(1) In the experiment procedure, HCl solution was added twice. What's the function of HCl? Write relative reaction formula.

(2) Explain the reason for diluting the solution before titration with $Na_2S_2O_3$.

(3) Why is the starch solution added till the solution turns yellow? Is there any disadvantages if the starch solution is added too early?

Determination of glucose in glucose oral solution

Experiment 3.17　Determination of Copper Content in Copper Alloys

Objectives

(1) Learn the pretreatment of copper alloy.

(2) Master the principle and operation process of determining copper content by indirect iodometry.

Principles

Copper alloy is an important basic material for manufacturing industry, which mainly includes copper tin alloy (bronze), copper zinc alloy (brass) and copper nickel alloy (white copper) for producing bearing gears, water pipe valves, corrosion-resistant components, respectively. Pure copper is mainly used to produce cables and wires. The content of copper has an important influence on the mechanical and electrical properties of the alloy. Therefore, the determination of copper content in copper alloy is very important to determine the grade of copper.

Solid copper alloy should be dissolved into solution first. Generally, HNO_3 is used to dissolve the alloy samples, and then excess nitric acid or other nitrogen oxides are removed by adding concentrated H_2SO_4 and heating the solution to avoid errors induced from oxidation of I^- by nitrogen oxides. Alternatively, the usage of HCl and H_2O_2 makes it easier to remove excess acid and H_2O_2.

$$Cu + 2HCl + H_2O_2 = CuCl_2 + 2H_2O \tag{3.50}$$

Generally, indirect iodometry is used to determine the copper content in laboratory. Copper ion reacts with potassium iodide and produces iodine quantitatively, which can be analyzed by titration with standard $Na_2S_2O_3$ solution. The reaction formulas are as follows.

$$2Cu^{2+} + 4I^- = 2CuI \downarrow + I_2 \tag{3.51}$$

$$I_2 + 2S_2O_3^{2-} = 2I^- + S_4O_6^{2-} \tag{3.52}$$

The reaction between Cu^{2+} and I^- is reversible. Therefore, excess KI is needed to facilitate the complete forward reaction. However, the precipitate CuI will absorb I_2 which induces an incomplete reaction between $Na_2S_2O_3$ and I_2 and brings out negative errors. To avoid this error, thiocyanate is usually added to produce CuSCN precipitate ($K_{sp} = 4.8 \times 10^{-15}$) with smaller solubility than CuI ($K_{sp} = 1.1 \times 10^{-12}$) right before stoichiometric point.

$$CuI + SCN^- = CuSCN \downarrow + I^- \tag{3.53}$$

This step can release the absorbed I_2 by CuI, and the produced I^- can react with free Cu^{2+} which makes amore complete forward reaction. It should be noted that thiocyanate can only be added right before the stoichiometric point. If it is added too early, SCN^- will consume part of Cu^{2+} and result in negative errors.

$$6Cu^{2+} + 7SCN^- + 4H_2O = 6CuSCN \downarrow + SO_4^{2-} + CN^- + 8H^+ \tag{3.54}$$

An acid solution circumstance is vital for the reaction between Cu^{2+} and I^-. Copper ions are easily hydrolyzed in weak acid solution. I^- ions are easily oxidized by oxygen in strong acid solution. Generally, NH_4HF_2 buffer solution is used to adjust the pH value in the range of 3.0-4.0 to avoid the hydrolysis of Cu^{2+} and the interference of As(V), Sb(V) or other ions. Moreover, F^- ions can complex with Fe^{3+} and thus avoid the interference from Fe^{3+}.

Materials

Instruments: burette, volumetric flask, pipette, conical bottle, weighing bottle, cylinder, beaker.

Reagents: solid $Na_2S_2O_3 \cdot 5H_2O$(AR), pure copper(s), KI(s, AR), HCl solution (6 mol/L), $NH_3 \cdot H_2O$(1:1), HAc(1:1), NH_4HF_2(20%), NH_4SCN(10%), starch solution, H_2O_2(30%), copper alloy sample.

Procedures

(1) Preparation and standardization of the standard $Na_2S_2O_3$ solution (0.1 mol/L) Please see Experiment 3.15.

(2) Determination of copper in alloy   Copper alloy (containing 65%-70% of copper) is accurately weighed in the range of( )-( )g and placed in a conical bottle. Then, 10 mL of HCl solution (6 mol/L) is added in and the solution is heated, followed by addition of 2 mL of H_2O_2 solution (30%) drop by drop till the copper is dissolved completely. Then the solution is heated fiercely to remove excess H_2O_2. After it is cooled down to room temperature, 20 mL of deionization water is added in.

Ammonia solution (1:1) is added in the above solution drop by drop till there are precipitates. Then, 8 mL of HAc (1:1), 10 mL of NH_4HF_2 (20%) and 1 g of KI are added in the mixed solution. When the solution turns light yellow after titration with $Na_2S_2O_3$, 3 mL of starch solution is added. The titration is continued till the solution becomes light blue. After that, 10 mL of NH_4SCN solution (10%) is added in and the titration is continued till the blue color fades. At this time, the solution shows grayish white or light complexion rather than colorless because of the precipitates produced by the impurities in the alloy. The above experiment is repeated three times. The consumed volume of $Na_2S_2O_3$ should be recorded and the content of copper in alloy can be derived.

Data Recording and Processing

(1) Preparation and standardization of the standard $Na_2S_2O_3$ solution (Table 3.34)

Table 3.34   Preparation and calibration of the standard $Na_2S_2O_3$ solution

Number	1	2	3
$m_{Na_2S_2O_3}$ (initial)/g			
$m_{Na_2S_2O_3}$ (final)/g			
$m_{Na_2S_2O_3}$/g			
$V_{initial}$/mL			
V_{final}/mL			
$V_{titrant}$/mL			
$c_{Na_2S_2O_3}$/(mol/L)			

(continued)

Number	1	2	3
$\bar{c}_{Na_2S_2O_3}$ /(mol/L)			
$|d_i|$			
$\bar{d}_r$/%			

(2) Determination of copper in alloy sample (Table 3.35)

Table 3.35 Determination of copper in alloy sample

Number	1	2	3
$m_{initial}$/g			
m_{final}/g			
m_{sample}/g			
Volume of $Na_2S_2O_3$			
$V_{initial}$/mL			
V_{final}/mL			
$V_{titrant}$/mL			
m_{Cu}/g			
w_{Cu}/%			
$\bar{w}_{Cu}$/%			
$|d_i|$			
$\bar{d}_r$/%			

Questions

(1) H_2O_2 is used to dissolve copper samples, how can the residual H_2O_2 impact on the final results if it is not eliminated totally?

(2) What reference materials can be used to calibrate the $Na_2S_2O_3$ solution? Why is pure copper used in this experiment?

(3) What's the function of excess KI?

(4) What's the function of NH_4HF_2? Why should NH_4HF_2 be added when the titration nears the end point?

Experiment 3.18 Determination of Chlorine in Tap Water by Mohr Method

Objectives

(1) Master the principle of Mohr method for determining trace of Cl^- in water.

(2) Learn the operation of precipitation titration.

Principles

The content of chlorine in some soluble chlorides can be determined by argentometry. Argentometry includes the Mohr method, Volhard method, and Fajans method with potassium chromate, ammonium iron sulfate, and special adsorbent as indicators respectively.

When titration is performed by Mohr method, the chloride is titrated with standard silver nitrate solution and a soluble chromate salt is added as the indicator. This produces a yellow solution. When the precipitation of the chloride is complete, the excessive Ag^+ reacts with the indicator to precipitate red silver chromate:

$$Ag^+ + Cl^- = AgCl \downarrow \text{(white)} \quad (3.55)$$
$$2Ag^+ + CrO_4^{2-} = Ag_2CrO_4 \downarrow \text{(red)} \quad (3.56)$$

The concentration of the indicator and the acidity of the solution are important factors that draw much attention.

(1) The concentration of Ag^+ at the equivalence point is 10^{-5} mol/L, which can be derived from K_{sp} of AgCl. By inserting this Ag^+ concentration in the K_{sp} equation of Ag_2CrO_4, the concentration of CrO_4^{2-} is calculated to be 0.011 mol/L. However, the intense yellow color of CrO_4^{2-} at this concentration obscures the red Ag_2CrO_4 precipitate color. Usually, the concentration of CrO_4^{2-} is kept in the range of 2.0×10^{-3}-5.0×10^{-3} mol/L. With K_2CrO_4 as the indicator, a blank titration should be performed to determine the blank titrant amount, which will be subtracted from the end point titration to correct the indicator error and obtain the real end point.

(2) The Mohr titration must be performed in slightly alkaline or neutral solution. If the solution is too acidic (pH<6), then part of the indicator is present as $HCrO_4^-$ and H_2CrO_4, and more Ag^+ will be required to form the Ag_2CrO_4 precipitate. Above pH 10, silver hydroxide may be precipitated. Generally, the pH value is kept in the range of 6-10.

It should be noted that only Cl^- and Br^- rather than I^- and SCN^- can be determined by Mohr method because AgI and AgSCN can absorb indicator and obscure the color change at the end point.

 Materials

Instruments: acid and basic burette, volumetric flask, pipette, conical bottle, weighing bottle, cylinder, beaker.

Reagents: (1) NaCl (s): NaCl is heated in a porcelain crucible and stirred continuously till there is no explosion sound. It is heated for another 15 min and cooled in a dryer before use. (2) 0.05 mol/L $AgNO_3$: 2.2 g $AgNO_3$ is dissolved in water with accurate volume of 250 mL and stored in a brown bottle away from light. (3) K_2CrO_4 solution (0.5%). (4) Water sample.

Procedures

(1) Standardization of standard $AgNO_3$ solution NaCl (0.7-0.8 g) is dissolved in deionized water and transferred into a 250 mL volumetric flask, followed by dilution till the liquid level reaches the mark.

Then, 25 mL of the NaCl solution is transferred into a conical bottle, followed by the addition of 25 mL deionized water and 1.0 mL K_2CrO_4 solution. Titration is conducted by using the standard $AgNO_3$ solution till there is red precipitate. The determination is repeated 3 times to calculate the average concentration of $AgNO_3$.

(2) Determination of chlorine in tap water Tap water (100 mL) is added in a conical bottle, followed by the addition of indicator K_2CrO_4 (1 mL). The mixed solution is then titrated by using $AgNO_3$ solution till red precipitate appears. The consumed volume of $AgNO_3$ should be recorded. And the titration is repeated 3 times. The content of chlorine in tap water and the RSD of the determination are calculated accordingly.

Data Recording and Processing

(1) Standardization of $AgNO_3$ solution (Table 3.36)

Table 3.36 Standardization of $AgNO_3$ solution

Number	1	2	3
m_{NaCl}(initial)/g			
m_{NaCl}(final)/g			
m_{NaCl}/g			
c_{NaCl}/(mol/L)			
V_{NaCl}/mL			
c_{AgNO_3}/(mol/L)			
$\bar{c}_{AgNO_3}$/(mol/L)			

(2) Determination of chlorine in tap water (Table 3.37)

Table 3.37 Determination of chlorine in tap water

Number	1	2	3
V(sample)/mL			
Volume of $AgNO_3$			
$V_{initial}$/mL			
V_{final}/mL			
$V_{titrant}$/mL			
c_{Cl^-}/(mol/L)			
$\bar{c}_{Cl^-}$/(mol/L)			
$|d_i|$			
$\bar{d}_r$/%			

Questions

(1) What is the suitable pH range for determination of Cl^- by using the Mohr method? Explain the reason.

(2) Why should the reference material NaCl be roasted at high temperature? Is there any disadvantages if NaCl is not roasted?

Experiment 3.19 Determination of Nickel in Alloy Steel by Dimethylglyoxime Gravimetric Method

Objectives

(1) Understand the application of organic precipitant, e.g., dimethylglyoxime, in gravimetric analysis for nickel.

(2) Learn the operation of drying in gravimetric analysis.

(3) Be familiar with the advantages of microwave ovens for drying samples.

Principles

Usually, there is a large amount of nickel in nickel-chromium alloy steel, which can be determined by using the dimethylglyoxime gravimetric method or EDTA complexometric titration method. The EDTA complexometric titration method suffers from complicated pre-separation procedure of iron. The dimethylglyoxime gravimetric method is easily conducted and is used widely.

Dimethylglyoxime exhibits high selectivity in the determination of nickel. It can react with nickel ion and form a red complex in ammonia solution. The complex precipitates from the solution and is stable ($K_{sp}=2.3\times10^{-25}$), which can be directly weighed. The structure of dimethylglyoxime nickel is illustrated in Fig. 3.1.

Fig. 3.1 The structure of dimethylglyoxime nickel

Dimethylglyoxime reacts with palladium/platinum and precipitates in an acidic solution. It can also react with nickel/ferrous and precipitates in an ammonia buffer solution. Therefore, when ferrous ions are present, they must be oxidized in advance to eliminate interference. Other ions, such as iron, chromium and titanium, don't react with dimethylglyoxime, but they will produce hydroxide precipitation in ammonia buffer solution and interfere with the determination. Usually, these interferences can be eliminated by adding a screening agent (tartaric acid or citric acid).

Dimethylglyoxime is binary acid (H_2D), and it can only react with Ni^{2+} in the form of HD^-. Therefore, it is critical to control the pH of the solution in the range of 7.0-8.0. If the pH value is higher than that, most of dimethylglyoxime will be present in the form of D^{2-}, which can hardly react with Ni^{2+}. And Ni^{2+} will react with ammonium and produce soluble complex. Thus, not all the nickel can be precipitated and errors are resulted.

Sample contains 50-80 mg of nickel is suitable for analysis. It is better to use excessive amount (40%-80%) of dimethylglyoxime to precipitate nickel completely. But if dimethylglyoxime is too much, it will precipitate itself, which results in a large positive error.

Dimethylglyoxime exhibits low solubility in water and thus ethanol is usually used to enhance the solubility. But too much ethanol will dissolve dimethylglyoxime nickel. Therefore, the solution should be diluted fully during the precipitating procedure and the concentration of ethanol is advised to be about 20% to avoid excessive precipitation of reagents.

Materials

Instruments: two glass crucibles (G4A or P16), vacuum pump and suction filter bottle, microwave oven, and desiccator.

Reagents: dimethylglyoxime (1%, ethanol), HCl solution (1 : 1), HNO₃ solution (2 mol/L), ammonium solution (1 : 1), tartaric acid solution (50%), ethanol (95%).

Procedures

(1) Make the weight of crucible constant (select one of the following methods)

① Drying with microwave oven. The crucible is washed with deionized water and dried with filtration equipment till the water mist disappears. Then, the crucible is heated in microwave oven at suitable power for 8 minutes (first time, 10 min for precipitates) and 3 minutes (second time). After each heating, the crucible should be cooled in a desiccator for 10-12 minutes and weighed with balance. If the difference between the two masses is smaller than 0.4 mg, the crucible can be used for next procedure. Otherwise, the above heating, cooling and weighing operation should be repeated.

② Drying with an electric heating oven. The crucible is washed with deionized water and dried in electric heating oven at (145±5)℃ for 1 hour first time and 30 minutes second time. After each heating, the crucible should be cooled in a desiccator for 30 minutes and weighed with a balance. If the difference between the two masses is smaller than 0.4 mg, the crucible can be used for next procedure. Otherwise, the above heating, cooling and weighing operation should be repeated.

(2) Preparation of sample solution   Nickel-chromium steel sample (0.35 g) is accurately weighed in a 250 mL beaker. The beaker is covered with a watch glass and then 20 mL HCl solution and 20 mL HNO₃ solution are added. Then the sample is heated till it is dissolved completely. The solution is boiled for another 10 minutes to eliminate nitrogen oxides. After cooling for a while, add in deionized water till the volume is 100 mL.

(3) Preparation of the precipitate   25 mL of the above solution is transferred into a 250 mL Erlenmeyer flask, and 100 mL of deionized water and 10 mL of tartaric acid solution are added. The mixed solution is heated and stirred in water-bath at 70 ℃, while ammonium solution is added drop by drop till the pH value is about 9. If there was precipitates, they should be removed by using slow filter paper. The filtrate is then collected in a 400 mL beaker, and the flask was washed three times with hot water, and then the filter paper was rinsed eight times with hot water. Finally the total volume of the solution was controlled at about 250-300 mL.

Subsequently, HCl solution (1 : 1) is added in drop by drop under continuous stirring till the pH value ranges in 3-4 and the color turns dark brownish green. Then the solution is heated in water-bath at 70 ℃, followed by addition of ethanol (20 mL) and dimethylglyoxime (35 mL). After the solution is mixed evenly, ammonia solution (1 : 1) is added in drop by drop till the pH value is about 8-9. Let the solution stand and age for 30 minutes. Complex of dimethylglyoxime and nickel will precipitate from the solution.

(4) Filtering, drying and weighing   The precipitates is transferred into the pre-treated crucible to remove liquid. Then, 10 mL of ethanol (20%) is used to wash the beaker and precipitates twice. After that, warm water is used to wash the beaker and precipitates for several times till there is no Cl⁻ in the filtrate. Finally, the precipitates are filtered till there

is no water mist in filter bottle.

Perform the procedure in Procedures (1) to make the weight of the precipitates constant. Record the mass of precipitates after weighing. The contents of nickel ($m/V, \%$) can be derived from the mass.

The molecular weight of dimethylglyoxime nickel is 288.91. The atom weight of nickel is 58.69.

Data Recording and Processing

Determination of nickel in alloy steel (Table 3.38)

Table 3.38 Determination of nickel in alloy steel

Number	1	2
m (empty crucible)/g		
m (sample)/g		
m (crucible and precipitates)/g		
m (precipitates)/g		
$w_{Ni}/\%$		
$\overline{w}_{Ni}/\%$		

Questions

(1) Why are the beaker and precipitates washed with ethanol solution first?

(2) How do you check whether Cl^- is removed completely?

Experiment 3.20   Determination of Trace Iron with Phenanthroline by UV-Vis Spectroscopy

Objectives

(1) Master the operation of a spectrophotometer.

(2) Understand the general procedure for investigation of experiment conditions and the methods for obtaining absorption curves and calibration curve.

(3) Master the principle of determining iron by using a spectrophotometer.

Principles

According to Lambert-Beer Law, absorbance (A) of substances is directly proportional to the concentration (C) of the substance within a certain concentration range when the incident light wavelength (λ) and the optical path (b) are constant. If we construct a calibration curve with absorbance as the ordinate and concentration as the abscissa, we can find the concentration of unknown sample from the calibration curve by measuring the absorbance.

Phenanthroline is usually used to determine trace iron in industrial products. It reacts with iron (II) and forms a red complex with stability constant ($\lg K$) of 21.3 and molar absorption coefficient of 1.1×10^4 L/(mol·cm) at 510 nm. Phenanthroline also reacts with iron (III) and forms light blue complex ($\lg K = 14.1$). To avoid interference, Fe (III) should be reduced to Fe(II) in advance by using hydroxylamine hydrochloride.

$$2Fe^{3+} + 2NH_2OH \cdot HCl \Longrightarrow 2Fe^{2+} + N_2 + 2H_2O + 4H^+ + 2Cl^- \qquad (3.57)$$

The acidity of solution affects the reaction between phenanthroline and iron. Too high acidity slows down the reaction rate, and too low acidity facilitates the hydrolysis of irons. A suggested pH value is about 5 which can be provided by using HAc-NaAc buffer solution.

This UV-vis spectroscopy method exhibits high selectivity. Sn^{2+}, Al^{3+}, Ca^{2+}, Mg^{2+}, Zn^{2+}, SiO_3^{2-}, with fortyfold iron equivalent, Cr^{3+}, Mn^{2+}, VO^{3-}, PO_4^{3-} with twentyfold, and Co^{2+}, Ni^{2+}, Cu^{2+} with fivefold don't interfere with the determination. Bi^{3+}, Cd^{2+}, Hg^{2+}, Zn^{2+}, Ag^+ can react with phenanthroline and form precipitates, which will interfere with the determination.

Materials

Instruments: spectrophotometer, pH meter, colorimetric cylinder (50 mL), pipette (1 mL, 2 mL, 5mL, 10 mL), cuvette, ear syringe.

Reagents: standard solution of iron (1.1×10^{-3} mol/L), standard solution of iron (100 μg/mL), hydrochloric acid, hydroxyammonium hydrochloride, sodium acetate (1 mol/L), phenanthroline (0.15%).

Procedures

(1) Preparatory work Open the UV-vis spectrometer and get it ready for the measurement.

(2) Measurement

① Absorption curve and selection of measurement wavelength. Measuring solution is first prepared by mixing 2.00 mL iron standard solution (1.0×10^{-3} mol/L), 1.00 mL hydroxyammonium hydrochloride (10%), 2.00 mL phenanthroline (0.15%) and 5.00 mL NaAc solution (1mol/L) in a colorimetric cylinder (50 mL), followed by dilution with deionized water till the water level reaches the mark. Reference solution is prepared by mixing the same solutions except iron standard solution. The absorbance of the measuring solution in the range of 440-560 nm is recorded on the UV-vis spectrometer by using a cuvette of 1 cm with interval of 10 nm. The absorbance curve is figured out with absorbance as the ordinate and wavelength as the abscissa, and the measuring wavelength is selected from the curve.

② Optimization of measurement conditions

a. Amount of phenanthroline. Two milliliters iron standard solution (1.0×10^{-3} mol/L) and 1.00mL hydroxyammonium hydrochloride (10%) are first added in 6 colorimetric cylinders (50 mL) and mixed well. Then 0.10 mL, 0.50 mL, 1.00 mL, 2.00 mL, 3.00 mL and 4.00 mL phenanthroline solution (0.15%) is added in the above 6 colorimetric cylinders, respectively. After that, 5.00 mL NaAc solution (1 mol/L) is added in and the solution is diluted with deionized water till the water level reaches the mark. Reference solution is prepared by mixing the above solutions except phenanthroline solution. Absorbance of the 6 solutions are recorded and a curve of absorbance-amount of phenanthroline is figured out with absorbance as the ordinate and the amount of phenanthroline as the abscissa. The optimized amount of phenanthroline can be derived from the curve.

b. pH value. Eight colorimetric cylinders (50 mL) are prepared, 2 mL iron standard solution (1.0×10^{-3} mol/L) and 1.00 mL hydroxyammonium hydrochloride (10%) are added in every cylinder. Two minutes later, optimized amount of phenanthroline solution (0.15%) is added in and mixed well. Then, different volumes, e. g., 0.00mL, 0.20mL, 0.50mL, 1.00mL, 1.50mL, 2.00mL and 3.00 mL of NaAc solution (1 mol/L) are added in the cylinders, respectively. The pH values of the 7 solutions are measured by using a pH meter. Then, the absorbance is recorded on UV-vis spectrometer. Reference solution is deionized water. Absorbance-pH curve is figured out with absorbance as the ordinate and the pH value as the abscissa. The optimized pH value (or the amount of NaAc) can be derived from the curve.

c. Reaction time. Measuring solution is prepared by mixing 2 mL iron standard solution (1.0×10^{-3} mol/L), 1.00 mL hydroxyammonium hydrochloride (10%), certain amount of phenanthroline solution (0.15%) and NaAc (1 mol/L). The mixed solution is diluted with deionized water till the water level reaches the mark. Absorbance-time curve is constructed with absorbance as the ordinate and the time as the abscissa. The optimized reaction time can be determined by the curve.

(3) Calibration curve Different volumes (0.00 mL, 2.00 mL, 4.00 mL, 6.00 mL, 8.00 mL, 10.00 mL) of iron standard solution (100 μg/mL) are added in six volumetric flasks (50 mL) respectively. Then, 1.00 mL of hydroxyammonium hydrochloride (10%) is added in each volumetric flask, followed by the addition of certain amount of phenanthroline solution (0.15%) and NaAc (1 mol/L). The mixed solution is diluted with deionized water till the water level reaches the mark. Reference solution is prepared by mixing all the above solutions except iron standard solution. After certain times, the absorbance are recorded with 1 cm cuvette at the optimized wavelength. Calibration curve is constructed with absorbance as the ordinate and the concentration as the abscissa.

(4) Determination of iron in samples The determination of iron in samples (water, industrial HCl, limestone sample) is conducted according to the above steps identically, except that iron standard solution is replaced by acertain amount of sample solution. The determination is repeated 3 times. The content of iron in samples (mg/L) can be derived from the calibration curve.

❊ Data Recording and Processing

(1) Absorption curve and selection of measurement wavelength (Table 3.39)

Table 3.39 Absorption Curve

Wavelength λ/nm											
Absorbance											
Wavelength λ/nm											
Absorbance											
Wavelength λ/nm											
Absorbance											

(2) Optimization of reaction conditions

① The amount of phenanthroline

Volume of phenanthroline/mL	0.10	0.50	1.00	2.00	3.00	4.00
Absorbance						

② pH value/volume of NaAc

Volume of NaAc/mL	0.00	0.20	0.50	1.00	1.50	2.00	3.00
Absorbance							

③ Reaction time

t/min	0	2	4	6	8	10	12	14	16	18	20
Absorbance											

(3) Calibration curve

$V_{100\,\mu g/mL}$/mL	0.00	2.00	4.00	6.00	8.00	10.00
Absorbance						

(4) Determination of iron in samples (Table 3.40)

Table 3.40 Determination of iron in samples

Sample	1	2	3
A			
$w_{Fe}/\%$			
$\overline{w}_{Fe}/\%$			

Questions

(1) Why should hydroxylamine hydrochloride solution be added in the determination of trace iron in this experiment?

(2) What's the function of the reference solution? Can deionized water be used as a reference solution in this experiment?

(3) What's the main conditions of the reaction between phenanthroline and iron?

Experiment 3.21 Potentiometric Titration of Phosphoric Acid

Objectives

(1) Master the operation of potentiometric titration.

(2) Master the methods for determining the stoichiometric point in potentiometric titration.

(3) Master the principle and method of determining pK_a of weak acid by potentiometric titration.

Principles

Potentiometric titration is a widely used analytical method for specific targets by determining the titration end point from the titration jump of electrode potential or pH value. Potentiometric titration exhibits advantages over other titration methods in reactions producing precipitates and solutions that are colored or turbid. Potentiometric titration can also be used to determine the ionization equilibrium constant of some specific targets.

For the potentiometric titration of phosphoric acid, standard solution of NaOH is used as titrant, pH meter is used to measure the change of pH value or electrode potential (Refer to section. 1.4.3 for the usage of pH meter). The titration curve for H_3PO_4 with NaOH is given in Fig. 3.2 with pH as the y-axis and the volume of NaOH as the x-axis. It can be seen from Fig. 3.2 that the first stoichiometric point and the second stoichiometric point are in the range of 4.0-5.0 and 9.0-10.0, respectively. The stoichiometric points can be derived from three tangent method, first derivative (dpH/dv-V) plot (Fig. 3.3) and second derivative plot (d^2pH/d^2v-V).

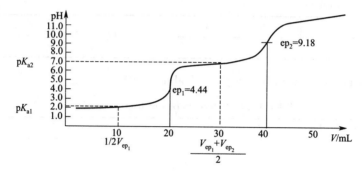

Fig. 3.2 The titration curve for H_3PO_4 with NaOH

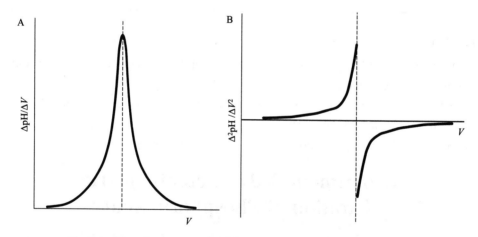

Fig. 3.3 First derivative (dpH/dv-V) plot and second derivative plot

Phosphoric acid is a polyfunctional acid, and the pK_a can be determined by the potentiometric titration method.

$$H_3PO_4 + H_2O \rightleftharpoons H_3O^+ + H_2PO_4^- \tag{3.58}$$

$$K_{a1} = [H_3O^+][H_2PO_4^-]/[H_3PO_4] \tag{3.59}$$

When the concentration of residual H_3PO_4 equals to that of produced $H_2PO_4^-$, e.g., $[H_3PO_4]=[H_2PO_4^-]$, $K_{a1}=[H_3O^+]$, $pK_{a1}=pH$.

$$H_2PO_4^- + H_2O \rightleftharpoons H_3O^+ + HPO_4^{2-} \tag{3.60}$$

$$K_{a2} = [H_3O^+][HPO_4^{2-}]/[H_2PO_4^-] \tag{3.61}$$

Similarly, when half of the H_3O^+ produced by the secondary dissociation are consumed by NaOH, $[H_2PO_4^-]=[HPO_4^{2-}]$, $K_{a2}=[H_3O^+]$, $pK_{a2}=pH$. The stoichiometric point can be determined from titration curve and the pK_a value can be derived from the stoichiometric point as illustrated in Fig. 3.2.

Materials

Instruments: pH meter (PHSJ-4A), magnetic stirrer, burette (50 mL), pipette, flask (100mL), volumetric flask (250 mL).

Reagents: phosphoric acid (0.1 mol/L), standard solution of NaOH (0.1 mol/L), standard buffer solution with pH values of 4.00, 6.68, and 9.18, methyl orange indicator (0.1%, in ethanol), phenolphthalein indicator (1%, in ethanol).

Procedures

(1) Preparation and standardization of NaOH solution (0.1000 mol/L) Refer to Exp. 3.4 for detailed procedures.

(2) Calibrate the pH meter with a standard buffer solution Refer to section. 1.4.3 for specific procedures.

(3) Assemble the titration device as shown in Fig. 3.4.

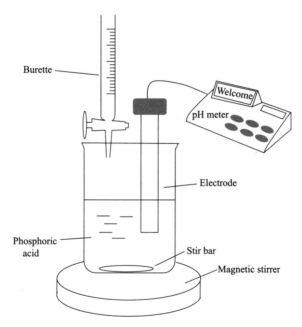

Fig. 3.4 Scheme of the titration device

(4) Titration process Phosphoric acid sample solution (20.00 mL, 0.1 mol/L) is added accurately into a 100 mL beaker from burette, stir bar and electrode of pH meter are then placed in solution, 2 or 3 drops of methyl orange and phenolphthalein are added subsequently. The titration is begun with 0.1000 mol/L NaOH after turning on the magnetic stirrer at an appropriate stirring speed. During the titration process, the pH values and volumes are recorded when every 2.00 mL of NaOH is added till the volume is close to 20.00 mL. After that, the pH value of the solution increases gradually and the volume interval is gradually decreased. When it is very close to the stoichiometric point, record the pH value after every drop (about 0.05 mL) of NaOH is added until the first titration jump is comple-

ted. At this time, the first stoichiometric point can also be observed by the color change of methyl orange. Please note that the pH meter should be calibrated by using alkaline standard solution (pH=9.86) when the pH value is about 7. The second titration jump and the second stoichiometric point are determined similarly to the first one. Continue the titration till the pH value is about 11.0.

Data Recording and Processing

(1) Preparation and standardization of NaOH solution (0.1000 mol/L) (Table 3.41)

Table 3.41 Preparation and standardization of NaOH solution (0.1000 mol/L)

Number	1	2	3
m_{NaOH}/g			
$m_{initial}$/g			
m_{final}/g			
$m_{standard}$/g			
$V_{initial}$/mL			
V_{final}/mL			
$V_{titrant}$/mL			
c_{NaOH}/(mol/L)			
$\bar{c}_{NaOH}$/(mol/L)			
$|d_i|$			
$\bar{d}_r$/%			

(2) Potentiometric titration of phosphoric acid (Table 3.42)

Table 3.42 Potentiometric titration of phosphoric acid

V_{NaOH}/mL							
pH value							
V_{NaOH}/mL							
pH value							
V_{NaOH}/mL							
pH value							
V_{NaOH}/mL							
pH value							
V_{NaOH}/mL							
pH value							

(3) Calculate the accurate concentration of NaOH and H_3PO_4. Detailed calculation procedures should be provided.

(4) Construct the curves of pH-V, $\Delta pH/\Delta V$-V, and $\Delta^2 pH/\Delta V^2$-V according to the data in Table 3.42 and determine the stoichiometric point from the curves.

(5) Find out the pH of the half neutralization point of the first stoichiometric point from the pH-V curve, and this is the pK_{a1} of H_3PO_4. Similarly, the pH of the half neutralization point from the first stoichiometric point to the second stoichiometric point is pK_{a2} of H_3PO_4.

Questions

(1) When H_3PO_4 is titrated with NaOH, the volume of NaOH consumed by the first stoichiometric point equals to that of the second stoichiometric point in theory. But they are

not the same actually. Try to explain the reason.

(2) Can the third dissociation constant of H_3PO_4 (K_{a3}) be determined from the titration curve? Illustrate the detailed method if yes. Explain the reason if not.

Tips

(1) Three tangent plot   Make two tangents AB and CD at the flat turning points of the titration curve, and make a tangent EF at the titration jump of the curve which intersects AB and CD at points Q and P. Then make two straight lines PG and QH parallel to the horizontal axis through points P and Q. Afterwards, make a parallel line at the very middle of PG and QH which intersects with the titration curve at point O. This point is the stoichiometric point from which we can find the pH value and the volume of titrant (Fig. 3.5).

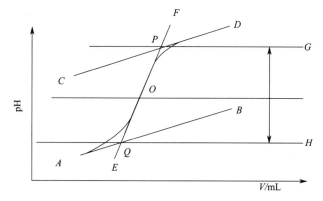

Fig. 3.5   Scheme for three tangent method

(2) First derivative (dpH/dv-V) plot and second derivative plot (d^2pH/d^2v-V)   Open an Excel sheet, input V in A1 and pH in B1, and input the recorded volume and pH into column A and column B from line 2. Select the data in the two columns, and click the "chart" button (scatter diagram with smooth line) in the "insertion" menu to obtain the potentiometric titration curve. Add the axis title to complete the graph.

Input ΔV in C1. Input "=A3−A2" in C3 to get the first value of ΔV. Select C3, move the cursor to the lower right corner of C3 till a black cross appears. Then click the left button of the mouse and pull down to the last line then obtain all ΔV values. Input ΔpH in D1 and obtain all ΔpH values by using the identical procedure. Input "=D3/C3" in E3 to obtain $\Delta pH/\Delta V$. Select the data in column A and column E to construct first derivative curve. Add the axis title to complete the graph.

Input $\Delta^2 pH$ in F1. Input "=D4−D3" in F4 to get the first value of $\Delta^2 pH$. Select F4, move the cursor to the lower right corner of F4 till a black cross appears. Then click the left button of the mouse and pull down to the last line obtain all $\Delta^2 pH$ values. Input "$\Delta^2 pH/\Delta^2 V$" in G1 and "=F4/C4" in G4 to obtain $\Delta^2 pH/\Delta^2 V$. Select the data in column A and column F to construct second derivative curve. Add axis title to complete the graph.

Chapter 4

Comprehensive Design Experiments

4.1 Chemistry

Experiment 4.1 Determination of Benzoic Acid Content

Objectives

(1) Master the titration principles of aromatic carboxylic acid (weak acid) in ethanol solvent and the selection of indicators.

(2) Master the preparation and detection method of neutral ethanol reagent.

Principles

Benzoic acid, also known as phenyl formic acid, is an aromatic acid organic compound and the simplest aromatic acid. Its molecular formula is $C_7H_6O_2$ ($M=122.12$ g/mol). See Fig. 4.1 for its structure. At the normal temperature, it is shiny, white, monoclinic flake or needle crystal, odorless or slight odor similar to benzoin or benzaldehyde. It can be slightly soluble in cold water and hexane, easily soluble in ethanol, ether, chloroform, benzene and other organic solvents. Benzoic acid exists widely in nature in the form of free acids, esters or its derivatives.

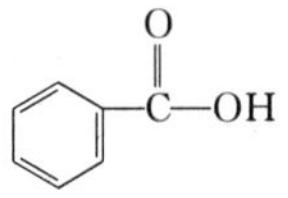

Fig. 4.1 The structure of benzoic acid

Benzoic acid can inhibit the growth of fungi, bacteria and mould. It is often used as a preservative in food or medicine. An appropriate amount of benzoic acid can inhibit the growth and reproduction of microorganisms, and prolong the storage life of food and drugs. However, excessive intake of benzoic acid will do great harm to the human body and cause certain irritation to the stomach, skin and mucous membrane. For sensitive people, more intake of benzoic acid and its salts will cause adverse reactions, such as drooling, abdominal pain, diarrhea, asthma, urticaria, metabolic acidosis and convulsions. Therefore, the determination of benzoic acid is very important.

At present, the main methods for the determination of benzoic acid are gas chromatography, thin layer chromatography, high performance liquid chromatography, UV spectrophotometry, acid-base titration and so on. Chromatography has high sensitivity and resolution and it is one of the important analytical methods for the detection of preservatives, but it has high requirements for equipment. In contrast, titration analysis is a common method for the determination of benzoic acid in the laboratory due to its low equipment requirements.

Benzoic acid is a monobasic weak acid ($k_a = 6.28 \times 10^{-5}$), NaOH standard solution can be used for direct titration[Formula(4.1)], and phenolphthalein is used as an indicator. At the stoichiometric point, the sodium benzoate hydrolysis solution is slightly alkaline, where phenolphthalein turns red and indicates the end point.

(4.1)

Since benzoic acid is slightly soluble in water and easily soluble in ethanol, the titration reaction must use ethanol as solvent. Ethanol is generally acidic, so in this acid-base titration, neutral ethanol solution should be used as solvent. The so-called "neutral" means that ethanol is neutral for the indicator, so as not to interfere with the discoloration of the indicator.

Materials

Instruments: analytical balance, platform scale, beaker, Erlenmeyer flask, graduated cylinder and burette.

Reagents: NaOH (solid), potassium hydrogen phthalate (KHP, G.R.), benzoic acid sample; 0.2% phenolphthalein ethanol indicator: 0.2 g phenolphthalein is dissolved in 100 mL 95% ethanol solution; neutral ethanol solution (neutral to phenolphthalein indicator): take 50 mL of absolute ethanol, add 50 mL of water, add 2-3 drops of phenolphthalein indicator, and titrate with 0.1 mol/L NaOH standard solution until the solution is light pink for later use.

Procedures

(1) Preparation and standardization of NaOH (0.1 mol/L) Preparation: weigh a certain amount of NaOH solid, fully dissolve it in deionized boiled water and cool to the room temperature, and transfer it into a polyethylene plastic bottle for later use.

Standardization: estimate the weighing mass of standard substance KHP according to the concentration of NaOH solution and titration error requirements. Use the decrement method to accurately weigh the calculated amount of KHP into a 250 mL Erlenmeyer flask, and add 20-30 mL of water to dissolve it. After that, 2-3 drops of phenolphthalein indicator were added, then titrate it with NaOH solution to light pink and keep the color for 30s, which is the end point. Record the consumed volume of NaOH solution. Repeat the titration 3 times (Table 4.1).

(2) Sample analysis Accurately weigh about 0.3 g of benzoic acid sample, put it into a 250 mL Erlenmeyer flask, add 25 mL of neutral ethanol solution to fully dissolve it, add 2-3 drops of phenolphthalein indicator, and titrate it with NaOH standard solution until the solution is light pink (does not fade within 30s) as the end point. Record the consumed volume of NaOH solution. Repeat the titration 3 times (Table 4.2).

Data Recording and Processing

According to the consumed volume of NaOH standard solution, calculate the mass percentage of benzoic acid by the following formula [Formula(4.2)]:

$$w = \frac{\bar{c}_{NaOH} V_{NaOH} M_{C_7H_6O_2}}{m_s \times 1000} \times 100\% \tag{4.2}$$

Where, $M_{C_7H_6O_2}$ is the molar mass of benzoic acid, 122.12 g/mol; m_s is the mass of sample, g.

Table 4.1 Standardization of NaOH solution

No.	1	2	3
m_{KHP} (initial)/g			
m_{KHP} (final)/g			
m_{KHP}/g			
V_{NaOH} (initial)/mL			
V_{NaOH} (final)/mL			
V_{NaOH} (consumed)/mL			
c_{NaOH}/(mol/L)			
$\bar{c}_{NaOH}$/(mol/L)			
$|d_i|$			
$\bar{d}_r$/%			

Table 4.2 Determination of benzoic acid mass fraction

No.	1	2	3
m_s/g			
V_{NaOH} (initial)/mL			
V_{NaOH} (final)/mL			
V_{NaOH} (consumed)/mL			
w/%			
$\bar{w}$/%			
$|d_i|$			
$\bar{d}_r$/%			

Questions

(1) Is neutral diluted ethanol really neutral? Why is this neutral diluted ethanol used instead of water?

(2) What kind of measuring instrument is used to measure neutral diluted ethanol in the experiment, and does it need to be measured accurately?

(3) In this experiment, benzoic acid was dumped too much (the weight exceeded 0.7 g) during the sample weighing by decrement method. Do you need to weigh again? Why?

(4) If NaOH standard solution absorbs CO_2 in the air, what is the impact on the determination results?

Experiment 4.2 Determination of Phenol Content

Objectives

(1) Be familiar with the principle, method and basic skills of $KBrO_3$ method for the de-

termination of phenol content.

(2) Understand the practical significance of "blank test" and learn the method and application of "blank test".

Principles

Phenol is an important chemical product. It is often used in the production of dyes, resins, drugs, synthetic fibers and other polymer materials. It is also widely used in disinfection, sterilization and so on. At the same time, phenol also has strong toxicity. It enters the human body mainly through the skin, digestive tract and respiratory tract, which corrodes, damages and inhibits the human mucosa, cardiovascular and central nervous system. The wastewater of many chemical industries and pharmaceutical enterprises contains phenol. If the wastewater is directly discharged into the environmental water, it will cause serious pollution to soil and water resources, and then do harm to humans and other organisms. Therefore, the monitoring of phenol content is very necessary in practical application.

At present, the determination methods of phenol content in water mainly include chromatography, electrochemical analysis, fluorescence analysis, spectrophotometry, chemiluminescence method and so on. Among them, titrimetric analysis is a common analytical method in the laboratory due to its low equipment requirements.

Br_2 can react with phenol to form a stable white precipitate of tribromophenol. The reaction is as follows:

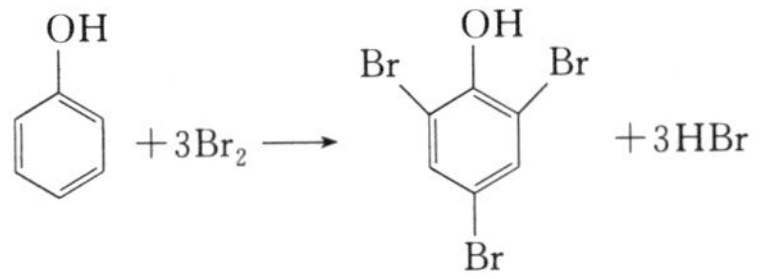

$$C_6H_5OH + 3Br_2 \longrightarrow C_6H_2Br_3OH + 3HBr \qquad (4.3)$$

Since the above reaction [Formula (4.3)] is slow and the Br_2 is very volatile, the Br_2 solution cannot be used directly for titrate reactions. $KBrO_3$ standard solution is generally used to react with excess KBr in acidic medium to produce a considerable amount of free Br_2:

$$BrO_3^- + 5Br^- + 6H^+ \rightleftharpoons 3Br_2 \text{(yellow)} + 3H_2O \qquad (4.4)$$

After the bromination reaction between the generated Br_2 and phenol, the remaining Br_2 can oxidize the excess KI to produce a certain amount of I_2. Then titrate the generated I_2 with $Na_2S_2O_3$ standard solution to calculate the content of phenol. The reaction is as follows:

$$Br_2\text{(left)} + 2I^- \rightleftharpoons 2Br^- + I_2 \qquad (4.5)$$

$$I_2 + 2S_2O_3^{2-} \rightleftharpoons 2I^- + S_4O_6^{2-} \qquad (4.6)$$

It can be seen that the reaction process has the following stoichiometric relationship, and then the content of phenol can be calculated.

$$KBrO_3 \approx 3\ Br_2 \approx C_6H_5OH \approx 3\ I_2 \approx 6\ Na_2S_2O_3 \qquad (4.7)$$

In this experiment, the blank test is carried out at the same time, and the determination of the sample and blank test are carried out under the same conditions, which can reduce the error.

Materials

Instruments: analytical balance, platform scale, pipette, beaker, volumetric flask, burette, iodine flask, graduated cylinder.

Reagents: $Na_2S_2O_3$ (0.05 mol/L, Table 4.3), 0.5% starch aqueous solution, KI solution (A.R.), 10% NaOH solution, $KBrO_3$ (standard substance), 1:1 HCl solution, KBr (A.R.) and industrial phenol sample.

Procedures

(1) Preparation of $KBrO_3$-KBr standard solution (about 0.02 mol/L) Accurately weigh 0.25-0.30 g of dry $KBrO_3$, add 1 g of KBr, dissolve it with an appropriate amount of deionized water, and fix its volume in a 100 mL volumetric flask for subsequent use.

(2) Determination of phenol content

① Determination of sample: accurately weigh about 0.2 g phenol sample, add 5 mL 10% NaOH solution, add a small amount of water to dissolve it, and fix the volume in a 250 mL volumetric flask. Accurately measure 25.00 mL of the phenol solution into a 250 mL iodine flask, add 10.00 mL $KBrO_3$-KBr standard solution and 10 mL 1:1 HCl solution, plug it immediately and add water to seal the bottle mouth, shake it fully for 1-2 min and let it stand for 10 min. Add 1 g KI, cap the stopper, shake and dissolve for 5 min, rinse the bottle stopper and the attachment on the bottleneck with a small amount of water, add 25 mL of water, and then titrate with $Na_2S_2O_3$ standard solution until the solution is light yellow, add 1-2 mL of starch indicator, continue titrating until the blue color fades out, which is the end point. Three samples are measured in parallel (Table 4.4).

② Blank test: accurately transfer 25.00 mL of deionized water instead of phenol test solution into a 250 mL iodine flask. Other operations are the same as those for the determination of phenol sample, and repeat the titration 3 times.

Data Recording and Processing

According to the relevant data in the experiment, calculate the percentage content of phenol through the following formula:

$$w = \frac{\left(c_{KBrO_3} V_{KBrO_3} - \frac{1}{6} \times \bar{c}_{Na_2S_2O_3} V_{Na_2S_2O_3}\right) M_{C_6H_5OH}}{m_s \times \frac{25}{250} \times 1000} \times 100\% \tag{4.8}$$

Where, m_s is the mass of phenol sample, g; $M_{C_6H_5OH}$ is the molar mass of phenol, 94.11 g/mol.

Table 4.3 Standardization of $Na_2S_2O_3$ solution

No.	1	2	3
m_{KBrO_3} (initial)/g			
m_{KBrO_3} (final)/g			
m_{KBrO_3}/g			
c_{KBrO_3}/(mol/L)			
V_{KBrO_3}/mL			
$V_{Na_2S_2O_3}$ (initial)/mL			

(continued)

No.	1	2	3
$V_{Na_2S_2O_3}$ (final)/mL			
$V_{Na_2S_2O_3}$ (consumed)/mL			
$c_{Na_2S_2O_3}$ /(mol/L)			
$\bar{c}_{Na_2S_2O_3}$ /(mol/L)			
$\|d_i\|$			
$\bar{d}_r$/%			

Table 4.4 Determination of phenol content

No.	1	2	3
m_s (sample)/g			
V (sample)/mL			
$V_{Na_2S_2O_3}$ (initial)/mL			
$V_{Na_2S_2O_3}$ (final)/mL			
$V_{Na_2S_2O_3}$ (consumed)/mL			
w/%			
$\bar{w}$/%			
$\|d_i\|$			
$\bar{d}_r$/%			

Questions

(1) Why add 10% NaOH solution when dissolving the phenol sample?

(2) After add 10 mL of 1 : 1 HCl solution and cover the plug tightly. What happens when let the solution stand? Why?

(3) Can the remaining Br_2 be titrated directly with $Na_2S_2O_3$ standard solution?

Experiment 4.3 Determination of Purity of Soluble Barium Salt by Gravimetric Method

Objectives

(1) To understand the principles and methods of determining the content of soluble barium salt by the barium sulfate gravimetric method.

(2) To understand the precipitation conditions, principles and methods of crystal precipitation.

(3) To master the basic operation techniques of filtration, washing and constant weight of crystal precipitation.

(4) To master the operation process of drying constant weight samples in a microwave oven.

Principles

Gravimetric method is to determine the component content by weighing the mass of the substance. In analysis, generally separate the component from the sample by appropriate methods, convert it into a certain weighing form and weigh it, and then calculate the content of the component from the mass.

Ba^{2+} ions can interact with some acid radical ions to form a series of slightly soluble or insoluble compounds, such as $BaCO_3$ ($K_{sp} = 2.58 \times 10^{-9}$, 298.15 K), BaC_2O_4 ($K_{sp} = 1.6 \times 10^{-7}$, 298.15 K) and $BaSO_4$ ($K_{sp} = 1.08 \times 10^{-10}$, 298.15 K). Among them, $BaSO_4$ has the lowest solubility and the most stable properties. Its chemical composition is consistent with the chemical formula, and its molar mass is large, which meets the requirements of gravimetric analysis for precipitation. $BaSO_4$ gravimetric method can be used not only to directly determine the content of Ba^{2+} and SO_4^{2-}, but also to indirectly determine the content of sulfur in sulfur-containing compounds that can be converted to SO_4^{2-}, such as sulfur trioxide in cement, sulfur in silicate, coal, and iron ore, etc. Due to the high accuracy of this method, the determination results of gravimetric method are also commonly used as a standard to check the accuracy of other analysis methods.

In the experiment, Ba^{2+} and SO_4^{2-} can form $BaSO_4$ precipitation, which is weighed in the form of $BaSO_4$ after aging, filtration, washing and drying, so as to obtain the content of Ba^{2+} or SO_4^{2-}. $BaSO_4$ precipitation is generally a fine crystal when it is initially formed, and it is easy to pass through the filter paper during filtration, resulting in the precipitation loss. In order to obtain the crystal precipitation of $BaSO_4$ as large and pure particles and prevent the generation of $BaCO_3$ and $Ba(OH)_2$, it is generally precipitated in HCl medium of 0.05 mol/L (appropriately increasing the acidity can increase the solubility of $BaSO_4$ in the precipitation process to reduce its relative supersaturation, which is conducive to obtain better crystalline precipitation). After acidification, heat the solution until it boils, add the diluted and hot H_2SO_4 solution slowly under continuous stirring, and Ba^{2+} reacts with SO_4^{2-} to form crystalline precipitation.

Compared with the traditional muffle furnace heating, the microwave oven heating is rapid and uniform, and can quickly reach a very high temperature. The sand core crucible does not need filter paper, so as to avoid the reduction of carbon on $BaSO_4$ precipitation at high temperature. However, if the precipitation contains high boiling point impurities such as H_2SO_4, it is difficult to decompose or volatilize during microwave heating and drying $BaSO_4$ precipitation. Therefore, the precipitation conditions and washing operations should be controlled. For example, the solution containing Ba^{2+} should be further diluted, and the excess of H_2SO_4 precipitant should be controlled to 20%-50%.

Materials

Instruments: analytical balance, hot plate, beaker (100 mL, 250 mL), glass rod, microwave oven, circulating water vacuum pump (equipped with suction bottle), measuring cylinder, watch glass, G4 sand core crucible, dryer.

Reagents: H_2SO_4 solution (1 mol/L, 0.1 mol/L); HCl solution (2 mol/L), $BaCl_2 \cdot 2H_2O$ (industrial product).

Procedures

(1) Preparation of $BaSO_4$ precipitation Accurately weigh 0.4-0.5 g of $BaCl_2 \cdot 2H_2O$

sample into a 250 mL beaker, add 100 mL of deionized water and 4 mL of HCl solution (2 mol/L), stir to dissolve and heat to near boiling. Take another 3 mL of H_2SO_4 solution (1 mol/L) into a 100 mL beaker, add 30 mL of deionized water, heat it to near boiling, and drop the diluted and hot H_2SO_4 solution (1 mol/L) into $BaCl_2$ solution under continuous stirring until the addition is completed. After $BaSO_4$ precipitation sinks, add 1-2 drops of 0.1 mol/L H_2SO_4 solution to the supernatant to determine whether $BaSO_4$ precipitates completely. After the precipitation is complete, cover the watch glass (do not take the glass rod out of the cup to avoid the loss of precipitation) and age for 12 h (tilt the beaker so that the precipitation is concentrated on one side of the beaker to facilitate separation and transfer). It can also be aged for 0.5 h in hot water bath or sand bath.

(2) Constant weight of crucible Wash the crucible with deionized water, do the suction filtration until the water mist disappears, and then heat the crucible in a microwave oven (10 min for the first time and 4 min for the second time). After that, transfer the crucible to the dryer to cool for 10-15 min and weigh it. If the difference in mass between the two times does not exceed 2 mg, which means that the weight is constant (otherwise, continue heating, cooling and weighing until constant weight). Record the mass of the crucible as m_1.

(3) Treatment of precipitation Transfer all the aged $BaSO_4$ precipitation to a dry crucible of constant weight, filter under reduced pressure, and wash the precipitation with warm water for several times. Put the crucible with precipitation into the microwave oven for drying, then cool and weigh until constant weight [perform the same operations as in Step (2) above]. At this time, record the mass of the crucible as m_2.

Data Recording and Processing

According to the content of barium in barium sulfate precipitation, calculate the mass of barium chloride, and then calculate the purity of $BaCl_2 \cdot 2H_2O$ sample (Table 4.5).

$$\eta = \frac{\dfrac{M_{BaCl_2 \cdot 2H_2O}}{M_{BaSO_4}} \times (m_2 - m_1)}{m_{BaCl_2 \cdot 2H_2O}} \times 100\% \tag{4.9}$$

Where, $M_{BaCl_2 \cdot 2H_2O} = 244.26$ g/mol; $M_{BaSO_4} = 233.39$ g/mol.

Table 4.5 Determination of purity of barium chloride

$m_{sBaCl_2 \cdot 2H_2O}$/g	m_1/g	m_2/g	(m_2-m_1)/g	η/%

Questions

(1) Why is the precipitation process carried out in a diluted HCl solution medium?

(2) In the process of precipitation when the heat H_2SO_4 solution is added drop by drop, why can't the speed be too fast and why does it need to be stirred constantly?

(3) In the whole process, why is it required that the glass rod cannot be taken out until the filtration and washing are completed?

(4) Why is it required that the glass rod should not touch the cup wall and bottom

when stirring?

(5) What is the purpose of tilting the beaker during aging?

Experiment 4.4 Determination of Methanol Content in Industrial Alcohol

 Objectives

(1) To master the principle and operation of determining methanol content by spectrophotometry.

(2) To master the basic operation of a spectrophotometer.

 Principles

Alcohol is an important raw material for making wine. However, some alcohol products contain a certain amount of methanol, which has a great impact on the human nervous system and blood system. Excessive intake through the digestive tract, respiratory tract or skin will produce toxic reactions. In recent years, the diseases caused by methanol poisoning and accumulation have become more and more frequent. Therefore, it is necessary to monitor the methanol content in alcohol products.

At present, the commonly used methanol detection methods mainly include spectrophotometry, gas chromatography, Raman spectroscopy, refraction method, etc. In contrast, spectrophotometry has the advantages of low equipment requirements and is suitable for large-scale product detection, but it has certain requirements for temperature, time and other conditions, which need to be strictly controlled. The magenta-sulfurous acid method is a common method to detect the content of methanol in industrial alcohol. The principle of the method is methanol in ethanol is oxidized to formaldehyde (HCHO) by $KMnO_4$ in H_3PO_4 medium:

$$5CH_3OH + 2KMnO_4 + 4H_3PO_4 = 5HCHO + 2KH_2PO_4 + 2MnHPO_4 + 8H_2O \quad (4.10)$$

Excessive $KMnO_4$ can be reduced by oxalic acid ($H_2C_2O_4$):

$$5H_2C_2O_4 + 2KMnO_4 + 3H_2SO_4 = 2MnSO_4 + K_2SO_4 + 10CO_2\uparrow + 8H_2O \quad (4.11)$$

Then, the colorless magenta-sulfurous acid was used as chromogenic agent to react with formaldehyde to form a quinone-type purple compound. Measure the absorbance at 590 nm and compare with the standard series.

 Materials

Instruments: spectrophotometer, colorimetric tube, cuvette, pipette, volumetric flask, balance, beaker and centrifuge.

Reagents

Potassium permanganate-phosphoric acid solution: Transfer 15 mL phosphoric acid solution (85%) into 70 mL water, and fully mix, then add 3 g potassium permanganate. After fully dissolved, its volume is fixed into 100 mL with distilled water, and stored in a brown reagent bottle for standing (note that the storage time should not be too long).

Oxalic acid-sulfuric acid solution: Dissolve 7 g oxalic acid dihydrate ($H_2C_2O_4 \cdot 2H_2O$) in 1:1 cold sulfuric acid, and determine the volume to 100 mL with the 1:1 sulfuric acid. Store the solution in a brown bottle for standing.

Magenta-sulfurous acid solution: Weigh 0.1 g of fine grinding alkaline fuchsin, add ~80℃ water several times (total of 60 mL), grind while adding water, centrifuge after full dissolution to obtain the supernatant in a 100 mL reagent bottle. After cooling, add 1 mL of concentrated hydrochloric acid and 10 mL of sodium sulfite solution (0.1 g/mL), add water to constant volume, and store it in a brown bottle (Note: if the solution is red, it needs to be prepared once more).

Methanol standard stock solution: Take 1.000 g methanol into a 100 mL volumetric flask with a small amount of distilled water, and add a constant volume of water to obtain a methanol standard solution (10 mg/mL), which is stored at low temperature for subsquent use.

Methanol standard solution: Accurately transfer 10.00 mL methanol standard stock solution and add water into a 100 mL volumetric flask to prepare 1 mg/mL methanol standard solution.

Ethanol solution (excluding methanol and formaldehyde): Add a small amount of $KMnO_4$ into 500 mL absolute ethanol, let stand for 24 hours after shaking and then distill it. Discard the initial and last 1/10 fraction, only collect the middle fraction.

Procedures

(1) Determination of maximum absorption wavelength Transfer 1.00 mL of methanol standard solution into a 25 mL colorimetric tube with a plug, add 0.5 mL of ethanol solution without methanol and formaldehyde, 2 mL of potassium permanganate-phosphoric acid solution and 3.5 mL of distilled water, fully mix and let stand for 15 min, then add 2 mL of oxalic acid-sulfuric acid solution to mix well. After its color fades, add 5 mL of magenta-sulfurous acid solution, fully mix and let stand at 25 ℃ for 30 min. Use a 1 cm cuvette on the spectrophotometer, take the reagent blank as the reference, measure the absorbance every 10 nm between 500-630 nm (between 580-600 nm, measure every 2 nm), draw the absorption curve with the wavelength as the abscissa and the absorbance as the ordinate, and select the appropriate wavelength for measurement (generally the maximum absorption wavelength λ_{max} is the measurement wavelength).

(2) Determination of experimental conditions Determine the optimum addition amounts of potassium permanganate-phosphoric acid solution, oxalic acid-sulfuric acid solution and magenta-sulfurous acid solution, respectively.

Transfer 1.00 mL methanol standard solutions (1 mg/mL) into several 25 mL colorimetric tubes with a plug, add 0.5 mL of ethanol solution without methanol and formaldehyde, add distilled water to 5 mL, mix well, add different amounts (0-3 mL) of potassium permanganate-phosphoric acid solution, mix well and let stand for 10 min. Add 2 mL of oxalic acid-sulfuric acid solution to each tube, mix well and let it stand for fading, then add 5 mL of magenta-sulfurous acid solution, mix well and let it stand at 25 ℃ for 0.5 h, take the reagent blank as the reference, measure the absorbance at the maximum absorption wavelength, and select the amount of potassium permanganate-phosphoric acid solution when the absorption value is large and stable.

Similarly, determine respectively the optimum addition amounts of oxalic acid-sulfuric acid solution and magenta-sulfurous acid solution.

(3) Plotting the standard curve   Accurately transfer 0.00 mL, 0.20 mL, 0.40 mL, 0.60 mL, 0.80 mL and 1.00 mL of methanol standard solution (1 mg/mL) into 25 mL colorimetric tube with a plug, add 0.5 mL of ethanol solution without methanol and formaldehyde, add water to the 5 mL, mix well, add 2 mL (based on the optimized amount) of potassium permanganate-phosphoric acid solution, mix well, let stand for 10 minutes, and add 2 mL (based on the optimized amount) of oxalic acid-sulfuric acid solution. After mixing, let stand for fading, add 5 mL (based on the optimized amount) of magenta-sulfurous acid solution, mix well and let stand at 25 ℃ for 0.5 h. Adjust the zero point (reagent blank) without methanol, measure the absorbance at the maximum absorption wavelength, and draw the standard curve with the methanol concentration as the abscissa and the absorbance as the ordinate.

(4) Determination of sample   Transfer 0.50 mL industrial ethanol sample into a 25 mL colorimetric tube with a plug, add water to 5 mL, and other operations are the same as the standard curve determination.

Data Recording and Processing

Methanol content calculation formula:

$$X = \frac{m}{V \times 1000} \times 100 \tag{4.12}$$

Where, X is the methanol content in samples, g/100 mL; m is the mass of methanol in samples, mg; V is the sample volume, mL.

Notes:

(1) The new prepared magenta-sulfurous acid solution should be placed in the refrigerator for 1-2 days before using. It should be re-prepared when the solution is red.

(2) In the above detection process, aldehydes in the industrial ethanol and others produced from other alcohols oxidized by $KMnO_4$, such as acetaldehyde and propionaldehyde, also react with the magenta-sulfurous acid solution. However, under certain acidic conditions, except that formaldehyde can form a long-lasting purple red, the color formed by other aldehydes will slowly fade, so there is no interference. Therefore, the time conditions in operation must be strictly controlled.

(3) In the experiment, heat will be released after adding oxalic acid-sulfuric acid solution, and the temperature will rise. Magenta-sulfurous acid solution should be added after cooling. Because the system temperature has an impact on the absorbance, the temperature of the sample tube and the standard solution tube should be kept as consistent as possible, and the temperature difference should be controlled within 1 ℃.

Experiment 4.5 Determination of Acetic Acid Content by Potentiometric Titration

Objectives

(1) To master the basic principle and method of potentiometric titration.

(2) To master the method of determination of acetic acid in edible vinegar and to use pH meter skillfully.

(3) To master the drawing of potentiometric titration curves and the calculation method of titration end point.

Principles

Acetic acid is an organic weak acid ($k_a = 1.8 \times 10^{-5}$) and could be titrated directly by NaOH standard solution [see Equation(4.13)]. Since weak acids satisfying $ck_a > 10^{-8}$ can be accurately titrated by NaOH, the total acid content is actually determined by this method and the result is expressed as acetic acid with the highest content ρ_{HAc} (g/L). The titration reaction product is NaAc, and the solution pH is about 8.7 at the stoichiometric point, so phenolphthalein can be used as an indicator. The color of the solution changes from colorless to slight pink at the end point.

$$NaOH + CH_3COOH = CH_3COONa + H_2O \tag{4.13}$$

Potentiometric titration is to determine the end point of titration by measuring the electromotive force change of the primary battery which is composed of two electrodes inserted into the solution, without the use of an indicator. Therefore, it can be used for the titration analysis of colored or turbid solution, and also for the titration reaction without a suitable indicator.

The experiment is an acid-base neutralization reaction, and the concentration of hydrogen ion in the solution (pH of the solution) changes continuously during the titration process, and a pH jump occurs near the stoichiometric point. So the end point of the titration can be determined by measuring the change of solution pH. A glass electrode with selective response to hydrogen ions is used as the indicator electrode, and a saturated calomel electrode is used as the reference electrode to form a working battery with the test solution, that is pH meter. The test solution is potentiometric titrated with NaOH standard solution to measure the change of pH value with the addition of NaOH solution. The experimental setup is shown in Fig. 4.2. The pH-V curve [Fig. 4.3(a)] is drawn with the volume of NaOH solution added as the horizontal coordinates and the pH value as the vertical coordinates to determine the titration end point. In addition, the titration end point can also be determined by the first-order derivative method [$\Delta pH/\Delta V$-V, Fig. 4.3(b)] or the second-order derivative method [$\Delta^2 pH/\Delta V^2$-V, Fig. 4.3(c)]. Then calculate the acetic acid content according to the concentration of NaOH solution, the volume of NaOH solution consumed at the measurement point and the amount of the test solution.

The glass electrode (GE), saturated calomel electrode (SCE) and the test solution form a galvanic battery, and the relationship between the electromotive force E of the bat-

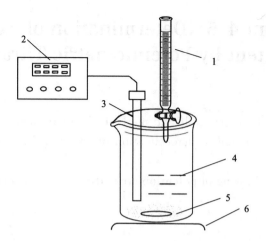

Fig. 4.2 Schematic diagram of the titration apparatus

1—burette; 2—pH meter; 3—composite pH electrode; 4—the test solution; 5—magneton; 6—magnetic stirrer

tery and the pH value of the solution to be tested is shown in formula (4.14), where E_M is membrane potential of GE and E_L is liguid junction potential

$$\begin{aligned} E &= E_{SCE} - E_{GE} + E_L \\ &= E_{SCE} - (E_{AgCl/Ag} + E_M) + E_L \\ &= E_{SCE} - \left(E_{AgCl/Ag} + K + \frac{2.303RT}{F}\lg a_{H^+}\right) + E_L \\ &= K' + \frac{2.303RT}{F}\text{pH} \end{aligned} \tag{4.14}$$

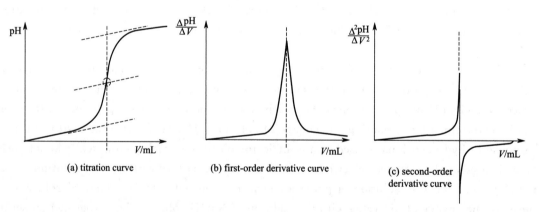

(a) titration curve　　(b) first-order derivative curve　　(c) second-order derivative curve

Fig. 4.3 Determination of titration endpoint

 Materials

Instruments: pHS-3C precision acidity meter, magnetic stirrer, analytical balance, volumetric flask, Erlenmeyer flask, pipette, burette, beaker, graduated cylinder.

Reagents: NaOH (solid), potassium hydrogen phthalate (KHP) primary standard substance, edible vinegar (3%-5%), 0.2% phenolphthalein ethanol indicator, deionized water, pH=4.00 (25 ℃) and pH=6.86 (25 ℃) standard buffer solution.

Procedures

(1) **Preparation and standardization of 0.1 mol/L NaOH solution** Weigh a certain amount of NaOH solid, fully dissolve it with deionized water which is boiled and cooled to room temperature, and transfer it into a polyethylene plastic bottle for subsequent use after cooling to room temperature.

According to the NaOH solution concentration and titration error requirements, estimate the weighing mass of the primary standard KHP. Accurately weigh the calculated amount of KHP in a 250 mL Erlenmeyer flask by subtraction method, add 20-30 mL of deionized water to dissolve it, and add 2-3 drops of phenolphthalein indicator. Titrate with NaOH solution to be standardized until the solution is light pink, and it will not fade for 30 s, which is the end point. Record the consumed volume of NaOH solution and repeat the titration 3 times (Table 4.6).

(2) **Determination of total acidity in edible vinegar** Calibration of pH meter: after preheating, the acidity meter is calibrated by the standard buffer solution two-point method, and the electrode is inserted into deionized water for standby.

The acetic acid content is measured roughly: accurately transfer 5.00 mL of edible vinegar into a 100 mL volumetric flask, dilute to the mark with freshly boiled and cooled deionized water, and the vinegar diluted solution is obtained. Accurately transfer 25.00 mL of the above diluted vinegar solution into a 250 mL beaker, add 20 mL of deionized water, put the beaker on the magnetic stirrer, insert the pH meter (dry the electrode head with filter paper), turn on stirring and start the titration. Measure the pH of the solution at each point after adding 0 mL, 1 mL, 2 mL up to 20 mL of NaOH standard solution, and the volume range of NaOH solution required for pH jump is preliminary determined.

The acidity of vinegar is measured precisely: repeat the experimental steps of the above rough measurement, take smaller volume increments of NaOH titrator near the chemical stoichiometric point judged in the rough measurement (such as recording the pH value once for each addition of 0.1 mL), and increase the density of the measure point near the stoichiometric point. Record the total volume V and the corresponding pH value after each addition of NaOH solution to draw the titration curve (Table 4.7).

Data Recording and Processing

Referring to the data recorded in Table 4.7, titration curves are drawn to determine the titration end point. As shown in Fig. 4.3(a), for the pH-V curve, the titration end point is the volume V corresponding to the turning point on the curve (at the maximum slope); as shown in Fig. 4.3(b), for the $\Delta pH/\Delta V$-V first-order derivative curve, the titration end point is the volume V corresponding to the peak of the curve (at the extreme value of $\Delta pH/\Delta V$); as shown in Fig. 4.3(c), for the $\Delta^2 pH/\Delta V^2$-V second-order derivative curve, the titration end point is the volume V corresponding to the intersection between the extreme positive value to the extreme negative value of $\Delta^2 pH/\Delta V^2$ and the zero line of the ordinate.

According to the consumed volume and the concentration of NaOH standard solution and the amount of tested solution at the end of titration, calculate the acetic acid content ρ_{HAc} (g/L) according to the following equation [Equation(4.15)].

$$\rho_{HAc} = \frac{\overline{c}_{NaOH} V_{NaOH} M_{HAc} \times \dfrac{100}{25}}{5.00} \qquad (4.15)$$

Where, M_{HAc}: the molar mass of HAc, 60.05 g/mol; $\overline{c}_{NaOH}$: actual concentration of NaOH standard titration solution, mol/L; V_{NaOH}: the consumed volume of NaOH standard solution at the end of titration, mL.

Table 4.6 Standardization of NaOH solution

No.	1	2	3
m_{KHP}(initial)/g			
m_{KHP}(final)/g			
m_{KHP}/g			
V_{NaOH}(initial)/mL			
V_{NaOH}(final)/mL			
V_{NaOH}(consumed)/mL			
c_{NaOH}/(mol/L)			
$\overline{c}_{NaOH}$/(mol/L)			
$|d_i|$			
$\overline{d}_r/\%$			

Table 4.7 Determination of acetic acid content

pH					
V_{NaOH}/mL					
pH					
V_{NaOH}/mL					
pH					
V_{NaOH}/mL					

 Questions

(1) Does CO_2 have any effect on this experiment? How do you eliminate it if there is any influence?

(2) What are the advantages and disadvantages of potentiometric titration analysis compared to volumetric analysis?

(3) If using the indicator method for this determination, how do you reduce the interference of the color of vinegar itself in the results?

Experiment 4.6 Determination of Iron, Aluminum, Calcium and Magnesium in Cement Clinker

 Objectives

(1) Learn sample dissolution and analysis methods for complex substances;

(2) Master the principle of precipitation separation of Fe and Al;

(3) Master the application principle of acid effect in cement analysis.

Principles

Cement is mainly composed of silicate. After the sample is decomposed by hydrochloric acid, iron, aluminum, calcium, magnesium and other groups in the form of Fe^{3+}, Al^{3+}, Ca^{2+}, Mg^{2+} exist in the solution, and they can form a stable complex with EDTA, as long as the appropriate acidity is controlled, they can be titrated with EDTA standard solution. Add ammonia water to the decomposed sample solution to precipitate Fe^{3+} and Al^{3+} into $Fe(OH)_3$ and $Al(OH)_3$, thus separating them from Ca^{2+} and Mg^{2+}. The precipitation was dissolved with hydrochloric acid, the pH value of the solution was adjusted to 2-2.5, and the Fe^{3+} was titrated by EDTA with sodium sulfosalicylate as an indicator. Add excess EDTA standard solution and heat to boiling to finish the reaction between Al^{3+} and EDTA completely at pH 4.2. Then use $CuSO_4$ standard solution to back-titrate with the PAN as indicator. Then calculate the percentage of Fe_2O_3 and Al_2O_3 respectively.

The filtrate containing Ca^{2+} and Mg^{2+} was separated and determined by referring to the analytical method of "Determination of total hardness of tap water" (Exp. 3.9).

Materials

Instruments: analytical balance, platform balance, pipette, beaker, volumetric flask, burette, graduated cylinder.

Reagents: 0.2% methyl red in ethanol solution; 10% sodium sulfosalicylate solution; 0.3% PAN in ethanol solution; cement sample; 0.02 mol/L EDTA standard solution; 0.02 mol/L $CuSO_4$ standard solution; 6 mol/L HCl solution; 20% NaOH solution; $CaCO_3$ primary standard substance; 1:1 $NH_3 \cdot H_2O$; NH_3-NH_4Cl buffer solution (pH=10.0, See Exp. 3.8, Preparation and Standardization of EDTA Solution); HAc-NaAc buffer solution (pH=4.2): 3.2 g anhydrous NaAc was dissolved in water, 5 mL glacial acetic acid was added, and diluted to 100 mL with water; K-B indicator: weigh 0.2 g acid chromic blue K and 0.4 g caophenol green B in a beaker, dissolve with water, and dilute to 100 mL.

Procedures

(1) Preparation and standardization of 0.02 mol/L EDTA standard solution (See Exp. 3.8, $CaCO_3$ as primary standard substance, Table 4.8).

(2) Preparation and determination of 0.02 mol/L $CuSO_4$ standard solution

Preparation: weigh 5.00 g $CuSO_4 \cdot 5H_2O$ and dissolve in water, add 4-5 drops of 1:1 H_2SO_4, and dilute to 1 L with water.

Measurement of volume ratio: accurately weigh 10.00 mL 0.02 mol/L EDTA solution, dilute it with water to about 150 mL, add 10 mL of HAc-NaAc buffer solution (pH=4.2), and heat it to 80-90 ℃. Add 5-6 drops of PAN indicator, titrate with $CuSO_4$ solution until the solution turns brown red and stable, that is, the end point. Calculate the volume of EDTA standard solution equivalent to 1 mL $CuSO_4$ solution (Table 4.9).

(3) Dissolution of sample Accurately weigh the 0.23-0.25 g sample into a 250 mL beaker, add a small amount of water to wet it, then add 15 mL of 6 mol/L HCl, cover the

watch glass, heat and boil until the sample is completely decomposed, and dilute with hot water to about 100 mL. Heat to boiling, add 2 drops of methyl red indicator, slowly add 1 : 1 $NH_3 \cdot H_2O$ under agitation until the solution is yellow, then heat to boiling. After the solution is clarified, filter with rapid quantitative filter paper while hot. The precipitate is fully washed with 0.1% $NH_3 \cdot H_2O$ hot solution until there is no Cl^- in the effluent. The filtrate was placed in a 250 mL volumetric flask, cooled to room temperature and diluted with water for determination of Ca^{2+} and Mg^{2+}.

(4) Determination of Fe_2O_3 Add 6 mol/L HCl to the filter paper to dissolve the hydroxide precipitate in the original beaker. Wash the filter paper with hot water several times and then discard it. Boil the solution to dissolve the possible hydroxide precipitate. Add 10 drops of sodium sulfosalicylate indicator, drop 1 : 1 $NH_3 \cdot H_2O$ until the pH value of the solution is 2-2.5 (the solution is purple), heat to 50-60 ℃, titrate with EDTA standard solution until the solution changes from dark purple red to light yellow as the end point. The amount of EDTA standard solution was recorded, and the mass percentage of Fe_2O_3 in the sample was calculated (Table 4.10). The solution after measuring Fe^{3+} was continued to be used for the determination of Al^{3+}.

(5) Determination of Al_2O_3 In the solution after titrating Fe^{3+}, accurately add 20 mL EDTA standard solution (0.02 mol/L), drop 1 : 1 $NH_3 \cdot H_2O$ to the solution pH value of ~4, add 10 mL HAc-NaAc buffer solution. Boil the solution for 1 min, take it off for slightly cold, and add 6-8 drops of PAN indicator. Titrate with $CuSO_4$ standard solution until the solution is red. Record the amount of $CuSO_4$ solution and calculate the mass percentage of Al_2O_3 in the sample (Table 4.11).

(6) Determination of CaO and MgO   Transfer 25.00 mL of the filtrate separated from Fe^{3+} and Al^{3+} into a conical bottle, add 15 mL of NH_3-NH_4Cl buffer solution, and then add 2-3 drops of K-B indicator. The solution was titrated by EDA until it changed from purple red to pure blue, and the consumed volume V_1 was recorded, which was the total volume of Ca^{2+} and Mg^{2+} consumption.

Another 25.00 mL of filtrate separated Fe^{3+} and Al^{3+} was transferred, followed by 5 mL of 20% NaOH solution, 20-30 mL H_2O, and 2-3 drops of K-B indicator, titrated with EDTA until the solution turned pure blue. The volume V_2 consumed by Ca^{2+} was recorded and the mass percentage of CaO was obtained.

The mass percentage of MgO is obtained by difference subtraction (Table 4.12).

 Data Recording and Processing

(1) Standardization of EDTA solution

Table 4.8 Standardization of EDTA solution

No.	1	2	3
m_{CaCO_3} (initial)/g			
m_{CaCO_3} (final)/g			
m_{CaCO_3}/g			

(continued)

No.	1	2	3
$c_{Ca^{2+}}$/(mol/L)			
$V_{Ca^{2+}}$/mL			
V_{EDTA}(initial)/mL			
V_{EDTA}(final)/mL			
V_{EDTA}(consumed)/mL			
c_{EDTA}/(mol/L)			
$\bar{c}_{EDTA}$/(mol/L)			
$|d_i|$			
$\bar{d}_r$/%			

(2) Determination of $CuSO_4$ solution

Table 4.9 Determination of $CuSO_4$ solution

No.	1	2	3
V_{EDTA}/mL			
V_{CuSO_4}(initial)/mL			
V_{CuSO_4}(final)/mL			
V_{CuSO_4}(consumed)/mL			
V_{EDTA}/V_{CuSO_4}			
$\overline{V_{EDTA}/V_{CuSO_4}}$			

(3) Determination of Fe_2O_3

Table 4.10 Determination of Fe_2O_3

No.	1	2	3
m_{sample}(initial)/g			
m_{sample}(final)/g			
m_{sample}/g			
V_{EDTA}(initial)/mL			
V_{EDTA}(final)/mL			
V_{EDTA}(consumed)/mL			
$m_{Fe_2O_3}$/g			
$w_{Fe_2O_3}$/%			
$\bar{w}_{Fe_2O_3}$/%			
$|d_i|$			
$\bar{d}_r$/%			

(4) Determination of Al_2O_3

Table 4.11   Determination of Al_2O_3

No.	1	2	3
V_{CuSO_4}(initial)/mL			
V_{CuSO_4}(final)/mL			
V_{CuSO_4}(consumed)/mL			
$w_{Al_2O_3}/\%$			
$\overline{w}_{Al_2O_3}/\%$			
$|d_i|$			
$\overline{d}_r/\%$			

(5) Determination of CaO and MgO

Table 4.12   Determination of CaO and MgO

No.	1	2	3
V_{sample}/g			
$V_{1,EDTA}$(initial)/mL			
$V_{1,EDTA}$(final)/mL			
$V_{1,EDTA}$(consumed)/mL			
$V_{2,EDTA}$(initial)/mL			
$V_{2,EDTA}$(final)/mL			
$V_{2,EDTA}$(consumed)/mL			
$w_{CaO}/\%$			
$\overline{w}_{CaO}/\%$			
$|d_i|$			
$\overline{d}_r/\%$			
$w_{MgO}/\%$			
$\overline{w}_{MgO}/\%$			
$|d_i|$			
$\overline{d}_r/\%$			

Questions

(1) Why do we use back-titration when titrating Al^{3+} with EDTA?

(2) In the determination of Fe^{3+}, Al^{3+}, Ca^{2+}, Mg^{2+}, why should the different pH values be strictly controlled?

(3) When determining Ca^{2+} and Mg^{2+}, what methods can be used to eliminate the interference of Fe^{3+} and Al^{3+}?

4.2 Environment

Experiment 4.7 Determination of Total Alkalinity of Water

Objectives

(1) To master the principle and method of acid-base titration to determine the water alkalinity.

(2) To grasp the expression and calculation of water alkalinity.

Principles

Alkalinity is the quantitative ability of aqueous media to react with hydrogen ions. Methyl orange was used as an indicator, and the end point pH of the titration was 4.3-4.5, which is called methyl orange basicity or M basicity. At this time, the hydroxide, carbonate and bicarbonate in the water are all neutralized, the sum of various weak acids in the water is measured, so it is also called the total basity, which can be converted to the corresponding basity in units mmol/L and represented by the amount of protons of 1 liter of water. mg/L ($CaCO_3$): the mass of $CaCO_3$ in the substance of 1 liter of water. M alkalinity= all HCO_3^{2-} + all CO_3^{2-} + all OH^-, and the determination results are expressed as a mass concentration equivalent to calcium carbonate in units mg/L.

Materials

Instruments: acid burette, pipette, 250 mL Erlenmeyer flask, analytical balance.

Reagents: 0.5 g/L methyl orange indicator, 0.05 mol/L hydrochloric acid, anhydrous sodium carbonate (reference reagent).

Procedures

(1) Standardization of 0.10 mol/L HCl solution Accurately weigh ( ) to ( ) g of anhydrous Na_2CO_3 using a weighing bottle with analytical balance (molar mass $M=105.99$ g/mol) into a 250 mL Erlenmeyer flask. Dissolve with 50 mL deionized water, add 1-2 drops methyl orange indicator, and titrate with HCl solution until the color of the solution changes from yellow to orange, which is the end point. Near the end point, in order to remove the dissolved CO_2 in the solution, it should be violently shaken or heated to drive CO_2, and then titrated to the end point after cooling. According to the mass of Na_2CO_3 and the volume of consumed hydrochloric acid, the concentration of HCl solution c_{HCl} is calculated as follows, and repeat the titration 3 times (Table 4.13).

$$c_{HCl} = \frac{2m \times 1000}{105.99V} \qquad (4.16)$$

Where, c_{HCl}: concentration of hydrochloric acid standard solution, mol/L; m: the mass of sodium carbonate, g; V: the volume of hydrochloric acid consumed.

(2) Determination of water samples Transfer 50.00 mL (V_1) water sample into a 250 mL Erlenmeyer flask, add 1-2 drops of methyl orange indicator, and titrate with hydrochloric acid standard solution until the color of the solution changes from yellow to orange. Record the consumed volume of hydrochloric acid standard solution V_2. Repeat the titration 3 times (Table 4.14).

The calculation equation of total alkalinity is as follows:

$$\rho_{CaCO_3} = \frac{c_{HCl} \times 100.1 V_2 \times 1000}{2V_1} \tag{4.17}$$

ρ_{CaCO_3}: the total alkalinity of water sample, mg/L; c_{HCl}: the concentration of hydrochloric acid standard solution, mol/L; V_1: the volume of water sample, mL; V_2: the volume of consumed standard hydrochloric acid solution, mL.

Data Recording and Processing

Table 4.13 Standardization of 0.1 mol/L HCl

No.	1	2	3
$m_{Na_2CO_3}$/g			
V_{HCl}(initial)/mL			
V_{HCl}(final)/mL			
V_{HCl}/mL			
c_{HCl}/(mol/L)			
$\bar{c}_{HCl}$/(mol/L)			
$|d_i|$			
$\bar{d}_r$/%			

Table 4.14 Determination of total alkalinity of water sample

No.	1	2	3
V_{sample}/g			
V_{HCl}(initial)/mL			
V_{HCl}(final)/mL			
V_{HCl}/mL			
ρ_{CaCO_3}/(mol/L)			
$\bar{\rho}_{CaCO_3}$/(mol/L)			
$|d_i|$			
$\bar{d}_r$/%			

Questions

What are the differences between acid-base titration and coordination titration?

Tips

Common water quality pH indicators

Alkalinity is the quantitative ability of an aqueous medium to react with hydrogen ions. Some common water quality standard pH indicators:

(1) Drinking water: Usually, the standard range of pH value of drinking water is 6.5-8.5, close to neutral.

(2) U.S. EPA (Environmental Protection Agency) drinking water quality standards: first-class water, no specific standards; secondary water, pH 6.5-8.5.

(3) European Union (EU) drinking water quality standards: pH 6.5-9.5. For bottled or barrel-loaded water, the minimum value of pH is reduced to 4-5.

(4) Japan's drinking water quality standards: pH 5.8-8.6.

(5) China's drinking water health standards: pH 6.5-8.5 (GB 5749—2022).

(6) Bottled drinking water: pH 5.0-7.0.

Experiment 4.8 Determination of Chemical Oxygen Demand (COD) in River Water

Objectives

(1) Understand the significance of measuring chemical oxygen demand (COD).

(2) Master the principle and method of determining COD in water samples by the acid $KMnO_4$ method.

 Principles

Chemical Oxygen Demand (COD) is a comprehensive indicator of the degree of pollution of water bodies by reducing substances (mainly organic), i.e. the reducing substances (inorganic or organic) in 1 L of water under certain conditions to be oxidised by the amount of oxygen consumed (mg/L). The higher the value of COD, the more serious the contamination of organic matter of the water body. In addition to inorganic reducing substances (such as NO_2^-, S^{2-}, Fe^{2+}, etc.), the water may also contain a small amount of organic matter. Organic matter decay prompted microbial reproduction in the water, contaminated water quality, affected human health. If the amount of COD in the water is high, it is yellow and has obvious acidity, endangering aquatic organisms. Industrial production with this water, the steam boiler will be corroded, the quality of printing and dyeing and other products will also be affected, so the determination of COD in the water body is very important.

Different conditions yield different oxygen demand, and therefore we must strictly control the reaction conditions. COD determination, according to the different oxidants, is divided into $KMnO_4$ method (COD_{Mn}) and $K_2Cr_2O_7$ method (COD_{Cr}). $KMnO_4$ method is easy to operate, and fast. To a certain extent, it can indicate the state of water body contaminated by organic matter, and is suitable for the determination of ground water, drinking water, river water and other water body of light pollution for $K_2Cr_2O_7$ method is suitable for the determination of the industrial and domestic sewage containing more components and complex, with more seriously polluted water quality. This method has a high oxidation rate and good reproducibility. In this experiment, the acidic potassium permanganate method was used to determine the chemical oxygen demand in water.

In the acidic solution, add an excess of $KMnO_4$ solution, heat it to make the organic and reducing substances in the solution fully oxidised. The remaining $KMnO_4$ is reduced with a certain amount of excess $Na_2C_2O_4$, and then use $KMnO_4$ standard solution to back-titrate the excess part of $Na_2C_2O_4$. The main reaction formula is as follows:

$$4MnO_4^- (excess) + 5C + 12H^+ = 4Mn^{2+} + 5CO_2 \uparrow + 6H_2O \quad (4.18)$$
$$2MnO_4^- (remaining) + 5C_2O_4^{2-} (excess) + 16H^+ = 2Mn^{2+} + 10CO_2 \uparrow + 8H_2O \quad (4.19)$$
$$5C_2O_4^{2-} (remaining) + 2MnO_4^- + 16H^+ = 2Mn^{2+} + 10CO_2 \uparrow + 8H_2O \quad (4.20)$$

Calculate the amount of $KMnO_4$ consumed by the organic and inorganic reducing substances contained in the water.

If the concentration of Cl^- in the water sample is greater than 300 mg/L, the result will be higher. Pure water can be added for appropriate dilution. Alternatively, Ag_2SO_4 can be added to precipitate Cl^-. Use alkaline potassium permanganate method or potassium dichromate method to eliminate interference. If Fe^{2+}, H_2S, NO_2^- and other reductive substances interfere with the determination of water samples, they can be oxidized by $KMnO_4$ at room temperature. Therefore, water samples are titrated with $KMnO_4$ solution at room temperature to remove the interfering ions, and the amount of $KMnO_4$ should not be counted. Water samples should be analyzed immediately after sampling. If there are special circumstances which need to be placed, a small amount of $CuSO_4$ can be added to inhibit the decomposition of organic matter by microorganisms. If the water sample is diluted with distilled water, the same amount of distilled water should be taken, and the blank value should be measured and corrected.

Materials

Instruments: acid burette, pipette, volumetric flask, Erlenmeyer flask, weighing bottle, analytical balance, graduated cylinder, beaker, electric hot plate.

Reagents: $KMnO_4$ (A.R.), 3 mol/L H_2SO_4, $Na_2C_2O_4$ (reference substance), water sample.

Procedures

(1) Preparation of 0.005 mol/L $Na_2C_2O_4$ standard solution Accurately weigh ( ) g reference substance $Na_2C_2O_4$ (dried) in a 100 mL beaker, add 40 mL distilled water to dissolve, then transfer it into a 500 mL volumetric flask, dilute it with water to the mark, and shake well for use.

(2) Preparation of 0.002 mol/L $KMnO_4$ Take 25.00 mL 0.02 mol/L $KMnO_4$ solution (prepared one week in advance) in a 250 mL volumetric flask, dilute it with water to the mark, and shake well for use (Table 4.15).

(3) Determination of COD in water samples Measure 100 mL of a well-stirred water sample in a conical flask with a measuring cylinder, add 5 mL of 1∶3 H_2SO_4 solution and a few glass beads (to prevent the solution from boiling violently), add 10.00 mL of $KMnO_4$

solution from a burette and heat immediately to boiling. Boil for 10.0 min from the first large bubble that rises (the red colour should not fade). Remove the conical flask, leave it for 0.5-1 min, accurately add 25.00 mL of $Na_2C_2O_4$ standard solution while it is still hot, shake well, and titrate immediately with $KMnO_4$ solution. Continue titration till the test solution is slightly red and does not fade for 0.5 min, that is the end point. The volume of consumption is V_1, the temperature of the test solution should not be less than 60 ℃.

Pipette 25.00 mL of water samples to be tested in a 250 mL conical flask, add 75.00 mL of distilled water (take 10.00 mL of more seriously contaminated water samples, and then add distilled water to 100 mL); add 8 mL of 3 mol/L H_2SO_4 solution, mixing well; add Ag_2SO_4 solution 2 mL; accurately add 0.002 mol/L $KMnO_4$ standard solution 10.00 mL, shake well; immediately put it into the boiling water bath heating for 30 min, the boiling water bath level should be higher than the liquid level of the reaction solution (or immediately heat it to boil); start timing from the first big bubbles, boil for 10 min, remove the conical flask for a cool a little; while it is still hot (70-80 ℃) accurately add 10.00 mL 0.005 mol/L $Na_2C_2O_4$ Standard solution, shake well, immediately titrate with $KMnO_4$ standard solution until the solution is stable reddish, that is the end point. The volume of consumption is V_2. Take 100.00 mL of distilled water instead of water samples for the experiment, perform the same operation as the above, and determine the blank value.

Three measurements were made in parallel and the chemical oxygen demand (mg/L) of the water samples was calculated according to the chemical reaction equation in the Principles, as well as the relative mean deviation (Table 4.16).

Data Recording and Processing

Table 4.15 Standardization of 0.002 mol/L $KMnO_4$ solution

No.	1	2	3
$m_{Na_2C_2O_4}$/g			
V_{KMnO_4} (initial)/mL			
V_{KMnO_4} (final)/mL			
V_{KMnO_4}/mL			
c_{KMnO_4}/(mol/L)			
$\bar{c}_{KMnO_4}$/(mol/L)			
$|d_i|$			
$\bar{d}_r$/%			

Table 4.16 Determination COD of water samples

No.	1	2	3
V_s/mL			
V_1 (initial)/mL			
V_1 (final)/mL			
V_1/mL			

No.	1	2	3
V_2(initial)/mL			
V_2(final)/mL			
V_2/mL			
COD/(mg/L)			
$\overline{\text{COD}}$/(mg/L)			
$|d_i|$			
$\overline{d}_r/\%$			

Questions

(1) What should be paid attention to in the collection and preservation of water samples?

(2) When Cl^- content in the water sample is high, can this method be used to determine? Why?

(3) What does it mean if the purple color disappears when $KMnO_4$ is added to the water sample and boiled? What measures should be taken?

(4) What are the methods for determining COD in water? Please express it by chemical equation.

Experiment 4.9　Determination of Sulfate Content in Urban Sewage

Objective

Master the principle and experimental operation of the gravimetric method for determining sulfate content in water.

Principles

Urban sewage mainly consists of industrial wastewater and domestic sewage, which contains heavy metal ions such as Fe^{3+}, Cu^{2+}, Pb^{2+}, Ag^+, as well as anions such as SO_4^{2-}, Cl^-, PO_4^{3-}, etc. Determination of the content of SO_4^{2-} can be completed by $BaSO_4$ precipitation generated quantitatively. Precipitation reaction is carried out at temperatures close to boiling, aging for a period of time. Filter out the precipitate, dry or burn precipitate, weigh and then according to the mass of $BaSO_4$ calculate the content of SO_4^{2-}, the reaction is as follows:

$$Ba^{2+} + SO_4^{2-} \Longrightarrow BaSO_4 \downarrow \tag{4.21}$$

However, any acid insoluble substance will interfere with precipitation, so before precipitation with $BaCl_2$, the sample should be acidified with hydrochloric acid and filtered to remove the interference.

Materials

Instruments: muffle furnace, electric heating plate, glass rod, beaker, watch glass, porcelain crucible, washing bottle, crucible clamp, dryer, slow quantitative filter paper.

Reagents: sewage sample, 1:1 hydrochloric acid solution, 100 g/L $BaCl_2$ solution, 0.1 mol/L $AgNO_3$ solution, methyl red.

Procedures

(1) Take 200 mL water sample (reduce the sample amount if SO_4^{2-} content exceeds 1000 mg/L), place it in a 400 mL beaker, add two drops of methyl red indicator, and use 1:1 hydrochloric acid to adjust the solution to red. Add another 2 mL, cover the watch glass, and put the beaker on the electric heating plate to boil. Continue boiling for at least 5 min, cool slightly, and rinse the watch glass 2-3 times with distilled water. Filter with qualitative filter paper, wash the beaker and filter paper 4-5 times with hot water, discard the precipitation and retain the filtrate.

(2) Heat the filtrate on an electric heating plate until boiling, and slowly add about 10 mL $BaCl_2$ under constant agitation until no precipitation occurs. Add another 2 mL $BaCl_2$. Keep at 80-90 ℃ for 2 h, cool and age overnight.

(3) Filter the aged precipitation with slow quantitative filter paper. Transfer and wash the precipitate with hot water, and repeatedly wash the precipitate with a small amount of warm water until the washing solution is free of chloride ions (test with 0.1 mol/L $AgNO_3$ solution).

(4) Wrap the precipitate with filter paper, put it into a porcelain crucible which was previously scorched to constant weight at 800 ℃. Dry and then make the sample ashed at low temperature on an electric furnace (do not let the filter paper burn flame). Then put the crucible into a muffle furnace, scorch it at 800 ℃ for 1 h and take it out, cover it with a slightly cold cover and put it into a desiccator and cool it to room temperature (about 30 min). Weigh until burn it to a constant weight (Table 4.17).

Data Recording and Processing

The formula for calculating sulfate content is as follows:

$$w = \frac{(m_2 - m_1) \times 411.6 \times 1000}{V} \qquad (4.22)$$

Where, w: the sulfate content, mg/L; m_2: the total mass of crucible and barium sulfate, g; m_1: the mass of crucible, g; V: the volume of sample, mL; 411.6: $BaSO_4$ factor of mass conversion to SO_4^{2-}.

Table 4.17 Determination of SO_4^{2-} content

No.	1	2	3
V_{sample}/mL			
m_1/g			
m_2/g			
m_{BaSO_4}/g			
$w_{SO_4^{2-}}$/(mg/L)			
$\overline{w}_{SO_4^{2-}}$/(mg/L)			

(continued)

No.	1	2	3
$|d_i|$			
$\overline{d}_r/\%$			

Notes:

(1) Constant weight refers to the mass difference is less than 0.3 mg after two consecutive calcinations or drying.

(2) The ashing of barium sulfate should ensure sufficient air supply, otherwise the precipitation is easily reduced by the carbon burned by filter paper. The burned deposits will appear gray or black. At this time, 2-3 drops of concentrated sulfuric acid can be added to the cooled precipitation, and then carefully heated until SO_2 white smoke is no longer produced, and burned at 800 ℃ to the constant weight.

 Questions

(1) Why should hydrochloric acid solution be added to the sample solution?

(2) How do you check whether there are chloride ions in the filtrate?

 Tips

Wastewater discharge standards require a sulphate discharge concentration <1500 mg/L, above which wastewater is classified as high sulphate wastewater. Although the sulphate in sulphate-containing wastewater is harmless, it encounters an anaerobic environment and produces H_2S under the action of sulphate-reducing bacteria (SRB), which can seriously corrode the treatment facilities and drainage pipes. And the odour of it is malodourous, it can seriously pollute the atmosphere. In addition, sulphate wastewater discharged into the water body will make the receiving water body acidic, pH reduced, be harmful to aquatic organisms; it discharged into the farmland will destroy the soil structure, harden the soil, reduce crop yields and lower the quality of agricultural products. At present, groundwater in many cities of China are suffering various degrees of sulphate pollution, and seeking effective sulphate wastewater treatment process has long been a common concern of the environmental engineering community.

Experiment 4.10 Determination of Formaldehyde in Waterborne Coatings by Spectrophotometry

 Objective

To master the extraction of formaldehyde from waterborne paint and the detection technology of acetylacetone color spectrophotometry.

Principles

The formaldehyde in the sample was evaporated by distillation method, and the formaldehyde in the fraction reacted with acetylacetone to form a stable yellow complex under heating conditions in pH=6 acetic acid-ammonium acetate buffer solution, and the absorbance was tested at the wavelength of 413 nm after cooling. According to the standard working curve, the content of formaldehyde in the sample was calculated.

Materials

Instruments: UV-vis spectrophotometer, distillation unit (including 500 mL distillation bottle, condensing tube, beaker), iodine flask.

Reagents: ammonium acetate, glacial acetic acid, acetylacetone, iodine solution (0.1 mol/L), sodium hydroxide solution (0.1 mol/L), hydrochloric acid solution (0.1 mol/L), sodium thiosulphate ($Na_2S_2O_3$, 0.1 mol/L), formaldehyde solution (37%).

Acetylacetone solution: 0.25% (volume ratio), weigh 25 g of ammonium acetate, add the appropriate amount of water to dissolve, add 3 mL of glacial acetic acid and 0.25 mL of acetylacetone reagent that has been distilled, transfer to a 100 mL volumetric flask, dilute with water to the mark, and adjust the pH=6. This solution can be stable for one month if it is stored at 2-5 ℃.

Procedures

(1) Preparation and standardization of $Na_2S_2O_3$ standard solution (0.1 mol/L) (Experiment 3.15).

(2) Preparation and standardization of formaldehyde stock solution Preparation of formaldehyde standard solution (1 mg/mL): pipette 2.8 mL formaldehyde solution, place in a 1000 mL volumetric flask, dilute with water to the mark.

Standardization of formaldehyde standard solution: take 20 mL of formaldehyde standard solution to be calibrated in an iodine measuring flask, accurately add 25 mL of iodine solution, and then add 10 mL of sodium hydroxide solution, shake well, and let it stand in the dark for 15 min. Add 11 mL of hydrochloric acid solution, titrate it with sodium thiosulphate standard solution until it turns yellowish, add 1 mL of starch solution, continue titration until the blue colour just disappears at the end point, record the volume of sodium thiosulphate standard solution used V_2 (mL). At the same time, do a blank sample, record the volume of sodium thiosulfate standard solution V_1 (mL).

(3) Calibration curve Pipette 0.00 mL, 0.20 mL, 0.50 mL, 1.00 mL, 3.00 mL, 5.00 mL, 8.00 mL of formaldehyde standard working solution (10 μg/mL), respectively, dilute with water to the scale, add 2.5 mL of acetylacetone solution. The solution was heated in a constant temperature water bath at 60 ℃ for 30 min, removed and cooled to room temperature, and the absorbance was measured at 413 nm using a UV-visible spectrophotometer with a 1 cm absorption cell (with water as reference). The standard curve was plotted with the mass of formaldehyde as the horizontal coordinate and the corresponding absorbance (A) as the vertical coordinate.

(4) Determination of formaldehyde content Weigh about 2 g (accurate to 1 mg) of the stirred specimen, place it in a 50 mL volumetric flask, add water and shake well, dilute to the scale. Then pipette 10 mL of the aqueous solution of the test sample in the volumetric flask and place it in a distillation flask to which 10 mL of water and a small amount of zeolite have been pre-added, and a suitable amount of water has been pre-added to the distillation

receiver to submerge the distillation outlet, and the external part of the distillation receiver is cooled in an ice-water bath. Heat the distillation unit so that the sample is evaporated to near dry, remove the fraction receiver, dilute to scale with water and leave to test. If the specimen to be tested is not easily dispersed in water, the stirred specimen is weighed directly to approximately 0.4 g (to the nearest 1 mg), placed in a distillation flask to which 20 mL of water has been pre-added, and gently shaken to disperse the sample before proceeding with the distillation process.

2.5 mL of acetylacetone solution was added to the calibrated fraction receiver and shaken well. It was heated in a constant temperature water bath at 60 ℃ for 30 min, removed and cooled to room temperature, and the absorbance was tested at 413 nm on a UV-visible spectrophotometer using a 1 cm cuvette (with water as reference). At the same time, a blank sample (water) was made under the same conditions, and the absorbance of the blank sample was measured. Subtract the absorbance of the blank sample from the absorbance of the sample, and find the corresponding mass of formaldehyde on the standard working curve.

Data Recording and Processing

(1) Standardization of $Na_2S_2O_3$ standard solution (Table 4.18)

Table 4.18 Standardization of $Na_2S_2O_3$ standard solution

No.	1	2	3
$V_{Na_2S_2O_3}$ (initial)/mL			
$V_{Na_2S_2O_3}$ (final)/mL			
$V_{Na_2S_2O_3}$ /mL			
$c_{Na_2S_2O_3}$ /(mg/L)			
$\bar{c}_{Na_2S_2O_3}$ /(mg/L)			
$|d_i|$			
$\bar{d}_r$ /%			

(2) Sandardization of HCHO Solution (Table 4.19) The concentration of formaldehyde solution is calculated using equation (4.23):

$$\rho_{HCHO} = \frac{(V_1 - V_2)c_{Na_2S_2O_3} \times 15}{20} \quad (4.23)$$

In the formula, ρ_{HCHO}: the mass concentration of the standard solution of formaldehyde, g/L; V_1: volume of $Na_2S_2O_3$ standard solution used for titration of blank sample, mL; V_2: volume of $Na_2S_2O_3$ standard solution used for formaldehyde solution calibration, mL; $c_{Na_2S_2O_3}$ is concentration of sodium thiosulfate standard solution, mol/L; 15: 1/2 of the molar mass of formaldehyde; 20: volume of formaldehyde standard solution removed during calibration, mL.

Table 4.19 Standardization of HCHO solution

No.	1	2	3
V_2 (initial)/mL			
V_2 (final)/mL			

(continued)

No.	1	2	3
V_2/mL			
V_1(initial)/mL			
V_1(final)/mL			
V_1/mL			
ρ_{HCHO}/(g/L)			
$\bar{\rho}_{HCHO}$/(g/L)			
$|d_i|$			
$\bar{d}_r$/%			

(3) Calibration curve of for maldehyde and sampee analysis (Table 4.20) The formaldehyde content (c) is calculated using equation (4.24):

$$c = \frac{m}{W} f \tag{4.24}$$

In the formula, c: formaldehyde content, mg/kg; m: mass of formaldehyde calculated from the standard working curve, μg; W: mass of sample, g; f: dilution factor.

Table 4.20 Calibration curve of formaldehyde and sample analysis

No.	Concentration of HCHO/(mg/kg)	Absorbance	The content of HCHO in the sample/%
1			—
2			—
3			—
4			—
5			—
6			—
Sample			

  Questions

(1) Explain the principle of spectrophotometric determination of formaldehyde.
(2) What methods are used to standardize formaldehyde standard solution? Why?

  Tips

Formaldehyde (chemical formula HCHO) is a colourless, water-soluble liquid that is highly volatile and has an irritating odour. Formaldehyde causes great harm to human body. First of all, formaldehyde can make people acutely poisoned and cause great damage to the skin. When formaldehyde reaches a certain level, the human body will experience uncomfortable symptoms, such as hoarseness, itchy eyes, throat discomfort, dermatitis, respiratory difficulties, and in severe cases, it can cause pulmonary oedema, anaphylactic shock and liver failure. When people inhale excess formaldehyde, it can lead to death. For the newly decorated room indoor formaldehyde and other harmful substances can be reduced through the ventilation. You can also place green plants in the room through photosynthesis to reduce formaldehyde content. Active carbon, bamboo charcoal and other adsorbents can also be very good sorbent of indoor formaldehyde.

Experiment 4.11 Determination of Glyphosate Content in Roundup

Objectives

(1) To master the principle of spectrophotometric determination of glyphosate content.

(2) To master the use of a spectrophotometer and plot the standard curve.

Principles

Glyphosate, chemically known as N-phosphate methylglycine, is also known as Roundup. The chemical formula is $C_3H_8NO_5P$, and the structural formula is shown in Fig. 4.4, which is an organophosphine herbicide. The molecule of glyphosate contains carboxylic acid group, amino group, methyl phosphate group, etc. It can carry out a series of typical biochemical reactions such as esterification, hydroxyamidation, amination, nitrosation and dehydration. Glyphosate is a highly effective, low-toxic, broad-spectrum herbicide, and is by far the most widely produced herbicides. Glyphosate was once suspected to be a carcinogen, and most of them now consider it to have very little carcinogenicity. However, consuming too much glyphosate can cause allergic reactions, damage to the liver and kidney functions and the digestive tract, nausea, vomiting, abdominal pain, and serious damage to the respiratory and the cardiovascular system, and even be life-threatening. In addition, if pregnant women are exposed to glyphosate for a long period of time, it may also lead to teratogenicity and birth defects in babies.

Fig. 4.4 The structure of glyphosate

After dissolved in water, glyphosate can interact with sodium nitrite in acidic medium to produce glyphosate nitroso derivative, which has the maximum absorption peak at 243 nm, and the content of glyphosate in the sample can be quantitatively calculated according to Lambert's law $A = \varepsilon bc$.

Materials

Instruments: UV-vis spectrophotometer, quartz cuvette (1 cm), volumetric flask (100 mL, 250 mL), pipette (1.00 mL, 2.00 mL, 5.00 mL).

Reagents: 50% sulfuric acid solution (Volume ratio), 250 g/L potassium bromide solution, glyphosate standard (99.8%), Roundup, 14 g/L sodium nitrite solution (weigh 0.28 g of sodium nitrite, dissolve it in 20 mL of water, and get it ready for use), distilled water.

Procedures

(1) Preparation of glyphosate standard solution Weigh 0.3000 g of glyphosate standard (accurate to 0.0001 g) in a small beaker, add 50 mL of water to dissolve, quantitatively transfer to a 250 mL volumetric flask, dilute to the scale and shake well.

(2) Plotting of glyphosate standard curve Pipette 0.80 mL, 1.10 mL, 1.40 mL, 1.70 mL, 2.00 mL of the above glyphosate standard solution into five 100 mL volumetric flasks, and at the same time, take another 100 mL volumetric flask as a reagent blank. Add 5 mL of distilled water, 0.5 mL of sulphuric acid solution, 0.1 mL of potassium bromide

solution and 0.5 mL of sodium nitrite solution to each of the above volumetric flasks. After add sodium nitrite solution, the stopper should be tightly plugged immediately, shake well and leave it for 20 min. Then dilute with water to the scale, shake well, and finally open the stopper and leave it for 15 min. Note that the temperature of nitrosylation reaction should not be lower than 15 degrees Celsius.

The absorbance of each solution was measured at 243 nm with a 1 cm cuvette using a blank as reference. The standard curve was plotted with the absorbance A as the vertical coordinate and the volume V of the corresponding standard sample solution as the horizontal coordinate (Table 4.21).

(3) Determination of glyphosate content in Roundup Weigh about 0.20 g of sample (accurate to 0.0001 g) and place it in a 200 mL beaker. Add 60 mL of water, slowly heat to dissolve, filter with quick filter paper while hot, rinse the filter paper carefully, put the filtrate to a 250 mL volumetric flask, cool to room temperature, dilute to the scale, shake well. Pipette 2.00 mL of sample solution in a 100 mL volumetric flask, dilute to the scale, determine the absorbance.

Data Recording and Processing

Table 4.21 Calibration curve of glyphosate and sample analysis

	Concentration of glyphosate/(mg/kg)	Absorbance	The content of glyphosate in the sample/%
1			—
2			—
3			—
4			—
5			—
6			—
Sample			

Questions

(1) What is the role of the reference solution? Can distilled water be used as a reference in this experiment?

(2) Briefly describe the principle of spectrophotometric determination of glyphosate content.

4.3 Biology

Experiment 4.12 Determination of Total Nitrogen Content for Compound Fertilizers

Objectives

(1) To understand the preparation method of compound fertilizer samples.

(2) To grasp the principle and method of determining total nitrogen by titrimetric after distillation.

Principles

Compound fertilizer is a fertilizer containing two or three of the crop nutrient elements such as nitrogen, phosphorus and potassium, which is made by chemical method or physical mixing method. The determination of total nitrogen content in the compound fertilizer is by converting various forms of nitrogen (nitrate, amide nitrogen, cyanamide nitrogen) into ammonia. Then, the ammonia is absorbed by the excess sulfuric acid solution, and is back titrated by sodium hydroxide standard solution with methyl red-methylene blue mixed indicator. Under alkaline conditions, the nitrate nitrogen is reduced by a nitrogen-fixing alloy and ammonia is distilled directly. Under acidic conditions, nitrate is reduced to ammonium salt. In the presence of mixed catalysts (potassium sulfate and copper sulfate), it is digested with concentrated sulfuric acid to convert organic nitrogen or amide nitrogen and cyanamide nitrogen into ammonium salt, and ammonia is distilled from an alkaline solution.

Materials

Instruments: analytical balance, round bottom distillation flask (1000 mL), pear-shaped glass funnel, electric furnace, distillation equipment.

Reagents:

(1) H_2SO_4 (A.R.);

(2) HCl (A.R.);

(3) Chromium powder: fineness less than 250 μm;

(4) Alloy in determination of nitrogen: Cu 50%, Al 45%, Zn 5%, fineness less than 250 μm;

(5) K_2SO_4 (A.R.);

(6) $CuSO_4 \cdot 5H_2O$ (A.R.);

(7) Mixed catalyst: 1000 g K_2SO_4 and 50 g $CuSO_4 \cdot 5H_2O$;

(8) NaOH solution: 400 g/L;

(9) NaOH standard solution: 0.5000 mol/L;

(10) Methyl red-methylene blue mixed indicator;

(11) Extensive pH test paper;

(12) Silicone grease.

Procedures

(1) Sample preparation of compound fertilizer Take about 100 g of the compound fertilizer sample after several times reduction, grind it with a grinder or a mortar until it passes through a 0.50 mm sieve (for wet fertilizer, it can pass the 1.00 mm sieve), mix it evenly, and put it in a clean and dry bottle. Then, components analysis was conducted. The grinding operation should be fast to avoid water loss or moisture absorption during the grinding process, and to prevent the sample from overheating.

Weigh 0.5-2 g (accurate to 0.0002 g) of the above sample into a round bottom distillation flask, which contains no more than 235 mg of total nitrogen and no more than 60 mg of nitrate nitrogen.

(2) Sample treatment and distillation

① Sample containing only ammonium nitrogen. Add 300 mL of water into the round bottom distillation flask, shake it to dissolve the sample, put the explosion-proof boiling substance into the flask, and connect the distillation flask to the distillation equipment. Add 40 mL of sulfuric acid solution (0.25 mol/L), 4-5 drops of methyl red-methylene blue mixed indicator, and appropriate amount of water to ensure that the gas outlet is sealed, and connect the receiver to the distillation equipment. The ground joint of the distillation is sealed with silicone grease. Add 20 mL of NaOH solution (400 g/L) through the drip funnel of the distillation, add 20-30 mL of water to wash the funnel when the solution is about to run out, and close the piston when 3-5 mL of water is left. Turn on the cooling water and the electric furnace. According to the amount of foam produced when the solution is boiling, adjust the heat intensity to avoid foam overflowing or droplet carrying over. After at least 150 mL of distillate distilled, pH test paper is used to check the liquid drop at the outlet of the condenser tube, and end the distillation if the solution does not show alkalinity.

② Samples containing nitrate nitrogen and ammonium nitrogen. Add 300 mL of water into the distillation flask, shake it to dissolve the sample, add 3 g of alloy in determination of nitrogen and explosion-proof boiling substance into the flask, and connect the distillation flask to the distillation equipment. In the distillation process, except for adding 20 mL of NaOH solution (400 g/L) and standing for 10 min before heating, complete other procedures the same as in the determination of ammonium nitrogen sample.

③ Samples containing amide nitrogen, cyanamide nitrogen and ammonium nitrogen. Place the distillation flask in a fume hood, carefully add 25 mL of H_2SO_4, insert a pear-shaped glass funnel, put the flask on the digestion heating device (1500 W electric furnace), heat the flask until the sulfuric acid emits white smoke, and stop heating after 15 minutes. After the distillation flask is cooled to the room temperature, carefully add 250 mL of water. In addition to adding 100 mL of NaOH solution (400 g/L) in the distillation process, complete other procedures the same as in the determination of ammonium nitrogen sample.

④ Samples containing organic matter, amide nitrogen, cyanamide nitrogen and ammonium nitrogen. Place the distillation flask in a fume hood, add 22 g of mixed catalyst, carefully add 30 mL of H_2SO_4, insert a pear-shaped glass funnel, and put the flask on the digestion heating device (1500 W electric furnace). If there is a lot of foams, reduce the heating intensity until the foam disappears, and continue heating until white smoke is emitted for 60 minutes or the solution is transparent. After the flask is cooled to the room temperature, carefully add 250 mL of water. In addition to adding 120 mL of NaOH solution (400 g/L) in the distillation process, complete other steps the same as in the determination of ammonium nitrogen sample.

⑤ Samples containing nitrate nitrogen, amide nitrogen, cyanamide nitrogen and ammonium nitrogen. Add 35 mL of water into the distillation flask, shake it to dissolve the sample, add 1.2 g of chromium powder and 7 mL of HCl, let it stand for 5-10 min, and insert a pear-shaped glass funnel. Place the distillation flask on the heating device (1500 W electric furnace) in the fume hood, heat it to boiling and the foams rise for 1 min. Then, cool it to the room temperature, carefully add 25 mL of H_2SO_4, continue to heat it to emit white smoke for 15 min, and carefully add 400 mL of water after the distillation flask cools to the room temperature. In addition to adding 100 mL of NaOH solution (400 g/L) in the distillation process, complete other steps the same as in the determination of ammonium nitrogen sample.

⑥ Samples containing organic matter, nitrate nitrogen, amide nitrogen, cyanamide nitrogen and ammonium nitrogen or unknown samples. Add 35 mL of water into the distillation flask, shake it to dissolve the sample, add 1.2 g of chromium powder and 7 mL of HCl, let it stand for 5-10 min, and insert a pear-shaped glass funnel. Place the distillation flask on the heating device (1500 W electric furnace) in the fume hood, heat it to boiling and the foams rise for 1 min. Then, cool it to the room temperature, add 22 g mixed catalyst, carefully add 30 mL H_2SO_4, and continue heating. If there is a lot of foams, reduce the heating intensity until the foams disappears, and continue heating until white smoke is emitted for 60 minutes. After the distillation flask is cooled to the room temperature, carefully add 400 mL of water. In addition to adding 120 mL of NaOH solution (400 g/L) in the distillation process, complete other steps the same as in the determination of ammonium nitrogen sample.

(3) Titration Use NaOH standard solution (0.5000 mol/L) to back-titrate the excess H_2SO_4 until the mixed indicator turns grayish green. At the same time, the blank test adopts the same operation steps, uses the same reagent, but does not include the sample.

Data Recording and Processing

The total N content w, expressed in mass fraction (%), is calculated by the following formula:

$$w = \frac{(V_2 - V_1) c \times 0.01401}{m} \times 100\% \tag{4.25}$$

where, V_2 is the volume of NaOH standard solution consumed in the blank test, mL; V_1 is the volume of NaOH standard solution consumed during the determination of compound fertilizer samples, mL; c is the concentration of NaOH standard solution, mol/L; the millimolar mass value of nitrogen is 0.01401, g/mmol; m is the mass of the compound fertilizer sample, g.

The calculation results are accurate to two decimal places, and the arithmetic mean of the parallel results is taken as the measurement result.

Experiment 4.13   Determination of Soil Organic Matter Content

Objectives

To master the principle and calculation method of soil organic matter determination.

Principles

Soil organic matter generally refers to the organic compounds containing carbon, which is present in various forms in soil. Organic matter is the main source of various nutrients such as N, P and S needed by crops, and it can improve soil fertility, soil buffering, and biological and enzymatic activities. It is the foundation of soil fertility and one of the important indicators of measuring soil fertility.

A quantitative potassium dichromate-sulfuric acid solution is used to oxidize organic carbon in the soil heated on an electric sand bath, and the remaining potassium dichromate is titrated with ferrous sulfate standard solution, and silica used as an additive for reagent blank to calculate the soil organic matter content. The reaction equation for the standardization of ferrous sulfate is as follows:

$$Cr_2O_7^{2-} + 6Fe^{2+} + 14H^+ = 6Fe^{3+} + 2Cr^{3+} + 7H_2O \qquad (4.26)$$

Materials

Instruments: analytical balance, electric sand bath, thermometer (200-300℃), ground Erlenmeyer flask (150 mL), ground simple air condenser (ϕ0.9 cm × 19 cm), burette (10.00 mL and 25.00 mL), copper wire sieve (1 mm, 0.25 mm), porcelain mortar.

Reagents:

(1) Potassium dichromate (Guaranteed reagent, G.R.);

(2) Concentrated sulfuric acid (A.R.);

(3) Ferrous sulfate (A.R.);

(4) Silver sulfate: ground into powder;

(5) Silicon dioxide: in powder;

(6) O-phenanthroline indicator: weigh 1.490 g of o-phenanthroline, and dissolve it in 100 mL of aqueous solution containing 0.700 g of ferrous sulfate. To prevent deterioration, store it in a brown bottle for use;

(7) Potassium dichromate standard solution (0.03330 mol/L): weigh 9.807 g of potassium dichromate baked at 130℃ for 1.5 h, dissolve it with a small amount of water, then transfer into a 1 L volumetric flask and add water to fix the volume;

(8) Potassium dichromate - sulfuric acid solution (0.4 mol/L): weigh 39.23 g of potassium dichromate, dissolve it in 600-800 mL of distilled water, add water to dilute it to 1 L, transfer the solution into a 3 L beaker; take 1 L of concentrated sulfuric acid and slowly add it into the potassium dichromate solution, stirring continuously. To avoid rapid temperature rise, pause for a moment after each addition of about 100 mL of sulfuric acid, and put the beaker in a basin with cold water to cool down. When the temperature of the solution drops to a level that is not too hot to the hands, add another portion of sulfuric acid until all of these has been added.

(9) Ferrous sulfate standard solution: weigh 56 g of ferrous sulfate, dissolve it in 600-800 mL of water, add 20 mL of concentrated sulfuric acid, mix well, add water to 1 L (filter if necessary), store it in a brown bottle. This solution is easy to oxidation by air and should be standardized when in use.

Procedures

(1) Standardization of ferrous sulfate standard solution   Take 25.00 mL of potassium dichromate standard solution in a 150 mL Erlenmeyer flask, add 3 mL of concentrated sulfu-

ric acid and 3-5 drops of *o*-phenanthroline indicator, titrate with ferrous sulfate standard solution, and calculate the concentration of ferrous sulfate standard solution according to the consumption of ferrous sulfate solution.

(2) Preparation of the solution to be measured Select representative soil samples dried by air, use tweezers to remove organic residues such as roots and leaves of plant, and crush the soil block into fine pieces to pass through a 1 mm aperture sieve. After the sample fully mixed, 10-20 g of sample will be taken out, ground and passed through 0.25 mm aperture sieve, and put into a 150 mL Erlenmeyer flask for use. For newly collected paddy soil or soil that has been under waterlogging conditions for a long time, it must be dried and crushed, spread evenly into a thin layer, turned over once a day, and exposed to the air for about a week before grind.

(3) Determination of organic matter content According to the regulations for organic matter content in Table 4.22, weigh 0.05-0.5 g of the prepared sample and be accurate to 0.0001 g. Put it into a 150 mL Erlenmeyer flask, add 0.1 g of silver sulfate in powder, accurately add 10.00 mL of 0.4 mol/L potassium dichromate-sulfuric acid solution with a burette, and shake well.

Insert a simple air condenser above the Eerlenmeyer flask, and then move it to the electric sand bath that had been preheated to 200-230℃ (Fig. 4.5). When the first drop of condensate falls from the air condenser, start timing, and proceed digest for (5 ± 0.5) min. After it digesting, remove the Erlenmeyer flask from the electric sand bath, cool for a moment, rinse the inner wall and bottom outer wall of the condenser with water, so that the washing solution flows into the Erlenmeyer flask, the total volume of the solution in the flask should be controlled in 60-80 mL, add 3-5 drops of *o*-phenanthroline indicator, and titrate the remaining potassium dichromate with ferrous sulfate standard solution. The color of the solution changes from orange-yellow to blue-green and finally to brown-red, that is the end point. If the volume of ferrous sulfate standard solution used for sample is less than 1/3 of the volume of ferrous sulfate standard solution consumed by the blank, the mass of the soil should be reduced and re-determined.

2-3 blank calibrations must be performed simultaneously for each batch of sample determination. Take 0.500 g of silica in powder instead of the sample, and follow the same steps as the sample measurement, take the average value.

Data Recording and Processing

Soil organic matter content X (calculated based on dried) is calculated as follows:

$$X = \frac{(V_0 - V)c \times 0.003 \times 1.724}{m} \times 100\% \tag{4.27}$$

where, X is soil organic matter content, %; V_0 is the volume of ferrous sulfate standard solution consumed in the blank test, mL; V is the volume of ferrous sulfate standard solution consumed in the determination of the sample, mL; c is the concentration of ferrous sulfate standard solution, mol/L; molar mass number of 1/4 carbon atom is 0.003, g/mol; the coefficient of conversion from organic carbon to organic matter is 1.742; m is the mass of sample, g.

The results are expressed by the arithmetic mean for the parallel measurement with three significant figures reserved.

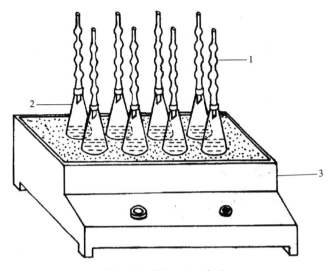

Fig. 4.5 Digestion device
1—simple air condenser; 2—ground Erlenmeyer flask; 3—electric sand bath

Table 4.22 Sample weight of different soil organic matter content

Organic matter content/%	Mass of sample/g
<2	0.4-0.5
2-7	0.2-0.3
7-10	0.1
10-15	0.05

Experiment 4.14 Determination of Chlorine Content for Compound Fertilizers

Objectives

(1) To understand the preparation method of compound fertilizer samples.

(2) To grasp the principle and method of determining chloride ion content using the back titration of the Volhard method.

Principles

The content of chloride ions in compound fertilizers directly affects the growth of crops, especially sweet potatoes and sugar beets that are sensitive to chloride ions. Therefore, the dosage of chloride ions should be strictly controlled during fertilization. In addition, long-term use of compound fertilizers with high chloride ion content can easily cause soil acidification and salinization, which will affect the germination and emergence of seeds, and inhibit the growth of crops. The experiment involves adding an excess of silver nitrate standard solution to a slightly acidic solution of a compound fertilizer sample to convert chloride ions into silver chloride precipitate, which is coated with dibutyl phthalate. The excess silver nitrate is back titrated with ammonium thiocyanate standard solution using ammonium ferric sulfate as an indicator. The titration reaction is as follows:

$$Ag^+ + Cl^- = AgCl \downarrow \tag{4.28}$$

$$Ag^+_{(excess)} + SCN^- = AgSCN \downarrow \tag{4.29}$$

Materials

Instruments: Erlenmeyer flask, volumetric flask, beaker, pipette, analytical balance, burette.

Reagents:

(1) Dibutyl phthalate (AR);

(2) Nitric acid: 1 : 1;

(3) Silver nitrate standard solution (0.05 mol/L): weigh 8.7 g of silver nitrate, dissolve in water, dilute to 1000 mL, and store in a brown reagent bottle;

(4) Chlorine ion standard solution (1.000 mg/mL): accurately weigh 1.6487 g of reference sodium chloride, which is dried at 270-300 ℃ to the constant mass. Dissolve it with water in a beaker, and then transfer it into a 1000 mL volumetric flask. Dilute it to the scale, mix it well, and store it in a plastic reagent bottle;

(5) Potassium chromate indicator (50 g/L): weigh 5.0 g of analytical pure potassium chromate, which is dissolved in a small amount of distilled water. Then add the silver nitrate solution until the brick red does not fade. Stir well, leave overnight, and filter. Dilute the filtrate with distilled water to 100 mL;

(6) Ammonium ferric sulfate indicator (80 g/L): dissolve 8.0 g of ammonium ferric sulfate in 75 mL of water, filter, add a few drops of sulfuric acid to make the brown color disappear, and dilute to 100 mL;

(7) Preparation of ammonium thiocyanate standard solution (0.05 mol/L): weigh 3.8 g of ammonium thiocyanate and dissolve it in water, dilute to 1000 mL.

Procedures

(1) Standardization of silver nitrate standard solution (0.05 mol/L) Accurately transfer 25.00 mL of chloride ion standard solution into a 250 mL Erlenmeyer flask, add 1 mL of potassium chromate indicator, and titrate with silver nitrate standard solution until the color of the solution changes from yellow to brick red precipitate, which is the titration end point. At the same time, perform a blank test. Keep four significant digits of the calculation result.

(2) Standardization of ammonium thiocyanate standard solution (0.05 mol/L) Accurately transfer 25.00 mL of chloride ion standard solution into a 250 mL Erlenmeyer flask, add 5 mL of nitric acid solution and 25.00 mL of silver nitrate standard solution, shake to precipitate and separate, add 5 mL of dibutyl phthalate, and shake for a moment. Add water to make the total volume of the solution approximately 100 mL, add 2 mL of ammonium ferric sulfate indicator, and titrate the remaining silver nitrate with ammonium thiocyanate standard solution until a light brick red color appears. At the same time, perform a blank test. Keep four significant digits of the calculation result.

(3) Sample preparation of compound fertilizer Take out about 100 g of the compound

fertilizer sample after several times reduction, grind it with a grinder or a mortar until it passes through a 0.50 mm sieve (for wet fertilizer, it can pass the 1.00 mm sieve), mix it evenly, and put it in a clean and dry bottle. Then, components analysis was conducted. The grinding operation should be fast to avoid water loss or moisture absorption during the grinding process, and to prevent the sample from overheating.

Weigh about 1-10 g (accurate to 0.001 g) of the above treated sample (see the Table 4.23 for the sample weight range) into a 250 mL beaker, add 100 mL of water, slowly heat to boil, continue to slightly boil for 10 min, cool to room temperature, transfer the solution to a 250 mL volumetric flask, dilute to the scale, and mix well. Dry filter and discard the initial portion of the filtrate.

Table 4.23 Sample weight range

Chloride ion content(w_1)/%	$w_1 < 5$	$5 \leqslant w_1 \leqslant 25$	$w_1 \geqslant 25$
Sample weight/g	10-5	5-1	1

(4) Determination of chloride ion content in the sample Accurately tranfer a certain amount of filtrate (containing about 25 mg of chloride ions) into a 250 mL Erlenmeyer flask, add 5 mL of nitric acid solution and 25.00 mL of silver nitrate standard solution, shake to precipitate and separate, add 5 mL of dibutyl phthalate, and shake for a moment. Add water to make the total volume of the solution approximately 100 mL, add 2 mL of ammonium ferric sulfate indicator, and titrate the remaining silver nitrate with ammonium thiocyanate standard solution until a light brick red color appears.

Data Recording and Processing

The mass fraction w (%) of chloride ions is calculated as follows:

$$w = \frac{(c_1V_1 - c_2V_2) \times 0.03545}{mD} \times 100\% \tag{4.30}$$

Where, V_1 is the volume of silver nitrate standard solution (25.00 mL); V_2 is the volume of ammonium thiocyanate standard solution consumed during titration, mL; c_1 is the concentration of silver nitrate standard solution, mol/L; c_2 is the concentration of ammonium thiocyanate standard solution, mol/L; m is the mass of compound fertilizer, g; D is the ratio of the transferred volume of the solution to the total volume of the solution; the millimole mass of chloride ion is 0.03545 g/mmol.

Take the arithmetic mean of the parallel measurements as the results. The allowable deviation for the determination of chloride ion content should comply with the requirements in Table 4.24.

Table 4.24 The allowable deviation for the determination of chloride ion content

Chloride ion content(w)/%	Maximum allowable deviation in parallel	Maximum allowable deviation in different laboratories' determination
<5	0.20	0.30
5-25	0.30	0.40
>25	0.40	0.60

Experiment 4.15　Determination of Potassium Content for Compound Fertilizers

Objectives

(1) To understand the preparation method of compound fertilizer samples.

(2) To grasp the principle and method of determining the potassium content by the tetraphenylborate gravimetric method.

Principles

In the weak alkaline solution, potassium tetraphenylborate solution and potassium ion in the sample solution form potassium tetraphenylborate precipitate, which is filtered, dried and weighed. If the sample contains cyanamide or organic matter, it can be treated with bromine water and activated carbon first. In order to prevent the interference of cations, an appropriate amount of EDTA can be added in advance to complex the cations. The precipitation reaction is as follows:

$$Na[B(C_6H_5)_4] + K^+ \rightleftharpoons K[B(C_6H_5)_4] \downarrow + Na^+ \tag{4.31}$$

Materials

Instruments: Erlenmeyer flask, volumetric flask, beaker, pipette, analytical balance, electric hot plate, G4 glass crucible, drier.

Reagents:

(1) $Na[B(C_6H_5)_4]$ solution: 15 g/L;

(2) EDTA solution: 40 g/L;

(3) NaOH solution: 400 g/L;

(4) Bromine water: 5% (mass fraction);

(5) Potassium tetraphenylborate saturated solution;

(6) Phenolphthalein indicator: 5 g/L ethanol solution (95%);

(7) Activated carbon: it should not absorb or release potassium ions.

Procedures

(1) **Sample preparation of compound fertilizer**　Take out about 100 g the compound fertilizer sample after several times reduction, grind it with a grinder or a mortar until it passes through a 0.50 mm sieve (for wet fertilizer, it can pass the 1.00 mm sieve), mix it evenly, and put it in a clean and dry bottle. Then, components analysis was conducted. The grinding operation should be fast to avoid water loss or moisture absorption during the grinding process, and to prevent the sample from overheating.

Perform parallel measurements on two samples. Weigh 2-5 g (accurate to 0.0002 g) of the above sample containing about 400 mg of potassium oxide, which is put into a 250 mL Erlenmeyer flask, add about 150 mL of water, heat and boil for 30 minutes, cool, and quantitatively transfer to a 250 mL volumetric flask. Dilute to the mark with water, mix well, and dry filter. Discard the initial 50 mL of filtrate.

(2) Sample treatment

① Samples not containing cyanamide compounds or organic matter. Accurately transfer 25.00 mL of the above filtrate into a 200 mL beaker, add 20 mL of EDTA solution (40 mL may be added if there are many cations), add 2 to 3 drops of phenolphthalein, then drop NaOH solution until red color appears, and add 1 mL excess of NaOH solution. The solution is slowly heated to boiling in the airing chamber for 15 minutes. Then place it to cool to the room temperature. If the red color disappears, adjust it to red with the NaOH solution.

② Samples containing cyanamide compounds or organic matter. Accurately transfer 25.00 mL of the above filtrate into a 200-250 mL beaker, add 5 mL of bromine water, and boil the solution until all the bromine water is completely removed (without bromine color). If it contains other colors, evaporate the solution to less than 100 mL, cool and add 0.5 g of activated carbon. Fully stir, then filter and wash with water 3-5 times, using about 5 mL of water each time. Collect all the filtrate, add 20 mL of EDTA solution (40 mL if there are many cations), and the following steps are the same as the operation ①.

(3) Precipitation and filtration   Add 10 mL of $Na[B(C_6H_5)_4]$ solution dropwise to the above sample solution (① or ②) stirring constantly, and add about 7 mL in excess. After let it stand for more than 15 minutes, filter with constant weight G4 glass crucible by tilting filtration. The crucible is then washed with potassium tetraphenylborate saturated solution 5-7 times (about 5 mL each time), and distilled water 2 times (about 5 mL each time).

(4) Dry   The crucible containing precipitation is placed in a drying oven, dried at 120 ℃ ± 5 ℃ for 1.5 hours, and then transferred into a desiccator, cooled to the room temperature, and weighed.

(5) Blank test   The analysis steps and reagent dosage are the same as those above, except that no sample is added.

Note: If the precipitation is not easy to remove, acetone can be used for further washing the crucible.

Data Recording and Processing

According to the weight of potassium tetraphenylborate precipitate, the mass fraction of K_2O in the compound fertilizer is calculated by the following formula:

$$w = \frac{(m_2 - m_1) \times 0.1314}{m_0 \times 25.00/250} \times 100\% \tag{4.32}$$

where, m_2 is the mass of potassium tetraphenylborate precipitation, g; m_1 is the mass of the potassium tetraphenylborate precipitate obtained during the blank test, g; 0.1314 is the coefficient of the mass of potassium tetraphenylborate converted to the mass of potassium oxide; m_0 is the mass of the compound fertilizer sample, g.

The calculation results are accurate to two decimal places, and the arithmetic mean of the parallel results is taken as the measurement result.

The allowable deviation of parallel measurement and different laboratory measurement results should meet the requirements of Table 4.25.

Table 4.25 The allowable deviation of parallel measurement and different laboratory measurement results

Mass fraction of K (calculated as K_2O)/%	Allowable deviation of parallel determination/%	Allowable deviation of different laboratory determination/%
<10.0	0.20	0.40
10.0-20.0	0.30	0.60
>20.0	0.40	0.80

Experiment 4.16   Determination of Available Phosphorus Content for Compound Fertilizers

 Objectives

(1) To understand the preparation method of compound fertilizer samples.

(2) To grasp the principle and method of determining the available phosphorus content by the phosphomolybdic acid quinoline gravimetric method.

 Principles

After the available phosphorus in the compound fertilizer is extracted by shaking with disodium dihydrogen ethylenediamine tetraacetate ($Na_2H_2Y \cdot 2H_2O$, EDTA) or ultrasonic extraction with citric acid solution ($C_6H_8O_7$, CA), the orthophosphate ion in the acid medium reacts with the excessive quinolomolactone reagent to generate yellow phosphomolybdic acid quinoline precipitation, and the phosphorus content is determined by phosphomolybdic acid quinoline gravimetry. The precipitation reaction is as follows:

$$H_3PO_4 + 3C_9H_7N + 12Na_2MoO_4 + 24HNO_3 =\!=\!=$$
$$(C_9H_7N)_3H_3[P(Mo_3O_{10})_4] \cdot H_2O\downarrow + 11H_2O + 24NaNO_3 \qquad (4.33)$$

 Materials

Instruments: analytical balance, ultrasonic cleaner, water bath oscillator, electric heating constant temperature drying oven (180 ℃±2 ℃), glass crucible filter (No. 4, volume 30 mL), electric hot plate.

Reagents:

(1) HNO_3 solution: 1∶1;

(2) EDTA solution: 37.5 g/L;

(3) CA solution: 20 g/L;

(4) Quinolomolactone reagent:

Solution Ⅰ: dissolve 70 g sodium molybdate dihydrate ($Na_2MoO_4 \cdot 2H_2O$) in 150 mL water.

Solution Ⅱ: dissolve 60 g of citric acid monohydrate ($C_6H_8O_7 \cdot H_2O$) in a mixture of 150 mL of water and 85 mL of nitric acid.

Solution Ⅲ: add solution A to solution B by stirring.

Solution Ⅳ: dissolve 5 mL of quinoline in a mixture of 35 mL of nitric acid and 100 mL of water.

Slowly add solution Ⅳ to solution Ⅲ and mix well. Place the solution in a dark place in a polyethylene bottle for 24 hours and filter it with a glass crucible filter. Measure 280 mL of acetone and add it to the above filtrate, dilute with water to 1000 mL, mix well, and store in another clean polyethylene bottle. The solution should be stored in dark for no more than one week.

Procedures

(1) **Sample preparation of compound fertilizer** Take out about 100 g of the compound fertilizer sample after several times reduction, grind it with a grinder or a mortar until it passes through a 0.50 mm sieve (for wet fertilizer, it can pass the 1.00 mm sieve), mix it evenly, and put it in a clean and dry bottle. Then, components analysis was conducted. The grinding operation should be fast to avoid water loss or moisture absorption during the grinding process, and to prevent the sample from overheating.

Perform parallel measurements on two samples. Each sample contains 100-180 mg of phosphorus pentoxide (accurate to 0.0002 g).

(2) **Extraction of available phosphorus** Extraction method one (EDTA oscillation extraction): according to the requirements for the sample weight of parallel measurements in step (1), another sample is weighed and placed on a filter paper, which is wrapped in filter paper and stuffed into a 250 mL volumetric flask. 150 mL of EDTA solution is added to the volumetric flask, and the flask is shaken to break the filter paper. The sample is dispersed in the solution, and placed in a constant temperature water bath oscillator at 60 ℃±2 ℃ for 1 hour (the oscillation frequency is determined by the sample inside the flask being able to freely flip). Then take out the flask, cool it to the room temperature, dilute with water to the mark, and mix well. Dry filter and discard the initial portion of the filtrate to obtain the solution A for the determination of the available phosphorus.

Extraction method 2 (citric acid ultrasonic extraction): for compound fertilizers with $NH_4H_2PO_4$, KH_2PO_4, and nitrophosphate as the phosphorus source, weigh the sample according to the requirements in step (1), place it in a 250 mL volumetric flask, add 150 mL citric acid solution, the volumetric flask is placed in the ultrasonic cleaner to extract for 6-8 min (the liquid level of the ultrasonic cleaner is higher than the liquid level in the volumetric flask). Then dilute it with water to the scale, mix well, dry filter, and discard the initial part of the filtrate to obtain the solution B for the determination of the available phosphorus.

(Note: citric acid ultrasonic extraction method is only applicable to fertilizers with ammonium dihydrogen phosphate, monopotassium phosphate and nitrophosphate compound fertilizer as phosphorus source).

(3) **Determination of available phosphorus** Use a pipette to transfer 25.00 mL of solution A or solution B to a 500 mL beaker, add 10 mL of nitric acid solution, and dilute with water to 100 mL. Heat the beaker on a hot plate for 2 min to 3 min, then add 35 mL of quinolomolactone reagent, cover a watch glass, and gently boil on the hot plate for 1 min or place it in a boiling water bath for heat preservation until the precipitation is separated, take out the beaker and cool it to the room temperature. Filter with a glass crucible filter that has been pre-dried to the constant weight in a 180 ℃±2 ℃ drying oven. The first step is to use the pouring method to filter the supernatant; the second step is to wash the precipitate 1-2

times with 25 mL of water each time; and the third step is to transfer the precipitate into the filter and wash the precipitate with 125-150 mL water. Place the precipitate and filter in a 180 ℃±2 ℃ drying oven. When the temperature reaches 180 ℃, dry the precipitate for 45 min, transfer it into a dryer, cool to the room temperature, and weigh.

(4) Blank test　Except for not adding samples, the analysis steps and reagent dosage are the same as the above steps, and parallel measurements are carried out.

Data Recording and Processing

The available phosphorus content ($w/\%$) in the compound fertilizer expressed by phosphorus pentoxide (P_2O_5) mass fraction is calculated as follows:

$$w = \frac{(m_1 - m_2) \times 0.03207}{m \times \dfrac{25.00}{250}} \times 100\% \tag{4.34}$$

Where, m_1 is the mass of phosphomolybdic acid quinoline precipitation, g; m_2 is the mass of phosphomolybdic acid quinoline precipitation in the blank test, g; m is the mass of the compound fertilizer sample, g; 0.03207 is the conversion factor for converting the mass of phosphomolybdic acid quinoline into the mass of phosphorus pentoxide; the volume of the sample solution is 25.00 mL; the total volume of sample solution is 250 mL.

The calculation results are accurate to two decimal places, and the arithmetic mean of the parallel results is taken as the measurement result.

Experiment 4.17　Determination of Available Phosphorus in Soil

Objectives

(1) To understand the principle of determining the available phosphorus content in soil by spectrophotometric method.

(2) To master the use of spectrophotometer.

Principles

Phosphorus is one of the essential elements for plant growth. The available phosphorus in soil refers to the total amount of phosphorus that can be absorbed and utilized by plants in the soil, which includes all water-soluble phosphorus, some adsorbed phosphorus, some slightly soluble inorganic phosphorus, and easily mineralized organic phosphorus. In agricultural production, the available phosphorus is generally used to guide the application of phosphorus fertilizer.

According to the properties of the soil, different extractants are selected to extract the available phosphorus from the soil. Ammonium fluoride-hydrochloric acid solution are used to extract available phosphorus from the acidic soil, and sodium bicarbonate solution is used to extract available phosphorus from the neutral and calcareous soil. The extracted available phosphorus is determined by Mo-Sbanti spectrophotometric method.

The principle of determining the available phosphorus by Mo-Sbanti spectrophotometric method is as follows: under acidic conditions, orthophosphate reacts with ammonium molybdate and antimony potassium tartrate to form phosphomolybdate heteropolymetalate. Mo(Ⅵ) in phosphomolybdic heteropolymetalate is partly reduced to Mo(Ⅴ) by ascorbic acid, and phosphorus molybdenum blue ($H_3PO_4 \cdot 10MoO_3 \cdot Mo_2O_5$ or $H_3PO_4 \cdot 8MoO_3 \cdot 2Mo_2O_5$) is formed.

Materials

Instruments: electronic balance, spectrophotometer, pH meter, volumetric flask (25 mL), colorimetric tube (50 mL), pipette (5 mL), measuring cylinder (50 mL), cuvette (1 cm), funnel, filter paper.

Reagents:

(1) Sulfuric acid (5%, volume ratio): take 5 mL of concentrated sulfuric acid and slowly add it to 90 mL of water. After cooling, dilute with water to 100 mL;

(2) Antimony potassium tartrate solution (5 g/L or 3 g/L): weigh 0.5 g (or 0.3 g) of antimony potassium tartrate ($KSbOC_4H_4O_6 \cdot 0.5H_2O$) and dissolve it in 100 mL water;

(3) Molybdenum antimony sulfate stock solution: weigh 10.0 g of ammonium molybdate and dissolve it in 300 mL of water at approximately 60 ℃, and cool it. Take another 126 mL of concentrated sulfuric acid, slowly pour it into approximately 400 mL of water, stir, and cool. Then slowly add the prepared sulfuric acid solution (5%) into the ammonium molybdate solution. Add 100 mL of antimony potassium tartrate solution, cool it, dilute it to 1 L with water, shake it well, and store it in a brown reagent bottle;

(4) Mo-Sbanti chromogenic reagent (15 g/L or 5 g/L): weigh 1.5 g (0.5 g) of ascorbic acid and dissolve it in 100 mL of molybdenum antimony sulfate stock solution. This solution is prepared and used immediately;

(5) Dinitrophenol indicator: weigh 0.2 g of 2,4-dinitrophenol or 2,6-dinitrophenol and dissolve it in 100 mL of water;

(6) Ammonia solution (1∶3): prepare it in a 1∶3 volume ratio of ammonia to water;

(7) Ammonium fluoride-hydrochloric acid extractant: weigh 1.11 g of ammonium fluoride and dissolve it in 400 mL of water, add 2.1 mL of concentrated hydrochloric acid, dilute it with water to 1 L, and store it in a plastic bottle;

(8) Boric acid solution (30 g/L): weigh 30.0 g of boric acid and dissolve it in hot water at about 60 ℃. After cooling, dilute it to 1 L with water;

(9) Phosphorus standard stock solution (100 mg/L): accurately weigh 0.4394 g of monopotassium phosphate (guaranteed reagent) dried at 105 ℃ for 2 hours, dissolve it in water, add 5 mL of 5% sulfuric acid, and dilute it to 1 L with water;

(10) Phosphorus standard solution (5 mg/L): suck 5.00 mL of phosphorus standard stock solution into a 100 mL volumetric flask, dilute it with water, shake it well for use;

(11) Sodium hydroxide solution (100 g/L): weigh 10 g sodium hydroxide and dissolve it in 100 mL water;

(12) Sodium bicarbonate extractant: weigh 42.0 g of sodium bicarbonate, dissolve it in about 950 mL of water, adjust the pH to 8.5 with sodium hydroxide solution (100 g/L), dilute it with water to 1 L, and store it in a polyethylene bottle for use. If the storage period

exceeds 20 days, pH must be checked and calibrated;

(13) Molybdenum antimony storage solution: weigh 10.0 g of ammonium molybdate and dissolve it in 300 mL of water at about 60 ℃, and cool it. Take another 181 mL of concentrated sulfuric acid, slowly add it into approximately 800 mL of water, stir, and cool. Then slowly add the prepared sulfuric acid solution (5%) into the ammonium molybdate solution. Add 100 mL of antimony potassium tartrate solution (3 g/L), cool it, dilute it to 2 L with water, mix well, and store it in a brown reagent bottle.

Procedures

(1) Soil sample pretreatment

① Acidic soil: weigh 5.00 g of air dried sample and place it in a 200 mL plastic bottle (pre-sieved through 2 mm sieve). Add 50.00 mL of NH_4F-HCl extractant and place it in an oscillator at a frequency of (180±20) r/min and (25±1)℃ for 30 min. Immediately filter with phosphorus-free filter paper. The filtrate is ready for the determination.

② Neutral or calcareous soil: weigh 2.50 g of air dried sample into a 200 mL plastic bottle (pre-sieved through 2 mm sieve). Add 50.00 mL of $NaHCO_3$ extractant. The other steps are the same as above.

(2) Plotting the standard curve

① Acidic soil: respectively and accurately transfer 0.00 mL, 1.00 mL, 2.00 mL, 4.00 mL, 6.00 mL, 8.00 mL and 10.00 mL of 5 mg/L phosphorus standard solution into a 50 mL volumetric flask, add 10 mL NH_4F-HCl extractant, then add 10 mL boric acid solution, shake well, add water to 30 mL, then add 2 drops of dinitrophenol indicator, adjust the solution to light yellow color with sulfuric acid solution or ammonia solution, add 5.00 mL Mo-Sbanti chromogenic reagent, and fill each volumetric flask to the mark with water, shaking thoroughly to obtain the phosphorus standard series solution. After let it stand at the room temperature higher than 20 ℃ for 30 min, adjust the zero point with 0.00 mL standard solution, use a 1 cm cuvette at 700 nm for the determination, and plot the standard curve.

② Neutral or calcareous soil: respectively and accurately transfer 0.00 mL, 0.50 mL, 1.00 mL, 2.00 mL, 3.00 mL, 4.00 mL and 5.00 mL of 5 mg/L phosphorus standard solution into a 25 mL volumetric flask, add 10 mL $NaHCO_3$ extractant, 5.00 mL Mo-Sbanti chromogenic reagent, slowly shake, discharge CO_2, and then fill each volumetric flask to the mark with water to obtain phosphorus standard solutions. After let it stand at the room temperature higher than 20 ℃ for 30 min, adjust the zero point with 0.00 mL standard solution, use a 1 cm cuvette at 880 nm for the determination, and plot the standard curve.

(3) Determination of available phosphorus

① Acidic soil: accurately transfer 10.00 mL of the above acidic soil filtrate into a 50 mL volumetric flask, add 10 mL of boric acid solution, shake well, add water to about 30 mL, then add 2 drops of dinitrophenol indicator, adjust the solution to light yellow with sulfuric acid solution or ammonia solution, add 5.00 mL of Mo-Sbanti chromogenic reagent, and fix the volume with water. After let it stand at the room temperature higher than 20 ℃ for 30 min, adjust the zero point with 0.00 mL standard solution, use a 1 cm cuvette at 700 nm for the determination.

② Neutral or calcareous soil: accurately transfer 10.00 mL of the above neutral or calcareous soil filtrate into a 50 mL volumetric flask or conical flask, slowly add 5.00 mL of Mo-Sbanti chromogenic reagent, slowly shake it, discharge CO_2, then add 10.00 mL of water, shake thoroughly, and gradually clean up CO_2. After let it stand at the room temperature higher than 20 ℃ for 30 min, adjust the zero point, use a 1 cm cuvette at 880 nm for the determination.

Data Recording and Processing

The available phosphorus content in soil samples, expressed as mass fraction w with values in milligrams per kilogram (mg/kg), was calculated according to the following formula:

$$w = \frac{(\rho - \rho_0)VD}{m \times 1000} \times 100\% \tag{4.35}$$

Where, ρ is the concentration of phosphorus, mg/L; ρ_0 is the concentration of phosphorus in the blank sample, mg/L; V is the volume of the sample extract solution, mL; D is the fractionation multiple, which is the ratio of the volume of the sample extracted to the volume of the fraction taken; m is the mass of the sample, g.

The results of the parallel measurements are expressed in the arithmetic mean with one decimal place.

Experiment 4.18 Macroporous Resin Adsorption and Spectrophotometric Determination of Flavonoids from Gingko Leaves

Objectives

(1) To master the application method of macroporous resin.
(2) To master the static and dynamic adsorption operation of macroporous resin.
(3) To master the use of UV-vis spectrophotometer.

Principles

China is the origin and the main production area of ginkgo biloba in the world. Ginkgo flavonoids are one of the most effective medicinal components extracted from Ginkgo leaves with about 20 types. Flavonoids mainly refer to a class of compounds whose mother nucleus is 2-phenylchromone (Fig. 4.6), and now they generally refer to a large class of natural compounds with a C_6—C_3—C_6 basic structure skeleton (Fig. 4.7), which is formed by the interconnection of two benzene rings through the central triple carbon. Due to the limitations of standard substances, it is difficult to directly determine the content of more than 20 components. At present, rutin, which is rich in ginkgo leaves, is generally used as the standard substance to analyze the content of flavonoids in ginkgo leaves and their extracts. The molecular formula of rutin is $C_{27}H_{30}O_{16}$ and the molecular weight is 610.51. Its structural formula is shown in Figure 4.8.

Fig. 4.6 2-Phenylchromone

Fig. 4.7 C_6—C_3—C_6 structural unit

Fig. 4.8 The structure of rutin

The main methods of separating and enriching flavonoids from the leaching solution of ginkgo leaves are the adsorption and extraction. The macroporous resin adsorption method has a relatively high selectivity for flavonoids, so the purity of flavonoids is high. The macroporous resin is a kind of organic polymer. Its adsorption principle is to use the van der Waals force between molecules to adsorb flavonoids from the solvent through a huge specific surface area, and then elute to obtain flavonoids with high purity. D101 is a macroporous weak acid cation exchange resin, whose functional group carboxyl forms hydrogen bonds with the hydroxyl groups of the flavonoids to adsorb well.

The principle of UV-vis spectrophotometry for the determination of flavonoid is that flavonoids in an alkaline solution containing sodium nitrite are complexed with Al^{3+} to produce yellow complex. Rutin is used as the reference to determine the content of flavonoids in the solution at 400-600 nm wavelength by UV-vis spectrophotometer. According to Lambert-Beer law, the content of flavonoids in the eluate is determined.

Materials

Instruments: ultraviolet visible spectrophotometer, pulverizer, electronic balance, electrothermal constant temperature water bath, pH meter, volumetric flask, colorimetric tube, cuvette, oscillation box, electric blast drying oven, glass chromatography column.

Reagents: ginkgo leaves, absolute ethanol (A.R.), sodium hydroxide (A.R.), aluminum nitrate (A.R.), sodium nitrite (A.R.), hydrochloric acid (A.R.), anhydrous methanol (A.R.), rutin reference (G.R.), D 101 macroporous resin.

Procedures

(1) Plotting the standard curve of rutin Accurately weigh 6.2 mg of rutin reference, dissolve it in 70% ethanol and fix the volume to 25 mL, shake well to obtain a rutin standard solution with a concentration of 0.248 mg/mL. Respectively and accurately transfer 0.00 mL, 0.25 mL, 0.50 mL, 0.75 mL, 1.00 mL, 1.25 mL and 1.50 mL of rutin standard solution (0.248 mg/mL) into a 10 mL volumetric flask, then add 0.30 mL of 5% sodium nitrite solution, shake well and let it stand for 6 min; add 0.30 mL of 10% aluminum nitrate solution, shake well and let it stand for 6 min, add 4.00 mL of 4% sodium hydrox-

ide solution, dilute it to the scale with 70% ethanol, shake well and let it stand for 15 min. The absorbance is measured at 510 nm with 70% ethanol as blank. The rutin standard curve is plotted with the concentration C of rutin as the x-axis and the absorbance value A as the y-axis.

(2) Extraction of flavonoids from ginkgo leaves　Take 5 g of ginkgo leaves and dry them at 50 ℃, crush them, and pass them through a 40 mesh (~0.42mm) sieve for later use. Weigh 0.5 g of ginkgo leaves powder, add 100 mL of 70% ethanol solution, heat it for 2 h in a closed chamber at 70 ℃, extract the supernatant by suction filtration, extract the filter residue in the same way, mix the two supernatants, concentrate the extract solution and transfer it into a 50 mL volumetric flask, dilute it to the scale with water, and shake well. The pH of the extraction solution needs to be adjusted before inject into the column.

(3) Pretreatment of D101 macroporous resin　Clean the glass chromatography column first. Add anhydrous ethanol equivalent to 0.4-0.5 times the volume of the resin into the column, and then add D101 macroporous resin into the column through wet filling. The liquid level above the resin layer is about 30 cm, and immerse the resin for 24 h. At a certain flow rate, the macroporous adsorption resin is hydrated and removed from impurities in the order of anhydrous ethanol, water, 5% hydrochloric acid, water, 2% sodium hydroxide and water.

Note that during loading column, add the resin into the chromatography column at a uniform speed along the glass rod at one time, and open the piston at the bottom of the column at the same time, so that the resin will naturally settle without bubbles.

(4) Static adsorption and desorption experiments of macroporous resin　Accurately weigh 1g of the treated macroporous resin, place it in a conical flask with a stopper, add 100 mL of flavonoid extract, and place it in an oscillation box for the static adsorption for 24h. After filtration, wash the saturated macroporous adsorption resin with distilled water until the eluent is colorless. Use the filter paper to absorb the residual solution on the resin surface, put the resin in a conical flask, add 80 mL of 70% ethanol solution, put it in the oscillation box for the static desorption for 24 h, and then filter out the resin. The concentration of flavonoids in the filtrate is determined with a spectrophotometer.

(5) Dynamic adsorption and desorption experiments of macroporous resin　Weigh 5 g of macroporous resin into a glass chromatography column (φ1.5 cm × 30 cm), with a diameter to height ratio of 1 : 8. Add 50.00 mL of flavonoid extraction solution to the column, pass through the column at a flow speed of 3 mL/min, and determine the concentration of flavonoids in the effluent with a spectrophotometer. Add a certain amount of distilled water to wash the residual extract in the column, and finally use 70% ethanol as the eluent to desorb at a flow speed of 3 mL/min, collect the effluent, and also use a spectrophotometer to determine the concentration of flavonoids in it.

(6) Regeneration of macroporous resin　Immerse the macroporous resin in 95% ethanol for 8 h, then wash with distilled water until there is no ethanol odor. Then it is immersed in 5% HCl solution for 3 h and washed with distilled water until the pH value of the effluent is neutral. At last, it is immersed in 5% NaOH solution for 3 h and washed with distilled water until the pH value of the effluent is neutral.

(7) Determination of flavonoid   Accurately transfer 2.00 mL of the sample solution into a 25 mL colorimetric tube, follow the steps of plotting the standard curve of rutin in (1) with 70% ethanol as the blank, and determine the absorbance at 510 nm. Using the standard curve and the absorbance value of the flavonoid in the eluent, obtain the concentration of rutin (mg/mL), and calculate the total flavonoid yield in ginkgo leaves.

Data Recording and Processing

The yield of total flavonoids w, expressed in mass fraction (%), is calculated by the following formula:

$$w = \frac{y \times 25 V_1}{mV \times 1000} \times 100\% \tag{4.36}$$

Where, V_1 is the volume of the leaching solution, mL; V is the volume of the test solution, mL; y is the concentration of rutin according to the standard curve, mg/mL; m is the mass of the sample, g.

Experiment 4.19   Determination of Mixed Amino Acids Content

Objectives

(1) To master the basic principle of separation of amino acids by ion exchange resin.

(2) To master the basic operation of ion exchange column chromatography.

(3) To master the principle of the ninhydrin color method for determination of amino acid content.

Principles

The separation of mixed amino acids by ion exchange column chromatography is based on the different charge behavior of each amino acid. Amino acids are amphoteric electrolytes and the net charge on the molecule depends on the isoelectric point (pI) of the amino acid and the pH of the solution. The pI of acid amino acid aspartate is 2.97, and the pI of basic amino acid lysine is 9.74. At pH 5.3, lysine can be dissociated into cation and bound to 732 resin because the pH is lower than the pI of lysine. On the contrary, aspartate can be dissociated into anion and flow out of the chromatographic column without being adsorbed by the resin. At pH 12, lysine can be dissociated into anions and be exchanged off the resin because the pH is higher than the pI of lysine. By the different pH of the eluent, the mixed amino acids can be respectively eluated to achieve separation.

The free amino group of amino acid and ninhydrin form a purple compound. In the certain extent, the color depth of the compound is correlated with the amino acid content. Therefore, the content of amino acids can be determined using a spectrophotometer.

Materials

Instruments: chromatography column (20 cm × ϕ1 cm), electronic balance, constant flow pump, collection device, thermostat water bath, spectrophotometer, pH meter, electronic magnetic stirrer, pipette, colorimetric tube with plug, rubber tipped dropper, graduated cylinder, beakers.

Reagents:

(1) 732 type cation exchange resin [100-200 mesh (0.0740-0.150 mm)];

(2) Hydrochloric acid (A.R.);

(3) Sodium hydroxide (A.R.);

(4) Standard amino acid solution: aspartate and lysine are respectively prepared into 2 mg/mL of 0.1 mol/L hydrochloric acid solution;

(5) Citric acid-sodium hydroxide-hydrochloric acid buffer solution (pH 5.3): weigh 14.25 g of citric acid, 9.30 g of sodium hydroxide, and transfer 5.25 mL of concentrated hydrochloric acid, dissolve them in a small amount of water, dilute it to 500 mL, and store it in the refrigerator;

(6) Ninhydrin chromogenic agent: take 0.5 g of ninhydrin dissolved in 75 mL of ethylene glycol monomethyl ether and add water to 100 mL;

(7) Amino acid mixed sample: 2 mg/mL aspartate and lysine solution were mixed in the ratio of 1 : 2.5, and then diluted with an equal volume of 0.1 mol/L hydrochloric acid solution and water.

Procedures

(1) Pretreatment of ion exchange resin For commercially available dry resin, it is necessary to first fully swell in water and obtain resin with appropriate particle size through flotation. Weigh 2 g of 732 type cation exchange resin and add it into the chromatographic column with water. Add 50 mL of 5% NaOH solution, control the flow speed to 1 drop/second, and then add 100 mL of deionized water to wash until neutral. Add another 50 mL of 5% HCl solution, control the flow speed to 1 drop/second, and then add 100 mL of deionized water to wash until neutral. Repeat this process three times.

(2) Packing column Take the chromatography column and fix it vertically on the iron stand, add 2-3 cm high citric acid-sodium hydroxide-hydrochloric acid buffer solution inside the column. Stir the pretreated ion exchange resin into suspension, slowly add the resin along the inner wall of the column into the chromatography column. When the resins gradually settle at the bottom of the column, slowly open the cock at the bottom of the column and continue to add the suspension until the resin layer reaches 3/4 of the column height.

(3) Balance After the chromatography column is packed, then slowly add the appropriate amount of buffer solution along the inner wall of the column to 2-3 cm above the resin, connect the constant flow pump, and balance with the buffer solution at a flow speed of 0.5 mL/min. Test the pH of the effluent solution with pH paper until it is equal to that of the buffer solution.

(4) Adding samples Close the constant flow pump, open the upper port of the column, slowly open the bottom outlet of the column, carefully release the solution inside the column until the concave surface of the liquid inside the column is exactly aligned with the upper level of the resin, and immediately close the bottom outlet (note: do not let the liquid level drop below the surface of resin). Transfer 0.5 mL of the amino acid mixture sample

and add it slowly along the wall of the column near the resin surface, taking care not to wash out the resin surface. After adding the sample, open the column bottom cock to let the liquid flow down as slowly as possible to the liquid concave surface, and close the cock immediately when the liquid concave level is exactly aligned with the resin surface. Then rinse the inner wall of the column with 0.5 mL of buffer solution using a rubber-tipped dropper, open the cock to release the liquid so that the lower surface of the liquid is aligned with the surface of the resin. Follow this method to clean twice. Then carefully add the buffer solution until it is 1 cm away from the top of the column and connect the chromatography column to the constant flow pump and collector.

(5) Elution

① Eluent with buffer: start elution with citric acid-sodium hydroxide-hydrochloric acid buffer solution at a flow rate of 6 s/drop, and collect eluent in test tubes from 1 to 10 with each tube collecting 1mL.

② Eluent with 0.01 mol/L sodium hydroxide solution (pH 12): close the constant flow pump and the cock, replace the above buffer solution with 0.01 mol/L sodium hydroxide solution, open the cock for eluting, and collect in tubes from 11 to 40 in the same way. After the collection, close the cock and constant flow pump.

(6) Plotting the standard curve and the determination of sample

① Plotting the standard curve. Take 2 mg/mL of standard amino acid solutions of 0.0 mL, 0.2 mL, 0.4 mL, 0.6 mL, 0.8 mL, and 1.0 mL into colorimetric tubes, and dilute to 1 mL with distilled water. Respectively add 1.5 mL citric acid-sodium hydroxide-hydrochloric acid buffer solution and 1 mL ninhydrin chromogenic agent, fully mix them, cover the tube, heat them in a 100 ℃ water bath for 15 min, cool them to the room temperature, and measure the absorbance at 570 nm with a spectrophotometer.

② Determination of amino acid samples. Respectively add 1.5 mL citric acid-sodium hydroxide-hydrochloric acid buffer solution to the 60 tubes collection solution, mix well, then add 1 mL ninhydrin chromogenic agent, and mix well. After heating in a 100 ℃ water bath for 15 min, cool to the room temperature. Colorimetry was performed at 570 nm using distilled water as blank. Draw an elution curve with the absorbance value as the vertical axis and the number of collected tubes as the horizontal axis.

(7) Resin regeneration   After the column used several times, the resin should be washed with 1 mol/L sodium hydroxide solution and then washed with distilled water until the eluent is neutral. The resin can be used again.

Experiment 4.20   Determination of Reducing Sugar in Tobacco Leaf

 Objectives

(1) To understand the preparation and processing methods of tobacco leaf samples.

(2) To master the principle and method of determining reducing sugar by arsenomolybdic acid colorimetry.

Principles

Heated under alkaline conditions, the reducing sugar can quantitatively reduce divalent copper ions to monovalent copper ions, which produces brick red cuprous oxide precipitation. Under acidic conditions, cuprous oxide can reduce ammonium molybdate, and the reduced ammonium molybdate reacts with disodium hydrogen arsenate to form a blue complex called arsenic molybdenum blue, the color of which is proportional to the content of reducing sugar. The content of reducing sugar in the sample can be determined by the spectrophotometry.

Materials

Instruments: 722 spectrophotometer, electronic analytical balance, crusher, water bath, volumetric flask, Erlenmeyer flask, beaker, graduated cylinder, colorimetric tube, pipette.

Reagents:

(1) Alcohol (80%);

(2) Copper reagent: weigh 12 g of sodium potassium tartrate, 24 g of anhydrous sodium carbonate and 16 g of sodium bicarbonate, which is respectively dissolved in 200 mL distilled water. Then mix the three parts of solution. Weigh 4 g of copper sulfate and dissolve it in about 200 mL of water. After it is completely dissolved, slowly add it to the above mixture while stirring. Then weigh 180 g of anhydrous sodium sulfate and dissolve it into the above solution. Heat the mixture in a boiling water bath for 20 min. If precipitation occurs after cooling, it should be filtered and removed. Finally, dilute the solution to 1000 mL. This solution is prone to crystallization and should be stored above 20 ℃;

(3) Arsenic molybdate reagent: weigh 25 g of ammonium molybdate [$(NH_4)_6Mo_7O_{24} \cdot 4H_2O$], dissolve it in 450 mL of distilled water, add 22 mL of concentrated sulfuric acid, and mix well. Take another 3 g of sodium hydrogen arsenate [$Na_2HAsO_4 \cdot 7H_2O$], dissolve it in 25 mL of water, and mix this solution with the above acidic ammonium molybdate solution. Keep the mixture at 37 ℃ for 24-48 h for use and should be stored in a brown bottle;

(4) Standard glucose solution (150 μg/mL): accurately weigh 150 mg of glucose and dissolve it in 1000 mL of water.

Procedures

(1) Plotting the standard curve   Take six 10 mL colorimetric tubes, add 0.0 mL, 0.2 mL, 0.4 mL, 0.6 mL, 0.8 mL and 1.0 mL 150 μg/mL glucose solution respectively, and add water to 1.0 mL. Then accurately add 1.0 mL of copper reagent and cover with a glass stopper. Put it in a boiling water bath, boil for 20 min, then take it out immediately, and cool it in cold water for 5 min. Finally, add 1.0 mL arsenic molybdate reagent, then add 7.0 mL distilled water to the scale, shake well and measure the absorbance at 570 nm with a spectrophotometer. Plot the standard curve and get the linear regression equation.

(2) Preparation of tobacco leaf samples   Accurately weigh 0.5 g of crushed tobacco leaf sample, put it into a conical flask, add 15 mL of 80% alcohol, extract it in a 70 ℃ water

bath for 30 min, and filter the supernatant into a 100 mL volumetric flask. Extract the residue twice with 10 mL of alcohol, filter, and rinse the residue with hot alcohol. Finally, dilute the extraction solution with distilled water into a 100 mL volumetric flask.

(3) Determination of reducing sugar Accurately transfer 5.0 mL of the above extraction solution and put it into a 50 mL volumetric flask, add water to the scale. Accurately measure 1.0 mL of diluent, measure the absorbance at 560 nm according to the standard curve steps, and calculate the reducing sugar content.

Data Recording and Processing

The reducing sugar content w, expressed in mass fraction (%), is calculated by the following formula:

$$w = \frac{y \times 10 \times V_1}{mV \times 10^6} \tag{4.37}$$

Where, V_1 is the volume of the extracting solution, mL; V is the volume of the test solution, mL; y is the concentration of the reducing sugar according to the standard curve, μg/mL; m is the mass of the sample, g.

4.4 Pharmacy

Experiment 4.21 Determination of NH_4^+ in Chinese Medicine White Sal-Ammoniac

Objectives

(1) To master the principle and procedure of determining the content of ammonium nitrogen in ammonium salt by the formaldehyde method.

(2) To be familiar with the titration operation and titration end point determination.

Principles

The Chinese medicine white sal-ammoniac (light sal-ammoniac) is a white crystal with irregular block or granular texture. It is uneven in size, salty and prickly. It has the effects of diuresis, dissipating blood stasis, detumescence, expelling phlegm, preventing ulcers, detoxifiying, eliminate corneal nebula and contraction of the uterus. The drug for external use Ma Ying Long Eye Ointment and the Mongolian medicine Li Niao Ba Wei San also contain white sal-ammoniac. Its main component is NH_4Cl, in addition, and it also contains Fe^{3+}, Ca^{2+}, Mg^{2+} and so on. The content of NH_4^+ in white sal-ammoniac is an important index to measure the quality of mineral medicine, and it is very important to determine its content accurately.

At present, the main methods for determining NH_4^+ include ion chromatography, colorimetry, electrochemical analysis and acid-base titration. Among them, the acid-base titration analysis method is a common analysis method in the laboratory because of its low requirements for equipment. The acidity of NH_4^+ in white sal-ammoniac is weak ($K_a = 5.6 \times$

10^{-10}), which cannot be titrated directly with standard base solution, but two indirect methods can generally be used for acid-base titration. The first is the distillation method, which needs to add excess alkali to the sample, and the ammonia distilled by heating is absorbed in the known excess acid standard solution, and then the excess acid is back titrated with the alkali standard solution to obtain the content of NH_4^+ in the sample. This method is more accurate, but the operation is more troublesome. The second formaldehyde method is used in this experiment, which is based on the reaction of formaldehyde with a certain amount of ammonium salt to produce a considerable amount of strong acid (H^+) and protonated hexamethylene tetramine $(CH_2)_6N_4H^+$ ($K_a = 7.1 \times 10^{-6}$), the reaction equation is as follows:

$$4NH_4^+ + 6HCHO \Longrightarrow (CH_2)_6N_4H^+ + 6H_2O + 3H^+ \quad (4.38)$$

From the above equation, 4 mol NH_4^+ can form 3 mol H^+ and 1 mol protonated hexamethylene tetramine, both of which can be accurately titrated with NaOH standard solution. It is equivalent to convert 1 mol of NH_4^+ into 1 mol of strong acid completely reacting with 1 mol of NaOH.

Therefore, NaOH standard solution is used to titrate the acid generated in the above reaction. At the stoichiometric point, the hexamethyl tetramine $(CH_2)_6N_4$ in the solution is a weak base ($K_b = 1.49 \times 10^{-9}$) and the pH of the solution is about 8.7, so phenolphthalein is selected as the indicator.

The reaction between ammonium salt and formaldehyde is slow at room temperature, and it needs to be left for a few minutes to complete the reaction after adding formaldehyde. Formaldehyde often contains a small amount of formic acid, so before use, it is necessary to first use phenolphthalein as an indicator and neutralize it with NaOH solution, otherwise the results will be too high. If the ammonium salt contains free acid, it also needs to be neutralized, that is, methyl red is used as an indicator, titrated with NaOH solution to orange yellow.

Materials

Instruments: analytical balance, platform scale, weighing bottle, beaker, volumetric flask (250 mL), pipette (25.00 mL), Erlenmeyer flask, burette.

Reagents: white sal-ammoniac sample, sodium hydroxide (A.R.), 40% of formaldehyde solution (A.R.), 0.2% of phenolphthalein ethanol solution, 0.1% of methyl red ethanol solution, potassium hydrogen phthalate ($KHC_8H_4O_4$, KHP, $M = 204.23$ g/mol).

Procedures

(1) Preparation and standardization of 0.1 mol/L NaOH solution Weigh about 1 g of sodium hydroxide with platform scale into a 100 mL beaker, then dissolve it with deionized water without CO_2. Transfer the solution into a 250 mL volumetric flask after cooling, add deionized water to the mark and shake well, and insert the rubber stopper for use.

Standardize 0.1 mol/L NaOH solution with KHP ($M = 204.23$ g/mol) as the primary standard substance. Accurately weigh 0.4-0.6 g of KHP into a 250 mL Erlenmeyer flask, add 20-30 mL of deionized water to dissolve, and add 2-3 drops of phenolphthalein solution. Titrate with NaOH solution until the solution changes from colorless to light pink, and it will not fade for 30 s, which is the end point. Calculate the exact concentration of the NaOH solution according to the mass of the primary standard KHP and the consumed volume of NaOH solution (Table 4.26).

(2) **Treatment of formaldehyde solution** Formaldehyde is often present in a white polymer state (paraformaldehyde). Take the original supernatant of formaldehyde (40%) in a beaker and dilute with water to obtain 20% of formaldehyde solution. Formaldehyde often contains a small amount of formic acid because it is easily air oxidized and the formic acid should be removed, otherwise a positive error will occur. The processing method is as follows: add 1-2 drops of 0.2% phenolphthalein indicator into 20% of formaldehyde solution, then neutralize it with 0.1 mol/L NaOH solution until the formaldehyde solution is light reddish (it is neutral to phenolphthalein).

(3) **Determination of NH_4^+ ions in the white sal-ammoniac** Because the traditional Chinese medicine, white sal-ammoniac contains insoluble impurities, it is necessary to take the supernatant for determination after full dissolution. Accurately weigh 1.5-2.0 g of white sal-ammoniac sample in a 100 mL beaker, add an appropriate amount of deionized water to dissolve, then transfer the solution into a 250 mL volumetric flask, add deionized water to the mark and shake well, insert the stopper. After let it stand for 5 min, accurately transfer 25.00 mL supernatant into a 250 mL Erlenmeyer flask, add 10 mL of neutralized 20% of formaldehyde solution and 2 drops of phenolphthalein indicator, and shake well. After standing for 1 min, the solution is titrated with 0.1 mol/L NaOH standard solution to pale pink. Parallel measurements are made 3 times (Table 4.27).

Data Recording and Processing

Calculate the content of NH_4Cl in white sal-ammoniac according to the volume of the consumed NaOH standard solution and the following formula (4.39):

$$w = \frac{\bar{c}_{NaOH} V_{NaOH} M_{NH_4Cl}}{m_s \times \frac{25}{250} \times 1000} \times 100\% \tag{4.39}$$

Where, M_{NH_4Cl} is molar mass of NH_4Cl, 53.49 g/mol; m_s is the mass of white sal-ammoniac sample, g.

Table 4.26 Standardization of NaOH solution

No.	1	2	3
m_{KHP}(initial)/g			
m_{KHP}(final)/g			
m_{KHP}/g			
V_{NaOH}(initial)/mL			
V_{NaOH}(final)/mL			
V_{NaOH}(consumed)/mL			
c_{NaOH}/(mol/L)			
$\bar{c}_{NaOH}$/(mol/L)			
$|d_i|$			
$\bar{d}_r$/%			

Table 4.27 Determination of NH_4^+ in white sal-ammoniac sample

No.	1	2	3
m_s/g			
V_{NaOH}(initial)/mL			
V_{NaOH}(final)/mL			
V_{NaOH}(consumed)/mL			
w			
$\bar{w}$			
$|d_i|$			
$\bar{d}_r$/%			

Questions

(1) Why cannot we determine the NH_4^+ by direct titration method with the NaOH standard solution in the experiment?

(2) What is the effect of adding formaldehyde in this experiment?

(3) Why should the added 20% of formaldehyde solution be neutralized by NaOH standard solution with phenolphthalein as an indicator in advance?

Experiment 4.22 Determination of Calcium in Calcium Supplement by $KMnO_4$ Method

Objectives

(1) To master the principle and method of determining calcium by the $KMnO_4$ method.

(2) To master the basic operational techniques of precipitation separation.

Principles

Calcium is one of the essential nutrients and the most abundant minerals in human body. It is also one of the most deficient elements in the body. At present, calcium intake in China is seriously insufficient, especially in children. In order to supplement calcium deficiency, calcium supplement products have developed rapidly at home and abroad. Therefore, calcium is a quality index that must be detected for the routine nutritional analysis of health foods, calcium products and dairy products. Accurately providing calcium content in calcium products is also the main basis for assessing the quality of calcium products. The main methods for determining calcium include the $KMnO_4$ method, spectrophotometry, EDTA complexometric titration, flame atomic absorption spectrometry, ion selection electrode and inductively coupled plasma mass spectrometry, etc. Among them, the $KMnO_4$ method has the advantages of simplicity, accuracy, fast and low error. It is a commonly used method to determine the content of calcium in calcium supplements.

Ca^{2+} can combine with $C_2O_4^{2-}$ to form insoluble white CaC_2O_4 precipitates under certain conditions. The precipitate is filtered, washed and dissolved with hot H_2SO_4 solution, and then $H_2C_2O_4$ can be indirectly titrated with $KMnO_4$ standard solution. The reactions are as follows:

$$Ca^{2+} + C_2O_4^{2-} \longrightarrow CaC_2O_4 \downarrow \qquad (4.40)$$

$$CaC_2O_4 + H_2SO_4 = CaSO_4 + H_2C_2O_4 \quad (4.41)$$
$$5H_2C_2O_4 + 2MnO_4^- + 6H^+ = 2Mn^{2+} + 10CO_2\uparrow + 8H_2O \quad (4.42)$$

When Ca^{2+} is precipitated, in order to obtain large crystalline precipitation which is easy to filter and wash, a sufficient amount of $(NH_4)_2C_2O_4$ precipitator is usually added into the acidic solution containing Ca^{2+} ($C_2O_4^{2-}$ is mainly in the form of $HC_2O_4^-$), and then ammonia is slowly added to neutralize H^+ in the solution. The concentration of $C_2O_4^{2-}$ slowly increases to obtain large crystalline precipitation of CaC_2O_4. However, when the precipitation is completed, the pH of the solution should be controlled between 3.5 and 4.5 (prevent CaC_2O_4 from dissolving too much and forming other insoluble calcium salts), and then keep heating and aging the solution for 30 min (during the process, small crystals dissolve and large crystals grow). After filtration, the $C_2O_4^{2-}$ adsorbed on the precipitation surface needs to be washed, but in order to reduce the washing loss of CaC_2O_4, first washed with diluted $(NH_4)_2C_2O_4$ solution, and then washed with slightly hot distilled water until the washing solution does not contain $C_2O_4^{2-}$. Finally, the washed CaC_2O_4 precipitate is dissolved in dilute H_2SO_4, heated to 70-80℃ and titrated with $KMnO_4$ standard solution.

The saled $KMnO_4$ often contains a small amount of impurities, such as sulfate, halide, nitrate and MnO_2, etc. Therefore, the $KMnO_4$ standard solution with accurate concentration cannot be directly prepared, and it needs to be standardized. $KMnO_4$ standard solution is often standardized with the reductant $Na_2C_2O_4$ as reference material ($Na_2C_2O_4$ does not contain crystal water and is easy to purify and stable). In the medium of H_2SO_4, MnO_4^- and $C_2O_4^{2-}$ react according to the equation (4.43), and the purple red color of MnO_4^- itself is used to indicate the end point during titration. To ensure that the reaction can proceed quantitatively and quickly, the reaction conditions should be strictly controlled: the temperature of the reaction should be controlled between 70-80℃ and the solution should be titrated while hot; the acidity of the solution is controlled in 0.5-1 mol/L in H_2SO_4 medium; the titration speed should be in a slow-fast-slow manner; the solution remains pale pink and does not fade within 0.5-1 min at the end point.

$$5C_2O_4^{2-} + 2MnO_4^- + 16H^+ = 2Mn^{2+} + 10CO_2\uparrow + 8H_2O \quad (4.43)$$

Materials

Instruments: burette (50.00 mL), beaker, funnel, graduated cylinder (10 mL, 50 mL), analytical balance, platform balance, watch glass, electric heating plate, water bath, Erlenmeyer flask, filter paper.

Reagents: $KMnO_4$ (solid), $Na_2C_2O_4$ (reference material), 3 mol/L H_2SO_4, 1 mol/L H_2SO_4, 6 mol/L HCl, 10% $NH_3 \cdot H_2O$, 0.25 mol/L $(NH_4)_2C_2O_4$, 10% ammonium citrate, 0.1% methyl orange, 0.1% $(NH_4)_2C_2O_4$, 0.5 mol/L $CaCl_2$.

Procedures

(1) Preparation and standardization of 0.02 mol/L $KMnO_4$ standard solution Weigh about 1.7 g of $KMnO_4$ and dissolve with 500 mL of distilled water, cover with a watch glass, heat to slightly boil and hold for 1 h, and replenish the evaporated water at any time. After cooling, let it stand for 2-3 days, then filter with a microporous glass funnel, and

store the filtrate in a brown bottle in the dark place for use.

Accurately weigh 0.2 g $Na_2C_2O_4$ in a 250 mL Erlenmeyer flask, add 50 mL water to dissolve it, then add 15 mL of H_2SO_4 (3 mol/L). Heat to 70-80 ℃ in a water bath (the solution begins to steam out), and titrate with $KMnO_4$ solution while it is hot until the solution is light pink and does not fade within 30 s. Perform parallel titration 3 times (Table 4.28).

(2) Determination of calcium in calcium supplement Accurately weigh three samples of calcium supplement (each containing about 0.05 g calcium) into a 250 mL beaker respectively. Add the appropriate amount of distilled water and cover with a watch glass, slowly drop 10 mL of HCl (6 mol/L) and heat to dissolve. After cooling, add 5 mL 10% ammonium citrate, 50 mL distilled water and 2-3 drops of methyl orange, and the solution is red. Add about 20 mL $(NH_4)_2C_2O_4$ (0.25 mol/L) into the solution, heat to 70-80 ℃ in a water bath and add 10% $NH_3 \cdot H_2O$ dropwise under continuous stirring until the color of the solution changes from red to yellow. Then it is aged in a hot bath for 30 min to form CaC_2O_4 big crystal precipitate.

Filter the precipitate after cooling, wash the precipitate with cold 0.1% $(NH_4)_2C_2O_4$ solution for several times, and then wash with water until the washing solution does not contain $C_2O_4^{2-}$ (add a few drops of 0.5 mol/L $CaCl_2$ to 1 mL of filtrate. If there is no turbidity in the solution, it indicates that the precipitate has been washed clean).

Unfold the filter paper with the precipitate and stick it onto the inner wall of the original beaker, wash the precipitate from the filter paper into the beaker several times with 50 mL of H_2SO_4 solution (1 mol/L). Then add distilled water to 100 mL, and heat to 70-80 ℃ in a water bath. The solution is titrated with $KMnO_4$ standard solution to light pink color, then dip the filter paper on the wall of the beaker into the solution and stir it gently. If the color fades, continue to add $KMnO_4$ dropwise until the light pink appears and it will not fade for 30 s, which is the end point. Record the consumed volume of $KMnO_4$ standard solution V_1. Repeat the titration 3 times (Table 4.29).

Data Recording and Processing

Calculate the content of calcium in some calcium products according to the consumed volume of $KMnO_4$ standard solution.

$$w_{Ca} = \frac{\frac{5}{2} \times \bar{c}_{KMnO_4} V_{KMnO_4} M_{Ca} \times 10^{-2}}{m_s} \times 100\% \tag{4.44}$$

Where, m_s is the mass of calcium products, g; M_{Ca} is the molar mass of calcium, 40 g/mol.

Table 4.28 Standardization of $KMnO_4$ solution concentration

No.	1	2	3
$m_{Na_2C_2O_4}$ (initial)/g			
$m_{Na_2C_2O_4}$ (final)/g			
$m_{Na_2C_2O_4}$ /g			
V_{KMnO_4} (initial)/mL			

(continued)

No.	1	2	3
V_{KMnO_4} (final)/mL			
V_{KMnO_4} (consumed)/mL			
c_{KMnO_4} /(mol/L)			
$\bar{c}_{KMnO_4}$ /(mol/L)			
$|d_i|$			
$\bar{d}_r/\%$			

Table 4.29 Determination of calcium content in calcium supplement

No.	1	2	3
m_s/g			
V/mL			
$w/\%$			
$\bar{w}/\%$			
$|d_i|$			
$\bar{d}_r/\%$			

Questions

(1) What are the advantages and disadvantages of the $KMnO_4$ method and the complexometric titration method for the determination of calcium content?

(2) Why do you put the filter paper from the beaker wall into the beaker for titration when the titration is close to the end point?

Experiment 4.23   Determination of Al^{3+} and Mg^{2+} in Compound Aluminium Hydroxide Tablets

Objectives

(1) To learn the pretreatment methods for determining the content of components in finished pharmaceutical products.

(2) To master the basic principles of back titration in coordination titration.

Principles

Compound Aluminum Hydroxide Tablet, a kind of stomach medicine, is used to relieve gastric pain, gastric burning sensation and acid regurgitation caused by excessive gastric acid. It can also be used for chronic gastritis. It is composed of aluminum hydroxide [$Al(OH)_3$] and magnesium trisilicate ($2MgO \cdot 3SiO_2 \cdot H_2O$), both of which can neutralize gastric acid, and combined with belladonna extract that can relieve spasm and pain. Among them, aluminium chloride, which is the product of aluminium hydroxide neutralising with gastric acid, can cause constipation, while magnesium ion, which is the product of magnesium trisilicate neutralising with gastric acid, has a light laxative effect. The combination of the two complements each other. However, aluminum is also a chronic neurotoxic substance, and excessive intake will deposit in neuronal fibers, causing chronic changes in the nervous system and inducing diseases such as Alzheimer's disease. Therefore, the determination of the aluminium and magnesium content in Compound Aluminium Hydroxide Tablet has important realistic meaning.

At present, the main methods for the determination of aluminium and magnesium include spectrophotometry, EDTA complexometric titration and inductively coupled plasma emission spectroscopy (ICP-AES). Among them, complexometric titration is a simple, rapid and widely used method for quantitative analysis.

Firstly, dissolve the tablets with acid, separate and remove the insoluble matter, and then take a test solution, adjust the pH to 3-4, add the excess and known amount of EDTA solution, boil for a few minutes to make the Al^{3+} react with EDTA completely.

$$Al^{3+} + H_2Y^{2-} \rightleftharpoons AlY^- + 2H^+ \tag{4.45}$$

After cooling, adjust the solution pH to 5-6. The excess of EDTA is back titrated with Zn^{2+} standard solution with xylenol orange as an indicator to calculate the Al^{3+} content.

$$Zn^{2+} + H_2Y^{2-} \rightleftharpoons ZnY^{2-} + 2H^+ \tag{4.46}$$

Take another test solution, adjust the pH to 8-9, separate Al^{3+} in the form of Al(OH)$_3$ precipitation. Then adjust the pH to 10, titrate Mg^{2+} in the filtrate with EDTA standard solution using EBT as an indicator, and calculate the content of Mg^{2+}.

$$Mg^{2+} + H_2Y^{2-} \rightleftharpoons MgY^{2-} + 2H^+ \tag{4.47}$$

Materials

Instruments: analytical balance, induction cooker, burette, pipette, volumetric flask, Erlenmeyer flask, graduated cylinder, beaker, weighing bottle, watch glass, mortar.

Reagents: Compound Aluminium Hydroxide Tablets, 0.02 mol/L EDTA standard solution, 0.02 mol/L Zn^{2+} standard solution, 0.2% xylenol orange indicator, 20% hexamethylene tetramine solution, 1:1 HCl solution, 1:1 ammonia water, 1:2 triethanolamine solution, $NH_3 \cdot H_2O$-NH_4Cl buffer solution (pH =10), 0.2% methyl red in ethanol solution, EBT indicator, NH_4Cl.

Procedures

(1) Preparation of 0.02 mol/L Zn^{2+} standard solution Accurately weigh 0.4 g of ZnO in a 250 mL beaker, cover with the watch glass, add 5-10 mL of 1:1 HCl from the beaker tip. After ZnO is dissolved completely, transfer the solution to a 250 mL volumetric flask and dilute to the mark with deionized water, shake well and calculate the exact concentration $c_{Zn^{2+}}$ (mol/L).

(2) Preparation and standardization of 0.02 mol/L EDTA standard solution Weigh 2 g of disodium dihydrogen ethylenediamine tetraacetate ($Na_2H_2Y \cdot 2H_2O$) into a 250 mL beaker, then dissolve it with 50 mL of water under heating and dilute to 250 mL, store it in a reagent bottle or polyethylene bottle.

Using xylenol orange as an indicator, accurately transfer 25.00 mL of Zn^{2+} standard solution in a 250 mL Erlenmeyer flask, add 2-3 drops of xylenol orange indicator, then add

20% hexamethylene tetramine solution until the color of the solution is stable purple-red and add an additional 5 mL. Titrate the above solution with EDTA standard solution until the color of the solution changes from purple-red to bright yellow as the end point. Repeat the titration 3 times (Table 4.30). Calculate c_{EDTA} (mol/L) according to the consumed volume of EDTA.

(3) **Treatment of samples** Take 10 tablets of Compound Aluminium Hydroxide to ground, accurately weigh about 2 g of powder in a 250 mL beaker, add 20 mL of 1∶1 HCl solution, dilute to 100 mL with deionized water and boil. After cooling, filter and wash the precipitate with deionized water. Collect the filtrate and washing liquid in a 250 mL volumetric flask, dilute to the mark with deionized water and shake well for use.

(4) **Determination of aluminium** Accurately transfer 10.00 mL of the above test solution in a 250 mL Erlenmeyer flask, dilute to about 25 mL with deionized water. Add 1∶1 $NH_3 \cdot H_2O$ until turbidity just appears, and then add 1∶1 HCl solution dropwise until the precipitate is completely dissolved. Add 25.00 mL of EDTA standard solution, heat and boil on an induction cooker for 3 min. After cooling to the room temperature, add 10 mL of 20% hexamethylene tetramine solution and 2-3 drops of xylenol orange indicator. Titrate with Zn^{2+} standard solution until the color of the solution changes from yellow to purple-red as the end point, and record the consumed volume V_1 of Zn^{2+} standard solution. Repeat the titration 3 times (Table 4.31).

(5) **Determination of magnesium** Accurately transfer 25.00 mL of the sample solution in a beaker, add 1∶1 $NH_3 \cdot H_2O$ until turbidity just appears, and then add 1∶1 HCl solution to dissolve the precipitate. After adding 2 g of NH_4Cl solid, add 20% hexamethylene tetramine solution until the precipitate appears, then add an additional 15 mL. Heat the test solution to 80 ℃ for 10 min, remove $Al(OH)_3$ by filtering after cooling, and wash the precipitate several times with a small amount of water. Collect the filtrate and washing liquid in a 250 mL Erlenmeyer flask. Then add 4 mL 1∶2 triethanolamine, 10 mL of NH_3-NH_4Cl buffer solution (pH=10), 1 drop of methyl red indicator and 3-5 drops of EBT indicator, and titrate with EDTA standard solution until the color of the solution changes from dark-red to blue-green as the end point. Record the consumed volume V_2 of EDTA standard solution. Repeat the titration 3 times (Table 4.31).

✺ Data Recording and Processing

Calculate the content of Al^{3+} in the Compound Aluminium Hydroxide Tablet according to the Eq. (4.48) by the quantitative addition of EDTA and the consumed volume V_1 of Zn^{2+} standard solution, expressed as Al_2O_3 (%).

$$w_{Al_2O_3} = \frac{(\overline{c}_{EDTA}V_{EDTA} - c_{Zn^{2+}}V_1)M_{Al_2O_3}}{m_s \times \frac{10}{250} \times 1000 \times 2} \tag{4.48}$$

Where, $w_{Al_2O_3}$ is the mass fraction of Al_2O_3; $\overline{c}_{EDTA}$ is the average concentration of EDTA standard solution, mol/L; V_{EDTA} is the volume of excess EDTA, mL; $c_{Zn^{2+}}$ is the concentration of Zn^{2+} standard solution, mol/L; V_1 is the volume of the consumed Zn^{2+} standard solution in the back titration, mL; $M_{Al_2O_3}$ is the molar mass of Al_2O_3, g/mol;

m_s is the mass of Compound Aluminium Hydroxide Tablets, g.

Calculate the content of Mg^{2+} in the Compound Aluminium Hydroxide Tablet according to the Eq. (4.49) by the volume V_2 of the consumed EDTA standard solution, expressed as MgO (%).

$$w_{MgO} = \frac{\bar{c}_{EDTA} V_2 M_{MgO}}{m_s \times \frac{25}{250} \times 1000} \times 100\% \qquad (4.49)$$

Where, w_{MgO} is the mass fraction of MgO; $\bar{c}_{EDTA}$: the average concentration of EDTA standard solution, mol/L; V_2 is the volume of the consumed EDTA standard solution in the titration of Mg^{2+}, mL; M_{MgO} is the molar mass of MgO, g/mol; m_s is the mass of Compound Aluminium Hydroxide Tablets, g.

Table 4.30　Standardization of EDTA solution

No.	1	2	3
m_{ZnO}(initial)/g			
m_{ZnO}(final)/g			
m_{ZnO}/g			
$c_{Zn^{2+}}$/(mol/L)			
V_{EDTA}(initial)/mL			
V_{EDTA}(final)/mL			
V_{EDTA}(consumed)/mL			
c_{EDTA}/(mol/L)			
$\bar{c}_{EDTA}$/(mol/L)			
$\lvert d_i \rvert$			
$\bar{d}_r$/%			

Table 4.31　Determination of Al^{3+} and Mg^{2+} content in tablets

	No.	1	2	3
Determination of Al^{3+}	m_s/g			
	V (test solution)/mL			
	$\bar{c}_{EDTA}$/(mol/L)			
	V_{EDTA}/mL			
	$c_{Zn^{2+}}$/(mol/L)			
	V_1/mL			
	$w_{Al_2O_3}$/%			
	$\bar{w}_{Al_2O_3}$/%			
	$\lvert d_i \rvert$			
	$\bar{d}_r$/%			
	No.	1	2	3
Determination of Mg^{2+}	V (test solution)/mL			
	$\bar{c}_{EDTA}$/(mol/L)			
	V_2/mL			
	w_{MgO}/%			
	$\bar{w}_{MgO}$/%			
	$\lvert d_i \rvert$			
	$\bar{d}_r$/%			

Questions

(1) Why do not you use the direct titration method to determine Al^{3+}?

(2) Why should the test solution be mixed with EDTA standard solution and then heated and boiled in the determination of Al^{3+}?

(3) Why do you use 20% hexamethyltetramine solution to adjust the pH for the precipitation of Al^{3+} instead of using ammonia in the determination of Mg^{2+}?

(4) What is the effect of triethanolamine in the determination of Mg^{2+}?

Experiment 4.24 Determination of the Content of Berberine Hydrochloride

 Objectives

(1) To understand the principle and method of determination of berberine hydrochloride by gravimetric method.

(2) To understand the precipitation conditions, principles and methods of crystal precipitation.

(3) To master the basic operating techniques of filtration, washing and constant weight for crystal precipitation.

 Principles

Berberine hydrochloride is the hydrochloride of quaternary ammonium berberine ($M_{C_{20}H_{18}O_4N \cdot Cl \cdot 2H_2O} = 407.85$ g/mol), which is slightly soluble in cold water and easily soluble in hot water. It is an isoquinoline alkaloid found in plants and a kind of broad-spectrum antibacterial drug, and has an inhibitory effect on pathogenic bacterium, etc. It is mainly used in clinic for intestinal infections such as gastroenteritis and bacterial dysentery caused by sensitive pathogens. The main component of Amaranth Berberine Capsule, Compound Berberine Tablet and Berberine Hydrochloride Tablet is berberine hydrochloride. The accurate determination of the effective component berberine hydrochloride content in drugs can effectively control the quality of the drug.

At present, the main methods for the determination of berberine hydrochloride in drugs are UV-vis spectrophotometry, HPLC, atomic absorption spectrometry, gravimetric method, capillary electrophoresis and so on. Among them, the accuracy of the gravimetric method is higher, and the determination results of the gravimetric method are often used as a standard in the analytical work to check the accuracy of other analytical methods. The gravimetric method determines the component content by weighing the mass of the substance. In the analyzing process, it is generally necessary to use appropriate methods to separate the tested component from the sample, convert it into a certain weighing form, and weigh it. The content of the component is calculated based on the weight. Under acidic conditions, trinitrophenol is used as a precipitator to form picric acid-berberine precipitation ($M_{C_{20}H_{17}O_4N \cdot C_6H_3O_7N_3} = 564.56$ g/mol) with berberine hydrochloride [Eq. (4.50)].

$$C_{20}H_{18}O_4N \cdot Cl + C_6H_3O_7N_3 = C_{20}H_{17}O_4N \cdot C_6H_3O_7N_3 \downarrow + HCl \quad (4.50)$$

The content of $C_{20}H_{18}O_4N \cdot Cl$ can be calculated by determining the mass of the precipitate after filtration, washing and drying.

Materials

Instruments: analytical balance, constant temperature drying oven, desiccator, weighing bottle, beaker, water circulation pump (equipped with suction filter bottle), G4 sand core crucible.

Reagents: concentrated hydrochloric acid (A. R.), trinitrophenol (A. R.), berberine hydrochloride tablets, 0.1 mol/L HCl, saturated aqueous solution of trinitrophenol, saturated solution of trinitrophenol berberine (prepare pure precipitate of trinitrophenol berberine and make a saturated solution with distilled water).

Procedures

(1) Preparation of precipitation Take 5 tablets of berberine hydrochloride, grind them finely, accurately weigh about 0.2 g, put it in a 250 mL beaker, dissolve it with 100 mL of heated distilled water, add 10 mL of 0.1 mol/L hydrochloric acid, and slowly add 30 mL of trinitrophenol saturated solution immediately. After the sediment sinks, add a small amount of trinitrophenol saturated solution to the supernatant to determine whether the precipitation is complete. After the precipitation is complete, heat it in a water bath for 15 minutes and let it stand for more than 2 hours.

(2) Constant weight of crucible Wash the crucible with deionized water, filter until the water mist disappears, and then heat the crucible in a constant temperature drying oven at 140 ℃ (60 min for the first time and 30 min for the second time). After that, transfer the crucible to the desiccator for cooling for 10-15 min and weigh it. If the difference between the two masses does not exceed 2 mg, it means that the weight is constant (otherwise, continue heating, cooling and weighing until constant weight). The mass of the crucible is recorded as m_1.

(3) Treatment of precipitation Transfer all the aged precipitation to a dry crucible of constant weight, filter under reduced pressure, wash with saturated trinitrophenol berberine aqueous solution several times, and then wash with water three times, 15 mL each time. Then, put the crucible with precipitation in a constant temperature drying oven, cool and weigh it until constant weight [the same as the operation of step (2)]. The mass of the crucible is recorded as m_2.

Data Recording and Processing

According to the mass of trinitrophenol berberine, calculate the content of berberine hydrochloride according to the following formula (Table 4.32):

$$w_{C_{20}H_{18}O_4N \cdot Cl} = \frac{(m_2 - m_1) \times \dfrac{M_{C_{20}H_{18}O_4N \cdot Cl}}{M_{C_{20}H_{17}O_4N \cdot C_6H_3O_7N_3}}}{m_s} \times 100\% \quad (4.51)$$

Where, $M_{C_{20}H_{18}O_4N \cdot Cl}$ is the molar mass of berberine hydrochloride, 371.85 g/mol; $M_{C_{20}H_{17}O_4N \cdot C_6H_3O_7N_3}$ is the molar mass in weighing form, 564.56 g/mol; m_s is the mass of sample, g.

Table 4.32 Determination of berberine hydrochloride content

m_s/g	m_1/g	m_2/g	(m_2-m_1)/g	$w_{C_{20}H_{18}O_4N \cdot Cl}$/%

Questions

(1) How do you determine the weighing amount of the sample? Should it be exactly 0.2 g?

(2) What is the purpose of adding HCl before adding the precipitant?

(3) Why should trinitrophenol be added slowly to the hot solution?

(4) Why should the precipitation be cooled to the room temperature in the desiccator after drying?

(5) Why should the precipitation be heated in a water bath and stand for more than 2 hours?

Experiment 4.25   Determination of Acetaminophen in Ankahuangmin Capsules

Objectives

(1) To master the principles of the determination of acetaminophen content.

(2) To master the method and calculation for the determination of drug content by UV-visible spectrophotometry.

Principles

Acetaminophen ($C_8H_9NO_2$, $M = 151.16$ g/mol, CAS: 103-90-2), also called paracetamol and p-hydroxyacetanilide (Fig. 4.9), is usually a white crystalline powder, odorless, slightly bitter, easily soluble in hot water or ethanol, moderately soluble in acetone and slightly soluble in water. Acetaminophen is the most commonly used as antipyretic and analgesic drug, and its antipyretic effect is slow and lasting, safe and effective, which is widely used in clinical practice and in the treatment of colds and fever, arthralgia, neuralgia, etc. It's the main component of paracetamol, quick-acting anti cold capsule, compound aminoglutethimide tablet, and chlorphenamine maleate capsule. However, acetaminophen has certain toxic effects on human. Excess acetaminophen can cause liver necrosis, so its content must be strictly controlled. Thus, determining the content of acetaminophen in drugs can better control the intake of acetaminophen.

At present, the methods for determining the content of acetaminophen include spectrophotometry, sodium nitrite method, electrochemical method and HPLC etc. Among them, the spectrophotometry uses the conjugated structure of acetaminophen, which has strong absorption in the UV region and a maximum absorption wavelength at 257 nm. The content of acetaminophen can be calculated by determination of absorbance at its maximum absorption wavelength.

Fig. 4.9 The structure of acetaminophen

Materials

Instruments: analytical balance, UV-vis spectrophotometer, pipette (5.00 mL, 10.00 mL), beaker, filter paper (the diameter of 10 cm), volumetric flask (100 mL, 250 mL).

Reagents: saled chlorphenamine maleate capsule (specifications: each capsule contains 250 mg of acetaminophen, 15 mg of caffeine, 1 mg of chlorpheniramine maleate and 10 mg of artificial bovis), acetaminophen standard substance, sodium hydroxide (A.R.), distilled water.

Procedures

(1) Preparation of acetaminophen standard solutions Accurately weigh 0.0400 g of standard acetaminophen in a beaker, add 50 mL of 0.4% sodium hydroxide solution and 50 mL of distilled water, shake for 15 min to dissolve. Transfer quantitatively to a 250 mL volumetric flask, add water to the scale, and shake up to obtain the standard stock solution (160 μg/mL).

(2) Plotting the standard curve Take the above standard diluent and determine the absorption spectrum of acetaminophen in the wavelength range of 200-400 nm (λ_{max}=257 nm).

Accurately transfer a certain amount of acetaminophen standard stock solution into a 100 mL volumetric flask respectively, add 10 mL of 0.4% sodium hydroxide solution, dilute to the mark with distilled water and mix well to get acetaminophen standard solution, and its concentrations are 2 μg/mL, 4 μg/mL, 6 μg/mL, 8 μg/mL and 10 μg/mL. Measure the absorbance A at λ_{max} with aqueous solution as a reference and plot the standard curve to determine the linear range and standard curve equation (Table 4.33).

(3) Preparation and determination of sample solution Take a few saled chlorphenamine maleate capsules and mix the contents evenly. Accurately weigh the appropriate amount of the contents (m_s, equivalent to about 40 mg of acetaminophen), add 50 mL of 0.4% sodium hydroxide solution and 50 mL of distilled water respectively, shake for 15 min to make the acetaminophen sample fully dissolved and transfer quantitatively to a 250 mL volumetric flask. Dilute to the mark with distilled water and mix well. After filtration, discard the initial filtrate. Accurately transfer 5.00 mL of filtrate into a 100 mL volumetric flask, add 10 mL of 0.4% sodium hydroxide solution, then add distilled water to the mark and shake well. Measure the absorbance A at λ_{max} with aqueous solution as a reference, and repeat the measurement three times in parallel. Determine the concentration of the acetaminophen based on the standard curve, and calculate the content of the acetaminophen in the chlorphenamine maleate capsule (Table 4.34).

Data Recording and Processing

The concentration of the acetaminophen c (μg/mL) in the sample solution was obtained based on the standard curve, then calculated the content of the acetaminophen in the chlorphenamine maleate capsule as follows.

The labeled amount of acetaminophen in the drugs:

$$X = \frac{\bar{c} \times 20 \times 250 \times 10^{-3} \times \bar{w}}{m_s} \tag{4.52}$$

where, $\bar{w}$ is the average mass of each chlorphenamine maleate capsule, mg; m_s is the mass of the contents in the chlorphenamine maleate capsule, mg.

Table 4.33 Plotting of standard curve

$c/(\mu g/mL)$	2	4	6	8	10
A					
Standard curve					

Table 4.34 Determination of acetaminophen content

No.	1	2	3
$\bar{w}$			
m_s/mg			
A			
$c/(\mu g/mL)$			
$\bar{c}/(\mu g/mL)$			
$\lvert d_i \rvert$			
$\bar{d}_r/\%$			
X/mg			

  Question

Besides the main component acetaminophen, the chlorphenamine maleate capsule also contains other components, such as caffeine, chlorpheniramine maleate, artificial bovis etc. Is there any interference with the determination of the ultraviolet spectrum?

4.5 Food

Experiment 4.26 Determination of Citric Acid Content in Commercially Available Citrus

  Objectives

(1) To master the method of preparation and standardization of the standard solution of NaOH.

(2) To master the principle and method of determination of citric acid content.

  Principles

China is one of the most important origins of citrus, with abundant citrus resources and a wide range of varieties, including oranges, pomelo, lemons, kumquats, and mandarin oranges, etc. In autumn and winter, citrus occupies half of the fruit market. Citrus fruit juice is rich, moderately sweet and sour, and contains a large amount of sugar, organic acids, vitamins, minerals and other nutrients. Citrus is also a very good Chinese medicine, with shunqi, anti-tussive, invigorating stomach, resolving phlegm, detumescence, relieving pain, dispersing liver and regulating qi, and other effects. The sugar and acid content of the fruit and the sugar-acid ratio or solid-acid ratio are the most important indicators for determining the quality of citrus.

Most organic acids are solid weak acids, such as citric acid, oxalic acid, tartaric acid, acetylsalicylic acid and benzoic acid. If the organic acid is soluble in water and the dissociation constant $K_a \geqslant 10^{-7}$, a certain amount of sample can be weighed, dissolved in water and titrated with NaOH standard solution. If titration jump is in the weak alkaline range, often use phenolphthalein as an indicator, and titrate till the solution changes from colourless to slightly red, that is the end point. According to the concentration c of NaOH standard solution and the volume V consumed in the titration, weigh the mass of organic acids, and calculate the content of organic acids.

Materials

Instruments: burette, Erlenmeyer flask, volumetric flask, pipette (25.00 mL), beaker.

Reagents: potassium hydrogen phthalate (reference material, dry at 100-125 ℃ for 1 h, then put in a desiccator to cool for use), NaOH (A.R.), citrus, phenolphthalein indicator.

Procedures

(1) Standardization of 0.10 mol/L NaOH solution   Accurately weigh 0.4-0.6 g of potassium hydrogen phthalate into a 250 mL of Erlenmeyer flask, add 20-30 mL of water, and gently heat to dissolve completely. After cooling, add 2-3 drops of 0.2% phenolphthalein indicator, and titrate with the NaOH solution until the solution turns reddish, and the color does not fade within half a minute, which is the end point (if the reddish color fades slowly in a long time, it is due to the absorption of carbon dioxide from the air). Record the volume of NaOH solution. Repeat the titration 3 times (Table 4.35).

(2) Determination of citric acid   The citrus samples were washed and dried, and the edible parts were homogenised in a tissue masher. Weigh 25 g of the slurry and dilute with water, and the volume is fixed to 250 mL, shake well, and filter through 0.45 μm filter membrane.

Pipette 25.00 mL of the above test solution into a 250 mL Erlenmeyer flask, add phenolphthalein indicator 1-2 drops, and drop the NaOH standard solution until the solution is slightly red, and keep it for 30 seconds without fading. That is the end point. Note down the volume of NaOH solution consumed and measure it 3 times in parallel to calculate the mass fraction of citric acid (Table 4.36).

$$w_{H_nA} = \frac{\frac{1}{n}cVM_{H_nA}}{m \times 1000} \times 100\% \qquad (4.53)$$

Where $n=3$ is the number of citric acid H^+, $M_{H_3A} = 210.14$ g/mol: molar mass of citric acid, and m is the sample mass.

Data Recording and Processing

Table 4.35 Standardization of NaOH solution

No.	1	2	3
$m_{KHC_8H_4O_4}/g$			
$V_{NaOH}(\text{initial})/mL$			
$V_{NaOH}(\text{final})/mL$			
V_{NaOH}/mL			
$c_{NaOH}/(mol/L)$			
$\bar{c}_{NaOH}/(mol/L)$			
$|d_i|$			
$\bar{d}_r/\%$			

Table 4.36 Determination of citric acid content

No.	1	2	3
m_{H_nA}/g			
$V_{NaOH}(\text{initial})/mL$			
$V_{NaOH}(\text{final})/mL$			
V_{NaOH}/mL			
$w/(mol/L)$			
$\bar{w}/(mol/L)$			
$|d_i|$			
$\bar{d}_r/\%$			

Question

Can oxalate acid, tartaric acid and other multiple acid be titrated with NaOH solution?

Tips

Citric acid is an edible acid that enhances normal metabolism in the body and is harmless in appropriate doses. The addition of citric acid to certain foods gives a good taste and promotes appetite, and its use in jams, beverages, canned goods and sweets is permitted in China.

Based on the fact that citric acid can affect the metabolism of calcium, people who often consume canned food, beverages, jams and sour candies, especially children, should pay attention to calcium supplementation, drink more milk, fish head and fish bone soup, eat more small shrimp, etc., so as not to lead to insufficient calcium in the blood and affect the health. Patients of the gastric ulcers, hyperacidity, caries, and diabetes should not consume citric acid regularly. Citric acid can not be added to pure milk, otherwise it will cause pure milk coagulation. In dairy industry, often prepare citric acid into a solution of about 10% added to a low concentration of milk solution, and add while quickly stirring.

Experiment 4.27 Determination of Calcium Content in Calcium Tablets/Milk Powder/Spinach

Objectives

(1) To master the method of standardizing EDTA.

(2) To master the principle and method of determining Ca^{2+} content by the EDTA method.

Principles

EDTA ($Na_2H_2Y \cdot 2H_2O$) standard solution is prepared by indirect method, which can be prepared of a rough concentration and then calibrated with reference substances such as Zn, ZnO, $CaCO_3$, $MgSO_4 \cdot 7H_2O$. After the sample has been processed, the calcium content in it can be determined by coordination titration, usually with EDTA as the titrant, and the end point is indicated by a calcium indicator under alkaline conditions. Calcium indicator is a purple-black powder, blue at pH=12-14, and forms a red complex with Ca^{2+}, used for the determination of calcium content in calcium-magnesium mixtures. At the end point the colour change is sharper than chromium black T.

$$Ca^{2+} + \underset{\text{(blue)}}{In} = \underset{\text{(fuchsia)}}{CaIn} \qquad (4.54)$$

$$\underset{\text{(fuchsia)}}{CaIn} + Y = CaY + \underset{\text{(blue)}}{In} \qquad (4.55)$$

Materials

Instruments: Erlenmeyer flask (250 mL), acid burette (50 mL), graduated cylinder (10 mL), beaker (250 mL), mortar, volumetric flask (250 mL), watch glass.

Reagents: $Na_2H_2Y \cdot 2H_2O$, $CaCO_3$ (primary standard), 20% NaOH solution, 6 mol/L HCl, calcium indicator, calcium tablets, milk powder, spinach.

Procedures

(1) Standardization of EDTA with $CaCO_3$

① Preparation of EDTA standard solution (0.02 mol/L): weigh 2.0 g $Na_2H_2Y \cdot 2H_2O$ in a 250 mL beaker, add 50 mL of water, heat to dissolve, dilute to 250 mL, and store in a reagent bottle or a polyethylene bottle.

② $CaCO_3$ standard solution preparation (0.02 mol/L): accurately weigh 120 ℃ dried $CaCO_3$ 0.4-0.6 g in a small beaker, add a few drops of water to moisten, cover it with watch glass, from the beaker mouth drop 6 mol/L HCl 5 mL until the $CaCO_3$ is completely dissolved, rinse with water to rinse the inside of the beaker, and then the solution was transferred to a 250 mL volumetric flask. Add water to the scale and shake well.

③ Standardization of EDTA standard solution: use a 25.00 mL pipette to suck $CaCO_3$ standard solution in a 250 mL conical flask, add water to 100 mL, add 5-6 mL of 20%

NaOH solution, add 2-3 drops of calcium indicator, titrate with the EDTA solution to be calibrated until the solution changes from violet to blue, which is the end point of the titration. The volume of EDTA solution consumed was recorded and measured three times in parallel. The data is recorded in Table 4.37.

(2) Determination of calcium content

① Sample pretreatment: put calcium tablets/milk powder into a mortar and grind it fine, accurately weigh 0.2-0.4 g of the finely ground sample, add 10 mL of distilled water, heat it up, add 6 mol/L HCl dropwise, stir it while heating it up, until it is completely dissolved. Transfer to a 250 mL volumetric flask, add distilled water to the mark, shake well.

Take a certain amount of fresh spinach washed with distilled water and chopped, put into the evaporation dish and weigh, put into the oven, bake for 2 h, and then put into the crucible, which will be placed in the muffle furnace and ashed for 10 h (note that the speed of temperature rise should not be too fast, and the highest temperature is 500 ℃). After cooling, add 2-3 drops of concentrated nitric acid, 1.0 mL of 6 mol/L HCl to dissolve, quantitatively transfer to a 250 mL volumetric flask, dilute to the scale with distilled water.

② Determination of calcium content: accurately remove 25.00 mL of the above solution in a 250 mL conical flask, add water to 100 mL, add 5-6 mL of 20% NaOH solution, add 2-3 drops of calcium indicator, titrate with EDTA solution to be calibrated until the solution changes from violet to blue, which is the end point of the titration. Titrate three times in parallel and calculate the calcium content in the sample. The data is recorded in Table 4.38.

✲ Data Recording and Processing

Table 4.37 Standardization of EDTA standard solution

No.	1	2	3
m_{CaCO_3}/g			
V_{EDTA}(initial)/mL			
V_{EDTA}(final)/mL			
V_{EDTA}/mL			
c_{EDTA}/(mol/L)			
$\bar{c}_{EDTA}$/(mol/L)			
$|d_i|$			
$\bar{d}_r$/%			

Table 4.38 Determination of calcium content

No.	1	2	3
m/g			
V_{EDTA}(initial)/mL			
V_{EDTA}(final)/mL			
V_{EDTA}/mL			
w_{Ca}/[mg/(100 g)]			
$\bar{w}_{Ca}$/[mg/(100 g)]			
$|d_i|$			
$\bar{d}_r$/%			

Tips

Calcium is the source of our life, plays a very important role in all stages of life growth, and is an essential and vital element for human health. Once calcium is insufficient, the body will not be able to function properly, which will easily lead to tooth decay, osteoporosis and osteochondrosis, poor development of young children, easily caus back pain and knee pain and so on. Many of us only know that children or the elderly should take calcium supplements—children for growth and the elderly to prevent osteoporosis. But in fact, all of us should take calcium supplements throughout our lives. After we are born, the calcium in our body has been in a continuous accumulation process, until about 35 years old when the calcium content of the human body reaches the peak in our life. After that, calcium loss begins to accelerate, and the amount of calcium lost is greater than the amount of calcium we normally accumulate in our bodies. If we store more calcium in our body before the age of 35, then we can maintain our body's various metabolic needs in the future, so calcium supplementation is essential for our healthy growth.

Experiment 4.28 Determination of NaCl Content in Soy Sauce

Objectives

(1) To learn the preparation and standardization of silver nitrate standard solution.

(2) To master the principle and method of determining chlorine content in chloride by the Mohr method.

Principles

The Mohr method is to titrate Cl^- directly with $AgNO_3$ standard solution in neutral or weak alkaline (pH=6.5-10) solution using K_2CrO_4 as an indicator. Due to the lower solubility of AgCl, according to the principle of step-by-step precipitation, AgCl white precipitation is precipitated first in the solution. When AgCl is quantitatively precipitated, slightly excessive Ag^+ and CrO_4^{2-} generate brick red Ag_2CrO_4 precipitates, which will make the solution slightly orange red and it is the end point.

The reaction is as follows,

$$Ag^+ + Cl^- \Longrightarrow AgCl\downarrow \text{ (white)} \qquad K_{sp}=1.8\times10^{-10} \qquad (4.56)$$

$$2Ag^+ + CrO_4^{2-} \Longrightarrow Ag_2CrO_4\downarrow\text{(brick red)} \quad K_{sp}=2.0\times10^{-12} \qquad (4.57)$$

Materials

Instruments: acid burette, Erlenmeyer flask (250 mL), volumetric flask (250 mL), dropper, graduated cylinder, pipette, brown reagent bottle.

Reagents: $AgNO_3$, NaCl (primary standard), soy sauce, 4 mol/L nitric acid, 4 mol/L NaOH solution, 5% $KMnO_4$ solution, K_2CrO_4 solution (50 g/L).

Procedures

(1) Preparation and standardization of 0.1 mol/L $AgNO_3$ Accurately weigh 8.5 g $AgNO_3$

in a small beaker, dissolve it with distilled water, transfer it to a brown reagent bottle, dilute it to 500 mL, cover the stopper and keep it in a dark place.

Accurately weigh 0.55-0.60 g NaCl reference material in a small beaker, dissolve with distilled water, transfer to a 100 mL volumetric flask, dilute with water to the scale, shake well.

Accurately pipette 25.00 mL NaCl solution in a 250 mL conical flask, add 20 mL of distilled water, 1 mL of K_2CrO_4 solution (50 g/L). Titrate with $AgNO_3$ solution until a brick red colour is obtained. The volume of $AgNO_3$ solution consumed was recorded, and three parallel measurements were made to calculate the average concentration of $AgNO_3$ solution (Table 4.39).

(2) Determination of NaCl content in soy sauce　Accurately pipette 10.0000 mL of soy sauce sample into a 250 mL volumetric flask, add water to the scale, and shake well. Pipette 25.00 mL of diluted solution in a 250 mL conical flask, add 20 mL of water and 1 mL of K_2CrO_4 solution, mix well. Titrate with $AgNO_3$ solution until the first appearance of brick red precipitate is the end point. Record the volume of $AgNO_3$ solution consumed, and measure three copies in parallel to calculate the NaCl content in soy sauce (Table 4.40).

 Data Recording and Processing

(1) Calculate the concentration of $AgNO_3$ standard solution

$$c_{AgNO_3} = \frac{m_{NaCl} \times \frac{25.00}{100.0}}{M_{NaCl} V_{AgNO_3}} \times 1000 \qquad (4.58)$$

Table 4.39　Standardization of 0.1 mol/L $AgNO_3$ solutiono

No.	1	2	3
m_{NaCl}/g			
V_{AgNO_3} (initial) /mL			
V_{AgNO_3} (final) /mL			
V_{AgNO_3} /mL			
c_{AgNO_3} /(mol/L)			
$\bar{c}_{AgNO_3}$ /(mol/L)			
$|d_i|$			
$\bar{d}_r/\%$			

(2) Calculate the concentration of NaCl in soy sauce

$$w_{NaCl} = \frac{c_{AgNO_3} V_{AgNO_3}}{m_s \times \frac{25.00}{250.0}} \times \frac{M_{NaCl}}{1000} \times 100\% \qquad (4.59)$$

(M_{NaCl} = 58.44 g/mol, M_{AgNO_3} = 169.88 g/mol)

Table 4.40　Determination of NaCl content in soy sauce

No.	1	2	3
m_s/g			
V_{AgNO_3} (initial)/mL			

(continued)

No.	1	2	3
V_{AgNO_3} (final)/mL			
V_{AgNO_3}/mL			
w_{NaCl}/%			
$\bar{w}_{NaCl}$/%			
$|d_i|$			
$\bar{d}_r$/%			

Questions

(1) When measuring chlorine by the Mohr method, why should the pH of the solution be controlled at 6.5-10.5? If there are ammonium ions, what range should the pH of the solution be controlled and why?

(2) When K_2CrO_4 is used as an indicator, what is the effect of excessive or insufficient indicator concentration on the determination?

(3) Can we directly titrate Ag^+ with NaCl standard solution by the Mohr method? Why?

Tips

Role and efficacy of sodium chloride: sodium chloride is indispensable for human beings. The total amount of sodium ions contained in an adult body is about 60 g, of which 80% exists in the extracellular fluid, that is, in the plasma and intercellular fluid; chloride ions also exist mainly in the extracellular fluid. The physiological functions of sodium and chloride ions are: (1) to maintain the osmotic pressure of the extracellular fluid, (2) to participate in the regulation of acid-base balance in the body, and (3) to participate in the generation of gastric acid in the body. In addition, sodium chloride plays a role in maintaining normal excitability of nerves and muscles.

Experiment 4.29 Determination of Anthocyanin Content in Blueberries

Objectives

(1) To master the technique of anthocyanin content determination.

(2) To master the separation and extraction technology of the actual sample with ultrasonic assisted extraction.

Principles

Anthocyanin (Anthocyanidin), also known as anthocyanin, is a benzopyran derivative, belonging to the phenolic compounds, and its structural formula is shown in Fig. 4.10. Anthocyanin is often formed with one or more glucose, rhamnose, galactose, arabinose, etc., through glycosidic bond, which is a class of water-soluble natural pigment widely existed in plants in nature, and is also the main colour-presenting substance in the leaves of the trees, and it presents different colours under different pH conditions in the plant cell vesicles. The content of cyanidin can be detected by UV-visible spectrophotometry using vanillin methanol solution mixed with hydrochloric acid methanol solution as the colour developer.

Blueberry is a species with high anthocyanin content in plants, and its highest anthocyanin content can reach 4200 mg/kg. Anthocyanin contained in blueberries is one of the anthocyanin species with the best performance and lowest side effect among all plant anthocyanins at present. In order to further develop and effectively use blueberry anthocyanins, the study of anthocyanin extraction and detection technology in blueberry is of great significance. Anthocyanins in plants are usually combined with proteins and cellulose in a bound state, so they are not easy to extract. Ultrasonic waves produce strong vibration, high acceleration, strong cavitation effect, stirring and other effects, can destroy the cell wall of the plant, so that the solvent penetrates into the cells and the chemical components are dissolved in the solvent, thus improving the extraction efficiency.

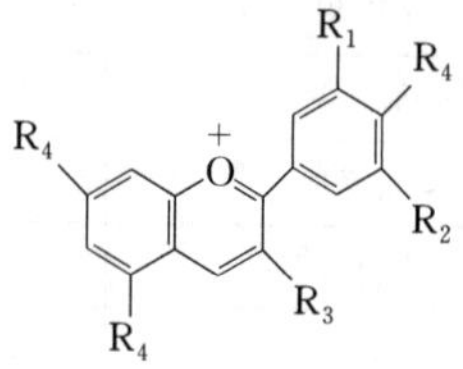

$R_1/R_2 = H$, OH or OCH_3;
$R_3 = H$ or glycone; $R_4 = OH$ or glycone

Fig. 4.10 Structure of anthocyanin

Materials

Instruments: analytical balance, spectrophotometer, beaker, cuvette, water bath, ultrasonic extractor, centrifuge, volumetric flask, measuring cylinder.

Reagents: blueberry, anthocyanin standard, anhydrous ethanol, methanol, vanillin, hydrochloric acid.

Procedures

(1) Ultrasound-assisted extraction of anthocyanins from blueberries Accurately weigh 1 g of fresh blueberry, mash blueberry into blueberry paste, then place in the extraction bottle, add 15 mL of 40% ethanol solution, extract under ultrasonic power 200 W, temperature 30 ℃ for 30 min. Repeat the extraction 3 times, combine the extract, centrifuge and filter, and fix the volume of extract to 100 mL.

(2) Selection of maximum absorption wavelength Take 1.0 mL of anthocyanin standard solution with a concentration of 0.1 mg/mL, add 5.0 mL of mixed colour developer (2% vanillin methanol solution : 8% hydrochloric acid methanol solution, 1 : 1), add methanol to 100 mL, and let react for 30 min while, avoiding light, and then measure the absorbance at the wavelength of 350-600 nm, plot the absorption curve to determine the maximum absorption wavelength of anthocyanin.

(3) Calibration curve Take 0.25 mL, 0.50 mL, 0.75 mL, 1.00 mL, 1.25 mL and 1.50 mL of anthocyanin standard solution with the concentration of 0.1 mg/mL, add 5.0 mL of mixed chromogenic agent, add methanol to 100 mL, let react for 30 min with the reagent blank as the reference solution, and then determine the absorbance at the maximum wavelength of absorption to draw the standard curve.

(4) Determination of anthocyanin content Take 10.00 mL of the above blueberry extract in a 100 mL volumetric flask, add 5.0 mL of mixed chromogen, add methanol to 100 mL, let react for 30 min away from light, use the reagent blank as the reference solution,

measure the absorbance at the maximum absorption wavelength, and calculate the anthocyanin extraction rate (Table 4.41).

Data Recording and Processing

Table 4.41 Absorbance of standard solution and determination of anthocyanins in samples

No.	Concentration of anthocyanin/(μg/mL)	Absorbance	The content of anthocyanins in the sample/%
1			—
2			—
3			—
4			—
5			—
6			—
Sample			

Tips

In recent years, anthocyanins as polyphenols have been widely noticed. Scientific research shows that the length of human life depends directly on the strength of people's antioxidant and anti-free radical ability, if the problem of free radical aggression is solved, then the body cells can be truly free to grow. The discovery of anthocyanins has found the simplest and most effective way for people all over the world to fight against oxidation and aging. Fundamentally, anthocyanin is a powerful antioxidant that protects the body from free radical damage. It also enhances blood vessel elasticity, improves the circulatory system and enhances the smoothness of the skin, improves vision, inhibits inflammation and allergies, improves joint flexibility, and helps to prevent a wide range of free-radical related diseases including cancer, heart disease, premature aging, and arthritis, so anthocyanin offers the human body a wide range of benefits.

Experiment 4.30 Determination of Melamine in Milk Products

Objectives

(1) To master the use of spectrophotometer.

(2) To master the principle and method of melamine detection in dairy products.

Principles

Melamine, with the chemical formula of $C_3H_6N_6$, commonly known as protein essence and named 1,3,5-triazine-2,4,6-triamine by IUPAC, is a triazine-containing nitrogen-containing heterocyclic organic compound, which is usually used as a chemical raw material. According to the structural characteristics of melamine molecule, the rapid detection of melamine was achieved by the complexation reaction between the colour rendering agent alizarin red and melamine (Fig. 4.11), and the absorbance was measured by UV-visible spectrophotometer.

Materials

Instruments: UV-vis spectrophotometer, analytical balance, weighing paper, volumetric flask, beaker, glass rod, measuring cylinder, ultrasonic extractor.

Reagents: milk powder, acetonitrile, 0.1% alizarin red solution (weigh 0.1 g of alizarin red dissolved in water and add water till the volume to 100 mL, mix well);

Melamine standard solution (1 mg/mL): accurately weigh 0.1000 g of melamine in a small beaker, add water to dissolve, transfer to a 100 mL volumetric flask, make volume reach the scale, shake well;

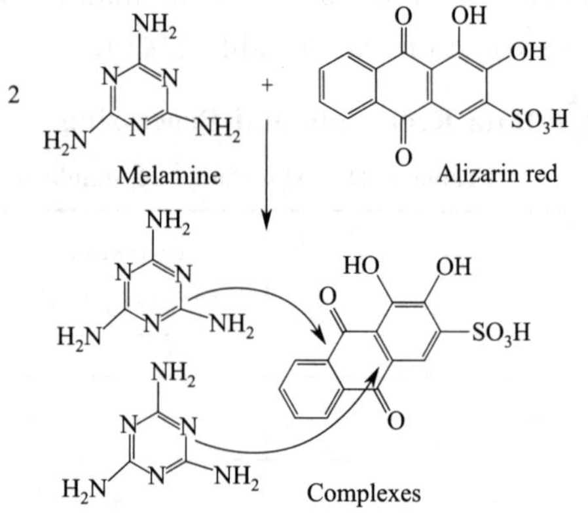

Fig. 4.11 Schematic diagram of the interaction between melamine and alizarin red solution

Melamine working solution (10 μg/mL): accurately pipette 1 mL of melamine standard solution (1 mg/mL) into a 100 mL volumetric flask, and then adjust the volume to the scale with water, shake well.

Procedures

(1) Determination of maximum absorption wavelength Add 1.0 mL of melamine working solution at 10 μg/mL and 0.1% alizarin red 1 mL into the cuvette, dilute to 10 mL with water, and let react for 10 min. Take the reagent blank as the reference solution, measure the absorbance at the wavelength of 400-800 nm, plot the absorption curve, and determine the optimum wavelength of the melamine colour reaction complex.

(2) Optimization of chromogenic conditions

① Selection of the chromogen concentration. Add 1.0 mL of melamine working solution at a concentration of 10 μg/mL and 0.5% alizarin red in six cuvettes, i.e., 0.005%, 0.010%, 0.020%, 0.030%, 0.040% and 0.050% of alizarin red, respectively, and dilute them to 10 mL with water, then let react for 10 min. The reagent blank was used as the reference solution, and the absorbance was measured at the maximum absorption wavelength to determine the optimal concentration of the colour developer.

② Selection of the color rendering time. Add 1.0 mL of melamine working solution with a concentration of 10 μg/mL and 0.1% alizarin red 1 mL into six colorimetric tubes, dilute to 10 mL with water, and then let react for 0 min, 5 min, 10 min, 15 min, 20 min and 30 min, respectively, and then take the reagent blank as the reference solution and measure the absorbance at the maximum absorption wavelength to determine the optimal time for the development of colour.

(3) Calibration curve Take 1 mL, 2 mL, 3 mL, 4 mL and 5 mL of melamine solution with the concentration of 10 μg/mL, add 0.1% alizarin red 1 mL, dilute it with water to 10 mL, let react for 10 min, and then use the reagent blank as the reference solution to measure the absorbance at the maximum absorption wavelength to draw the standard curve.

(4) Sample analysis Blank sample: take 1.0 g of milk powder, add 10 mL of water

to dissolve, add 10 mL of acetonitrile to precipitate proteins, treat it by ultrasonic extraction for 10 min, use centrifugation to deal with it for 10 min, filtrate, and then take the supernatant as the extraction solution for use.

Spiked samples: take 1.0 g of milk powder, add 25 μL of melamine standard solution with a concentration of 1 mg/mL, dissolve it in 10 mL of water, add 10 mL of acetonitrile to precipitate proteins, treat it by ultrasonic extraction for 10 min, use centrifugation to deal with it for 10 min, and then filter the supernatant as the extract for use.

Take 5 mL of the supernatant of blank and spiked samples, add 0.1% alizarin red 1 mL respectively, dilute to 10 mL with water, let react for 10 min, use the reagent blank as the reference solution, determine the absorbance at the maximum absorption wavelength, and use the working curve to find out the concentration of melamine. The concentration of melamine was determined three times in parallel, and the recoveries and relative standard deviations were calculated (Table 4.42).

Data Recording and Processing

Table 4.42 Absorbance of standard solution and determination of melamine in samples

No.	Concentration of melamine/(μg/mL)	Absorbance	Content of melamine in the sample/%
1			—
2			—
3			—
4			—
5			—
6			—
Sample			
1			
2			
3			

Questions

Do you know other methods for the determination of melamine? What are the advantages and disadvantages of each method?

Experiment 4.31 Determination of Vitamin B12 in Health Foods

Objectives

(1) To master the structure and basic operation of the ultraviolet-visible spectrophotometer.

(2) To master the determination of vitamin B12 in health food.

Principles

Vitamin B12 participates in many metabolic activities in the body and is an indispensable nutrient for maintaining health. The human body can not synthesise its own vitamin B12, and needs to obtain it through food or nutritional supplements with vitamin B12. Vitamin B12 is purified by adding sodium cyanide, which converts the natural form of vitamin B12 into the more stable cyanocobalamin. Cyanocobalamin is usually added as a food additive to health food multivitamin tablets for the purpose of vitamin B12 supplementation. The structure of vitamin B12, as shown in Fig. 4.12, is easily soluble in water and ethanol, and is easily decomposed in strong acid (pH<2) and alkaline solution, while it is the most stable in weak acidic conditions at pH 4.5-5.0.

UV spectrophotometry can be used for qualitative and quantitative analysis of organic compounds. The absorption spectrum of each substance generally has a maximum absorption peak, which is the qualitative basis of UV spectrophotometry. At the appropriate measurement wavelength, the same compound has different absorption intensities at different concentrations, which is the basis for quantitative analysis by UV absorption spectroscopy. According to the characteristic that vitamin B12 aqueous solution in 361 nm wavelength absorption is the largest, use ultraviolet spectrophotometry to determine the content of vitamin B12 complex tablets.

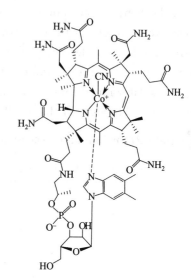

Fig. 4.12　Structure of vitamin B12

Materials

Instruments: UV-visible spectrophotometer, cuvette, measuring cylinder, volumetric flask, filter membrane (water system), ultrasonic cleaner, centrifuge.

Reagents: vitamin B12 standard substance, health products (multivitamin tablets), ethanol.

**Procedures**

(1) Preparation of vitamin B12 standard solution (1.0 mg/mL)　Accurately weigh 10.0 mg of vitamin B12 standard substance, dissolve it in 5% ethanol, and then dispense it into a 10 mL brown volumetric flask and mix well to obtain the standard reserve solution of vitamin B12.

(2) Selection of measurement wavelength　Using water as reference, the absorption spectrum of vitamin B12 standard solution was measured in the wavelength range of 200-500 nm to determine the characteristic absorption wavelength of vitamin B12 (λ_{max}).

(3) Calibration curve　Take 0.05 mL, 0.10 mL, 0.25 mL, 0.50 mL, 1.00 mL, 2.50 mL, 5.00 mL of standard reserve solution in a 10 mL brown volumetric flask, respectively, and dilute it with water to obtain the vitamin B12 standard solutions 1-7. Under the selected λ_{max}, measure the absorbance A of the vitamin B12 standard solutions from No. 1 to No. 7, and plot the standard curve. Plot the standard curve with absorbance A and mass concentration c to get the regression equation.

(4) Sample analysis　10 tablets were crushed and mixed well. Weigh 3.0000 g of the sample in a 50 mL centrifuge tube, add 10 mL of water, mix well, and place it in an ultra-

sonic cleaner, put it in ultrasonic extraction for 10 min, then centrifuge it at 4000 r/min for 5 min. The extraction was repeated twice, the extracts were combined, and the sample solution was filled in a 25 mL brown volumetric flask after passing through a 0.45 μm aqueous filter, and then the sample solution was obtained.

The absorbance A of the sample solution was measured at the selected λ_{max}, and the content of vitamin B12 in the sample was calculated according to the standard curve equation (Table 4.43).

Data Recording and Processing

Table 4.43 Absorbance of standard solution and determination of vitamin B12 content in the sample

No.	Concentration of vitamin B12/(μg/mL)	Absorbance	Content of vitamin B12 in the sample/%
1			—
2			—
3			—
4			—
5			—
6			—
Sample			
1			
2			
3			

Tips

The main source of vitamin B12 is animal food, which exists in animal liver, beef, pork, eggs, milk and cheese, and it needs to be combined with calcium when absorbed in order to benefit the body's functional activities. Vitamin B12 can promote the formation and regeneration of red blood cells, prevent anaemia; promote children's growth, improve appetite; enhance physical strength; maintain the normal function of the nervous system; promote concentration, enhance memory and sense of balance. It can make fats, carbohydrates and proteins suitable for use in the body. Vitamin B12 is necessary for cell division and maintaining the integrity of the myelin sheath of nerve tissue, and is mainly used in the treatment of pernicious anaemia. It can combine with folic acid to treat megaloblastic anaemia caused by antifolate drugs, steatorrhea and so on.

Chapter 5

Innovative Experiments

5.1 Chemistry Experiment Competition for College Students

Experiment 5.1 Preparation of Rare Earth Europium (Eu) Complex and Determination of Coordination RatFio

Objectives

(1) Prepare rare earth europium complex Eu(DBM)$_n$(H$_2$O)$_2$.
(2) Determine the content of Eu in the complex by compleximetry.
(3) Determine the content of dibutyl maleate (DBM) in the complex by UV-vis.
(4) Calculate the coordination number n of the complex.

Principles

Rare earth elements(REEs) are important strategic resources due to their unique physical and chemical properties such as photo, electricity, magnetism and catalysis, have become the target of research in the 21st century. China has the richest rare earth resources in the world.

As a typical metal element, REs can form coordination bonds with most non-metallic elements in the periodic table. As β-diketone ligands are good chelating ligands for rare earth ions, the complexes formed by REs and β-diketone ligands have attracted more attention. The reaction equation of rare earth Eu^{m+} with dibenzoyl methane (HDBM) to form the complex Eu(DBM)$_n$(H$_2$O)$_2$ is shown below:

(5.1)

The content of Europium in Eu(DBM)$_n$(H$_2$O)$_2$ (lgβ_1<11) can be determined by compleximetry, as the complex formed by ethylenediamine tetraacetic acid (EDTA) and Eu^{m+}

ions has a high stability constant (lg$K$ = 17.14). Due to the special electronic configuration of rare earth ions, the spectral intensity and peak shape of the complex are largely determined by the properties of ligands. Therefore, under certain conditions, the content of ligand in Eu(DBM)$_n$(H$_2$O)$_2$ can be determined by comparing the UV-Vis absorption spectra of ligand and complex, so the chemical composition of the complex can be determined.

Materials

Instruments: UV-vis spectrophotometer, portable UV lamp, pH test paper, balance, circulating pump, tool box (including weighing paper, filter paper, scissor, label paper, magnetic rod, marker, pH comparison card), nitrile gloves, eye shield, oven, beaker for products recovery (400 mL), analytical balance, paper strip, cotton gloves, magnetic stirrer (one set with one water bath), buret support, acid buret (50.00 mL), pipette (5.00 mL, 25.00 mL), Allihn condenser (with rubber tube), Buchner funnel and filter flask, eggplant type flask (100 mL), beaker (100 mL, 150 mL, 400 mL), volumetric flask (50 mL, 100 mL, 250 mL), Erlenmeyer flask (250 mL), graduated cylinder (10 mL, 50 mL), petri dish, watch glass, stainless steel scraper, dropper, glass rod, aurilave, washing bottle, reagent bottle (1 L), tweezer, white ceramic plate, stirring magnet.

Reagents: Eu^{m+} aqueous solution (0.96 mol/L), dibenzoyl methane (HDBM), HDBM stock solution (prepared with anhydrous ethanol, 5.00×10^{-4} mol/L), anhydrous ethanol, 95% ethanol, NaOH solution (2 mol/L, 0.5 mol/L), HCl solution (6 mol/L, 1 mol/L), ZnO, xylenol orange (0.2% aqueous solution), hexamethylenetetramine (20% aqueous solution), dimethyl sulfoxide (20% ethanol solution), disodium ethylenediamine tetraacetic acid (Na$_2$H$_2$Y·2H$_2$O), phosphate buffer solution (pH=8.0, 0.035 mol/L), deionized water.

Procedures

(1) Preparation of complex Eu(DBM)$_n$(H$_2$O)$_2$ Add 1.6 g dibenzoyl methane (HDBM, excess) and 25 mL anhydrous ethanol to a 100 mL eggplant type flask and heat in a 50 ℃ water bath to dissolve the solids. Add 1.80 mL solution into eggplant type flask from the burette containing 0.96 mol/L europium chloride aqueous solution (record the reading of burette before and after dropping, accurate to 0.01 mL). First, drop 2 mL 2 mol/L NaOH solution into the reaction bottle with a dropper (dip a small amount of solution on the filter paper with a glass rod, and observe the luminescence of the sample before and after dropping the NaOH solution under a portable UV lamp). Then add 0.5 mol/L NaOH solution dropwise to pH=8. Raise the temperature to 65 ℃ and let react for 10 min. Cool, filter and wash with suitable solvent. Put the product in an oven at 60 ℃ for dring and weighing.

(2) Determination of Eu in Eu(DBM)$_n$(H$_2$O)$_2$

① Preparation of zinc standard solution. Accurately weigh 0.2100 g (accurate to 0.0001 g) the reference material ZnO with the decrement method into a 150 mL beaker, add 5 mL 6 mol/L HCl solution. After dissolution, add an appropriate amount of water to dilute, quantitatively transfer it to a 250 mL volumetric flask and fix volume. Calculate the concentration of zinc standard solution (mol/L).

② Preparation and standardization of EDTA solution. Weigh 3.7 g Na$_2$H$_2$Y·2H$_2$O and dissolve it in 1000 mL distilled water to obtain an EDTA solution with a concentration of about 0.01 mol/L. Accurately transfer three 25.00 mL of zinc standard solutions to a 250 mL Erlenmeyer flask respectively, add a little amount of xylenol orange indicator, adjust it to purple-red with 20% hexamethylenetetramine solution, and then add 5 mL in excess. Titrated with 0.01 mol/L EDTA solution until the color of the solution turns from purplish-red to bright yellow, which is the end point. Record the consumed volume of EDTA and repeat the titration 3 times. Calculate the concentration of EDTA solution(mol/L) and the relative average deviation.

③ Determination of Eu in Eu $(DBM)_n(H_2O)_2$. Accurately weigh three self-made products(0.2000 g each, accurate to 0.0001 g) by the method of addition and put them in a 250 mL Erlenmeyer flask respectively(Note: Please moisten the Erlenmeyer flask with anhydrous ethanol first), and use 25 mL ethanol solution of dimethyl sulfoxide(20%, volume ratio) to dissolve. Take xylenol orange as the indicator, adjust it to purple-red with 1 mol/L HCl solution, add 2 mL 20% hexamethylenetetramine solution, and titrate it to the end point with EDTA standard solution. (Note: if turbidity occurs during titration, add a little amount of ethanol to dissolve it). Calculate the content of Eu in the product(mol/g) and the relative average deviation.

(3) Determination of DBM in Eu $(DBM)_n(H_2O)_2$

① Drawing the standard curve. Use HDBM stock solution(5.0×10^{-4} mol/L ethanol solution) to prepare 0-5.0×10^{-5} mol/L series of standard solutions. Each standard solution should contain 10 mL anhydrous ethanol and 5 mL pH 8.0 phosphate buffer(0.03 mol/L), and dilute with deionized water. Select the detection wavelength between 300-500 nm. Determine the absorbance and draw the standard curve.

② Determination of DBM in Eu $(DBM)_n(H_2O)_2$. Accurately weigh 0.0200 g(accurate to 0.0001 g) the self-made product, dissolve it with anhydrous ethanol, quantitatively transfer it to a 100 mL volumetric flask and fix volume with anhydrous ethanol. Transfer 5.00 mL of the above solution to a 100 mL volumetric flask and fix volume with deionized water [it needs to contain anhydrous ethanol and phosphate buffer in the same proportion as (3) ①]. Measure the absorbance of the solution. According to the above standard curve, calculate the content of DBM in the product (mol/g).

Data Recording and Processing

(1) According to the results of steps(2) and(3), calculate the coordination number n of Eu $(DBM)_n(H_2O)_2$.

(2) Calculate the molecular weight and the yield of the product(atomic weight of Eu: 151.96).

Questions

(1) In this experiment, NaOH is required to adjust the pH to 8 when preparing the complex. Why? If thenoyl trifluoroacetone (HTTA, structural formula is as Fig. 5.1) is used as ligand to synthesize the Eu complex, how much pH should be adjusted at least

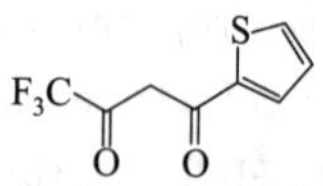

Fig. 5.1 Structure of HTTA

by calculation? (It is known that pK_a of HDBM and HTTA is 14.17 and 6.23, respectively.)

(2) What parameters are used to compare the absorption intensity of different solutions? The absorption intensity of DBM was quantitatively compared with that of Eu(DBM)$_n$(H$_2$O)$_2$ according to the data in the experiment.

(3) Both chemical analysis and instrumental analysis were used in this experiment. Please list the main differences between these two analysis methods.

Experiment 5.2 Preparation of Iron Oxide Nanoparticles and Its Application in the Determination of Melamine in Dairy Products

Objectives

(1) Prepare of iron oxide nanoparticles.

(2) Determine of melamine in dairy products.

Principles

Iron oxide magnetic nanoparticles (Nps) are a new type of nano materials with peroxide mimic enzyme properties. The research results show that compared with horseradish peroxidase and other peroxidase mimetic nanomaterials, Nps show better catalytic activity, stability, monodispersity and reusability.

Based on the catalytic activity of peroxide mimic enzyme of iron oxide magnetic nanoparticles, a rapid and simple spectrophotometric analysis method for the determination of melamine in dairy products can be established by using its catalytic effect on the redox reaction between hydrogen peroxide and 2,2′-azino-bis(3-ethylbenzothiazoline-6-sulfonic acid) diammonium salt(ABTS).

Hydrogen peroxide in the system is excessive for melamine. Hydrogen peroxide first reacts quantitatively with melamine to form a 1∶1 adduct. Iron oxide magnetic nanoparticles act as a catalyst to catalyze the redox reaction of the remaining hydrogen peroxide and ABTS to form colored compounds. The absorbance of colored compounds decreases with the increase of melamine content, and has a good linear relationship, which can determine the content of melamine.

(5.2)

The Eq. (5.2) is the reaction of hydrogen peroxide with melamine.

$$H_2O_2 \xrightarrow{Nps} HO^- + HO^+ \tag{5.3}$$

$$HO^+ + ABTS \xrightarrow{Nps} 2H_2O + ABTS^+ \tag{5.4}$$

The Eq. (5.3) and Eq. (5.4) are redox reaction of H$_2$O$_2$ and ABTS catalyzed by iron oxide magnetic nanoparticles(Nps).

 Materials

Instruments: analytical balance, 721G spectrophotometer, nitrogen gas bag, magnetic stirrer, magnetic-iron, thermostatic waterbath, flask with three necks, roundbottom flask, constant pressure funnel, sealing pipe, Erlenmeyer flask, volumetric flask, gloves, magneton, airway, exhaust, colorimetric tube, centrifuge tube, small test tube, cuvette, acid burette, washing bottle, pipette, weighing bottle, centrifuge.

Reagents: $K_2Cr_2O_7$(GR), 1 mol/L $FeCl_3$ solution(containing 2 mol/L HCl), 1 mol/L $FeCl_2$ solution(containing 2 mol/L HCl), 1.25 mol/L $NH_3 \cdot H_2O$, acetonitrile(CH_3CN), concentrated HCl, sulfur-phosphorus mixed acid solution, diphenylaminesulfonic acid sodium salt ($C_{12}H_{10}NSO_3Na$), anhydrous ethanol, acetone (CH_3COCH_3), milk [containing 0.25 mmol/L melamine($C_3H_6N_6$)], milk sample, 10% trichloroacetic acid solution, 6% $SnCl_2$ solution, 5% $HgCl_2$ solution, ABTS(0.03 mol/L) solution, HAc-NaAc buffer solution(pH=4.75), H_2O_2(0.01 mol/L) solution.

 Procedures

(1) Preparation of iron oxide nanoparticles—Coprecipitation method Mix 8 mL 1 mol/L $FeCl_3$ solution and 4 mL 1 mol/L $FeCl_2$ solution in a 100 mL flask with three necks, stirring with magnetic stirrer. Add 50 mL 1.25 mol/L $NH_3 \cdot H_2O$ to a constant pressure funnel. Fill the mixed solution with nitrogen for 10 min to remove the oxygen in the solution, then quickly add $NH_3 \cdot H_2O$, and continue stirring under the protection of nitrogen for 20 min. Transfer the solution to a 100 mL beaker, deposit the magnetic nanoparticles at the bottom of the beaker with a magnet, pour out the supernatant by decantation method, and wash repeatedly with distilled water until neutral. Add water to 60 mL, disperse evenly, transfer 10 mL solution to a 25 mL round-bottom flask, process it with electromagnetic stirring and disperse for use. Pour out the remaining supernatant in the beaker, wash the nanoparticles at the bottom with ethanol 2-3 times, and then wash them with acetone 2-3 times. During each separation, the magnetic particles can be sucked to the bottom of the beaker with the help of magnets, dried in the air(the drying can be accelerated with the help of aurilave), and weighed.

(2) Determination of melamine in dairy products—Spectrophotometric method

① Sample pretreatment. Transfer 2.00 mL milk containing 0.25 mmol/L melamine into a centrifuge tube, add 2.00 mL acetonitrile, 1.00 mL 10% trichloroacetic acid($C_2HCl_3O_2$) solution and 5.00 mL distilled water in sequence. Mix well, ultrasonic oscillate the solution for 10 min, then put it into a centrifuge at the speed of 4000 r/min for 10 min. Take the supernatant for the next test immediately.

② Plotting the working curve. Add(a) 0.00 mL, (b) 0.10 mL, (c) 0.20 mL, (d) 0.30 mL, (e) 0.50 mL and(f) 0.80 mL clear solution obtained from sample pretreatment in small test tubes, add distilled water to 1.00 mL, then add 1.00 mL HAc-NaAc buffer solution, 0.50 mL H_2O_2 solution respectively. After mixing well, add 0.50 mL ABTS solution, and finally transfer 50.0 μL ferric oxide magnetic nanoparticle solution with a pipette. Mix evenly, let react in 45 ℃ water bath for 15 min, and then put it into an ice water bath for 10 min to terminate the reaction.

Deposit iron oxide magnetic nanoparticles on the bottom of the test tube with a magnet, transfer 2.00 mL supernatant to a 10 mL colorimetric tube with a pipette, fill it to the mark with distilled water, mix well and take water as the reference solution, measure the absorbance with a spectrophotometer at 417 nm, draw the working curve.

③ Determination of melamine in the sample. Transfer 2.00 mL milk sample into a centrifuge tube. After treatment according to the above procedures, transfer two 0.50 mL supernatants in parallel for the same color reaction, determine the absorbance, and calculate the content of melamine in the sample according to the working curve.

(3) Determination of iron content in iron oxide nanoparticles—$K_2Cr_2O_7$ method

① Preparation of dichromate titration standard solution. Prepare 250 mL 0.01000 mol/L $K_2Cr_2O_7$ standard solution.

② Determination of iron content in iron oxide nanoparticles. Accurately weigh 0.10-0.12 g sample into a 250 mL Erlenmeyer flask, add a few drops of water to wet the sample and shake it to disperse, then add 2 mL concentrated HCl, cover with watch glass and heat to boil. After the sample is fully dissolved, 6% $SnCl_2$ solution is added dropwise while hot to reduce Fe^{3+} until the yellow color just disappears. Then in excess add 1-2 drops of $SnCl_2$ solution, quickly cool it to the room temperature with running water, immediately add 2-3 mL 5% $HgCl_2$ solution, mix well and stew for 3-5 min. The sample solution is diluted to 80-100 mL with distilled water, add 10 mL sulfur-phosphorus mixed acid and 5-6 drops of diphenylaminesulfonic acid sodium salt as indicator. Titrate immediately with $K_2Cr_2O_7$ standard solution until the solution is stable purple, and calculate the percentage content of iron in iron oxide nanoparticles.

Data Recording and Processing

According to procedures (2) and (3), calculate the content of melamine in the sample and the percentage content of iron in the iron oxide nanoparticles.

Experiment 5.3 Study on Synthesis of Complex [Ni(Me₃en)(acac)]BPh₄ and Its Solvent/Thermochromic Behavior

Objectives

(1) Prepare complex [Ni(Me₃en)(acac)]BPh₄.

(2) Determine Ni(II) content in complex [Ni(Me₃en)(acac)]BPh₄.

Principles

The red complex [Ni(Me₃en)(acac)]BPh₄ can be synthesized by acetylacetone and N,N,N'-trimethylethylenediamine ligands, and the complex has solvent and thermochromic effects. Organic solvent recognition materials (solid particles or thin film) can be prepared by mixing [Ni(Me₃en)(acac)]BPh₄ with layered structure saponite(SAP) or polymer Nafion (a perfluorinated sulfonic acid polymer). For example, the material is immersed in dichloromethane, ether and other solvents, the material remains red color. However, when the material is put into methanol, ethanol, acetone, acetonitrile, DMF (N,N-dimethylformamide) and other solvents, the material changes from red to blue-green. Put the blue-green material into the vacuum oven for heating and drying(80-100 ℃), and the material turns red again. In the rapid detection, the color of such materials changes significantly in some specific solvents which can be recognized by the naked eye, and the changes are reversible. They have certain application values in environment, industrial production and so on.

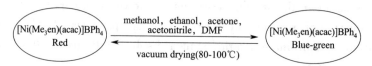

Materials

Instruments: pipette(1 mL), circulating water pump, rotary evaporator, low temperature bath, oven, magnetic rod, UV-visible spectrophotometer, magnetic stirrer, flask clip, condenser clip, buret support, bottle holder, white porcelain plate, round-bottom flask(50 mL), glass funnel(6 cm), solid funnel, filter flask(250 mL), Buchner funnel(6 cm), porcelain drip plate(white-6 holes), Erlenmeyer flask(150 mL, 250 mL), grinding plug, beaker(50 mL; 100 mL; 400 mL), Allihn condenser, graduated cylinder (50 mL; 100 mL), graduated cup(10 mL; 25 mL), petri dish(6 cm), glass-surface vessel(7 cm), volumetric flask(50 mL), colorimetric tube(10 mL, 25 mL), volumetric pipet(25 mL), burette(50 mL), weighing bottle(ϕ25 mm×40 mm), dryer(15 cm), dropper, glass rod, washing bottle, stirring magneton, nickel spoon, funnel plate, tool box [including aurilave, tweezers, scissors, label paper, marker, cotton, filter paper (medium speed, slow speed), latex gloves, gauze gloves, disposable gloves and goggles].

Reagents: nickel nitrate hexahydrate[$(Ni(NO_3)_2 \cdot 6H_2O$], acetylacetone ethanol solution(4.0 mol/L), N,N,N'-trimethylethylenediamine(Me_3en) ethanol solution(4.0 mol/L), Na_2CO_3, sodium tetraphenylboron($NaBPh_4$), zinc, disodium ethylenediamine tetraacetic acid ($Na_2H_2Y \cdot 2H_2O$, 0.01 mol/L), 0.2% xylenol orange solution, HCl solution(6 mol/L, 2 mol/L), 30% hexamethylenetetramine solution, anhydrous ethanol, dichloromethane, petroleum ether(60-90 ℃), acetonitrile, ethanol-water($V_{ethanol} : V_{water} = 1 : 1$), dichloromethane-petroleum ether(60-90 ℃, $V_{dichloromethane} : V_{petroleum\ ether} = 1 : 2$), pH test paper (1-14), precision pH test paper(5.4-7.0).

Procedures

(1) Synthesis of Ni (Ⅱ) complex[Ni(Me$_3$en)(acac)]BPh$_4$ In a 50 mL round bottom flask, add 0.87 g nickel nitrate hexahydrate [Ni(NO$_3$)$_2 \cdot$ 6H$_2$O, 3.0 mmol, dry the sample surface with filter paper as much as possible before weighing] and 15 mL anhydrous ethanol. After stirring and dissolving, successively add 0.75 mL acetylacetone(Hacac), ethanol solution (4.0 mol/L, 3.0 mmol, taken with a pipette), 0.16 g sodium carbonate (Na$_2$CO$_3$, 1.5 mmol), 0.75 mL N,N,N'-trimethylethylenediamine(Me$_3$en), ethanol solution(4.0 mol/L 3.0 mmol, taken with a pipette), continue stirring for 30 min. Filtrate under normal pressure, evaporate the solution under reduced pressure until it is nearly dry,

dissolve the residue in 20 mL dichloromethane, add 1.44 g sodium tetraphenylboron(NaBPh$_4$, 4.2 mmol), and let react for 30 min. Then, filtrate with slow qualitative folding filter paper, and add petroleum ether(60-90 ℃) to the filtrate until the product precipitates [$V_{dichloromethane}$: $V_{petroleum\ ether}$ = 1 : (1-2)]. After vacuum suction filtration, wash with ethanol-water mixed solvent($V_{ethanol}$: V_{water} = 1 : 1), anhydrous ethanol and dichloromethane-petroleum ether (60-90℃) mixed solvent ($V_{dichloromethane}$: $V_{petroleum\ ether}$ = 1 : 2), and drain. Transfer the product to a petri dish, dry it in an oven at 80 ℃ for 30 min, weigh and calculate the yield.

(2) Determination of Ni(II) content in Ni(II) complexe

① Preparation of zinc standard solution. Accurately weigh 0.15-0.17 g metal zinc, put it into a 100 mL beaker, add 5 mL 6 mol/L HCl(slight heat if necessary). After the zinc is completely dissolved, add an appropriate amount of deionized water for dilution, quantitatively transfer it to a 250 mL volumetric flask. Calculate the concentration of zinc standard solution(mol/L).

② Standardization of EDTA solution. Accurately transfer two 25.00 mL of the above zinc standard solution into 250 mL Erlenmeyer flask, add 50 mL deionized water, 3-4 drops 0.2% xylenol orange indicator, and 10 mL 30% hexamethylenetetramine solution. Titrate with 0.01 mol/L EDTA standard solution until the solution changes from purplishred to pure yellow, which is the end point, and record the data. Calculate the concentration of EDTA solution(mol/L) and relative mean deviation.

③ Determination of Ni(II) content. Accurately weigh 0.10-0.12 g two products into 250 mL Erlenmeyer flask, dissolve with 20 mL of ethanol(if it does not dissolve, heat slightly), accurately add 38-40 mL 0.01 mol/L EDTA standard solution and place it for 5 min. Then add 5 mL 30% hexamethylenetetramine solution and adjust the pH value of the solution to 5.8-6.2(check the pH value of the solution with pH test paper). Use 0.2% xylenol orange solution as indicator(3-4 drops), titrate with 0.01 mol/L zinc standard solution until the solution turns purple-red, which is the end point, and record the data. Calculate the percentage content of Ni(II) and relative mean deviation.

(3) Solvent discoloration of Ni(II) complex and characterization of UV-vis absorption spectra In a 10/25 mL colorimetric tube, use dichloromethane and acetonitrile as solvents respectively, prepare 10 mL of Ni(II) complex solution with 1 mg/mL(dichloromethane as solvent) and 15 mg/mL(acetonitrile as solvent). Observe the color of the solution and determine UV-vis absorption spectra(wavelength measurement range: 400-750 nm). Give the maximum absorption wavelength of each solution(λ_{max}).

(4) Thermochromism of Ni(II) complexes Take about 5 mg Ni(II) complex(about 1/3 nickel spoon) into a white porcelain dropping plate, drop 1-2 drops anhydrous acetonitrile, and observe the color of the sample after the solvent volatilizes. Then place it in an oven at 80℃ for 10 min and observe the color of the sample again.

✹ Data Recording and Processing

Calculate the yield of[Ni(Me$_3$en)(acac)]BPh$_4$ and the content of Ni(II) in the product according to procedures (1) and (2).

Questions

(1) Draw the structure of the coordination cation in the complex [Ni(Me₃en)(acac)]BPh₄, and explain why it has such a structure with the complex crystal field theory.

(2) Combining with the phenomena observed in procedures (3) and (4), explain the reason for solvent and thermochromism of the Ni(Ⅱ) complex.

(3) Complexometry is one of the common methods to determine the content of metal ions, try to list at least two other methods to quantitatively determine the content of Ni(Ⅱ).

Experiment 5.4   Determination of Copper Content in Polynuclear Copper(Ⅰ) Complex

Objectives

(1) Prepare $[Cu_x(dppy)_y(CH_3CN)_{x/2}](ClO_4)_x$ complex with heterogeneous luminescence and discoloration.

(2) Determine copper content in complex.

 Principles

Cu(Ⅰ) complexes have variable structure, excellent optical/electrical and catalytic physical and chemical properties. In recent years, they have shown a wide range of applications in luminescent materials, chemical sensors, biological probes and catalysis. Cu(Ⅰ) complexes have attracted much attention in chemical science research. For example, $[Cu_x(dppy)_y(CH_3CN)_{x/2}](ClO_4)_x$ complex has isomeric luminescence and discoloration characteristics. It can emit strong blue light under the excitation of 365 nm ultraviolet light. When it is recrystallized in methanol vapor or solvent containing methanol, the structure will change, resulting in the luminescence color becoming green. $[Cu(CH_3CN)_4]ClO_4$ is obtained by the reaction of $Cu(ClO_4)_2 \cdot 6H_2O$ with copper powder in acetonitrile. The product further reacts with diphenyl-2-pyridylphosphine(dppy) to obtain $[Cu_x(dppy)_y(CH_3CN)_{x/2}](ClO_4)_x$ complex.

There are many methods to determine the content of copper, such as inductively coupled plasma mass spectrometry, atomic absorption spectrometry, spectrophotometry, iodometry, complexometric titration and so on. Among them, titration analysis is a simple, rapid and widely used quantitative analysis method. Cu^{2+}, Ni^{2+}, Ca^{2+}, Mg^{2+}, Al^{3+} can be determined by complex titration with ethylenediamine tetraacetate(EDTA). However, copper in $[Cu_x(dppy)_y(CH_3CN)_{x/2}](ClO_4)_x$ complex exists in the form of Cu(Ⅰ), and its content is difficult to be determined by conventional chemical titration analysis method. Therefore, it is necessary to oxidize Cu(Ⅰ) to Cu(Ⅱ) before determination. Cu(Ⅱ) can form a relatively stable complex with EDTA($lgk = 18.8$). In the range of pH = 2-12, Cu(Ⅱ) can also form a stable complex with yellow 1-(2-pyridylazo)-2-naphthol (PAN) indicator($lgk = 6.70$).

In this experiment, $[Cu_x(dppy)_y(CH_3CN)_{x/2}](ClO_4)_x$ is dissolved in concentrated nitric acid. Under certain pH conditions, PAN is used as an indicator and EDTA standard solution is used for titration to determine the content of copper in the product.

Materials

Instruments: rotary evaporator, low temperature coolant circulating pump, circulating water pump, 365 nm UV flashlight, ultrasound equipment, magnetic stirrer, iron shelf, white ceramic plate, butterfly clip, safety bottle, spider clamp, 25.00 mL pipette, titration tube(50.00 mL), analytical balance, platform scale, round bottom flask(100 mL), glass plate funnel(30 mL), watch glass, beaker(100 mL), volumetric flask(250 mL), Erlenmeyer flask (250 mL), graduated cylinder(10 mL, 50 mL).

Reagents: copper(II) perchlorate hexahydrate, acetonitrile, diphenyl-2-pyridylphosphine(dppy, $M = 263.27$ g/mol), copper powder, anhydrous ether, concentrated nitric acid, 0.2% 1-(2-pyridylazo)-2-nitrophenol(PAN) ethanol solution, zinc oxide(ZnO, $M = 81.38$ g/mol), about 0.01 mol/L ethylenediamine tetraacetate(EDTA, $M = 372.24$ g/mol), 0.2% xylenol orange solution, 20% hexamethylene tetramine solution, HAc-NaAc buffer solution (pH=4.2).

 Procedures

(1) Preparation of $[Cu(CH_3CN)_4]ClO_4$ Add about 1.1g $Cu(ClO_4)_2 \cdot 6H_2O$, 40 mL acetonitrile and excess 2/3 copper powder into the dry 100 mL round bottom flask successively, plug the bottle, stir at room temperature until the solution is basically colorless. Quickly filter the reaction solution into a dry 100 mL round bottomed flask using a glass plate funnel, and the acetonitrile is evaporated under vacuum distillation with a rotary evaporator to obtain $[Cu(CH_3CN)_4]ClO_4$ solid.

(2) Preparation of $[Cu_x(dppy)_y(CH_3CN)_{x/2}](ClO_4)_x$ Add 20 mL acetonitrile, 1.46 g dppy, 1.17g $[Cu(CH_3CN)_4]ClO_4$ and 0.20 g copper powder into the dry 100 mL round bottom flask, plug the bottle, and stir at room temperature for 1.5 h. After the reaction, filter the reaction solution in the flask into a dry 100 mL round bottom flask with a glass plate funnel, rotate, evaporate and concentrate the solution to about 5 mL, add anhydrous ether dropwise until it is white and turbid. Allow the turbid solution to crystallize fully by standing, then discard the supernatant by decantation method, wash the crystal with appropriate solvent, dry it at room temperature, and weigh after the crystals were detached from the bottle wall by ultrasound.

(3) Analysis of copper(I) content in $[Cu_x(dppy)_y(CH_3CN)_{x/2}](ClO_4)_x$

① Preparation of zinc standard solution. Accurately weigh a certain amount(about 0.2 g) of reference reagent ZnO into a 100 mL small beaker, first add a small amount of water to wet, then drop hydrochloric acid solution(volume ratio 1∶1) to completely dissolve ZnO, then quantitatively transfer the solution to a 250 mL volumetric flask, dilute to the mark and shake well, and calculate the molar concentration of zinc standard solution.

② Standardization of 0.01000 mol/L EDTA. Accurately transfer 25.00 mL of the above zinc standard solution into a 250 mL Erlenmeyer flask, add 25 mL of deionized water, 2-3 drops of 0.2% xylenol orange indicator, drop 20% hexamethylene tetramine until the solution is stable purplish red, and then add 3 mL in excess hexamethylene tetramine solution, titrate with EDTA standard solution to bright yellow as the end point. The molar concentration of EDTA solution and relative average deviation are calculated. Repeat the titration 3 times.

③ Determination of copper in $[Cu_x(dppy)_y(CH_3CN)_{x/2}](ClO_4)_x$. Accurately weigh 0.14-0.16 g $[Cu_x(dppy)_y(CH_3CN)_{x/2}](ClO_4)_x$ into a 250 mL dry Erlenmeyer flask, add about 1 mL concentrated nitric acid to completely dissolve the sample(confirm with a 365 nm UV flashlight), add 20 mL deionized water and 20 mL HAc-NaAc buffer solution, drop 6-8 drops of 0.2% PAN indicator, and titrate with EDTA standard solution until the color of the solution changes from blue to yellow green, which is the end point. The mass percentage and relative average deviation of copper in the product are calculated by parallel titration 3 times.

Data Recording and Processing

According to the volume consumed by EDTA standard solution, calculate the content of Cu according to the following formula,

$$w = \frac{\overline{c}_{EDTA} V_{EDTA} M_{Cu}}{m_s \times 1000} \times 100\% \tag{5.5}$$

Where, m_s is the mass of $[Cu_x(dppy)_y(CH_3CN)_{x/2}](ClO_4)_x$ sample, g; M_{Cu} is the molar mass of copper, 63.5 g/mol.

Question

What is the function of copper powder when synthesizing $[Cu_x(dppy)_y(CH_3CN)_{x/2}](ClO_4)_x$?

5.2 Research Training Experiments

Experiment 5.5 Gemini Ionic Liquid Modified Graphene Oxide Membranes for ReO_4^-/TcO_4^- Adsorption

Objectives

(1) To learn the synthesis methods of click chemistry.
(2) To learn to use the adsorption method to separate and recover substances.
(3) To master the method of analyzing ReO_4^- concentration in solution using ICP-OES.

Principles

Technetium(^{99}Tc) is a dangerous radioactive isotope with a long half-life and high toxicity. It exists in the form of $^{99}TcO_4^-$ in aqueous solution. However, it was difficult to analyze high radioactive elements ^{99}Tc in the laboratory, ReO_4^- as a non-radioactive analogue with similar charge density and thermodynamic properties to TcO_4^- was often used to evaluate the removal properties of TcO_4^-.

Click chemistry is a new combinatorial chemistry method based on the synthesis of carbon heteroatom bonds(C—X—C). In this experiment, a functionalized graphene oxide flexible membrane (GO-C_6) was prepared using a hydrophobic olefin click reaction. The C—C bonds of the graphene network can further connect to the Gemini ionic liquid. As the number of alkyl chains C_1, C_4, and C_6 increases, its adsorption capacity increases. On the imidazole modified graphene oxide membrane, the ion exchange reaction between ReO_4^-/TcO_4^- and Cl^- plays a dominant role in the adsorption mechanism.

Procedures

GO was sonicated to prepare GO dispersion in N,N-dimethylformamide (DMF), and then pentaerythritol tetra (3-mercaptopropionate) (PETMP) and 10 mg of 2,2′-azobis (2-methylpropionitrile) (AIBN) was added into the dispersion. The resulting mixture was stirred at 343 K for 16 h under nitrogen atmosphere. Afterwards, the functional graphene dispersion was centrifuged. Subsequently, the sediment (GO-SH) was dispersed in MeOH, then DMF, PETMP, DMAP and Gemini C6 IL were added. After ultrasonic dispersion for 1 h, the mixture was immediately poured into the petri dish and exposed under UV light for 3 h. Finally, the membranes (GO-C_6) were dried at room temperature.

GO-C_6 membranes were cleaved into small pieces, then the required mass was weighed. Batch adsorption tests were carried out by shaking 20 mL of ReO_4^- solutions (20 mg/L) with 10 mg of GO-C_6 at various acid concentrations. Then use ICP-OES to determine the ReO_4^- concentration in the solution before and after the adsorption.

Experiment 5.6 Colorimetric Determination of Uric Acid with Gold Nanoparticles

Objectives

(1) Learn the gold nanoparticle colorimetry to detect uric acid content.

(2) Master the use of a UV-vis spectrophotometer.

Principles

Gold nanoparticle colorimetry is a newly developed colorimetric analysis method in recent years. It can realize high sensitivity detection by using the characteristics of high light absorption coefficient of gold nanoparticles due to surface plasmon resonance effect. Using this method, determination the content of uric acid in the solution is to use the hydrogen bond between uric acid and melamine to react with part of melamine and inhibit the aggregation of gold nanoparticles induced by melamine. With the increase of uric acid concentration, the solution gradually changes from blue to red, and its absorption curve and absorbance also change, which is easy to observe.

Procedures

(1) Preparation of gold nanoparticles   Gold nanoparticles were prepared by the classical Frens sodium citrate reducing chloroauric acid method. Add 35 mL ultrapure water and 0.5 mL 25 mmol/L $HAuCl_4$ solution a into 100 mL Erlenmeyer flask, shake well, heat to boiling. Quickly add a certain volume of 38.8 mmol/L sodium citrate solution, shake well,

after the solution changes color, continue to heat for 5 min, then close the heating plate, take down the solution, cool it to room temperature, and put it into a 4 ℃ refrigerator for subsequent use.

(2) Determination of uric acid Take a clean colorimetric tube, add a certain volume and concentration of melamine solution, pH=7.0 acetic acid buffer solution and uric acid solution, shake well, then add 1.0 mL of the prepared gold nanoparticles solution and use ultra-pure water to constant volume to 5 mL, shake well, take the solution into a quartz colorimetric plate, scan its UV-vis absorption spectrum.

Experiment 5.7 A Green Chemistry Technique for Dissolution and Recovery of Gold from Electronic Wastes

Objectives

(1) Learn the properties of complexes and master the principle of complexes reducing the oxidation electrode potential.

(2) Learn the basic knowledge of recycling.

Principles

The electrode potential of gold in aqueous solution is very high ($E^\ominus$ = 1.52-1.83 V), gold will not dissolve in nitric acid, sulfuric acid, hydrochloric acid, hydrofluoric acid and alkali. When gold forms a complex with ligands, it can effectively reduce its electrode potential. Au^+ can form a stable complex with CN^- and the electrode potential decreases sharply to -0.596 V:

$$Au \Longrightarrow Au^+ + e^- \quad E^\ominus = +1.83V \quad (5.6)$$

$$Au \Longrightarrow Au^{3+} + 3e^- \quad E^\ominus = +1.52V \quad (5.7)$$

$$4Au^+ + 8CN^- + O_2 + 2H_2O \Longrightarrow 4[Au(CN)_2]^- + 4OH^- \quad (5.8)$$

Due to $E^\ominus([Au(CN)_2]^-/Au) = -0.596V < E^\ominus(O_2/OH^-) = 0.401$ V, so gold can be dissolved, which is the principle of using complexes to dissolve gold. In the aqua regia method, hydrochloric acid actually acts as a ligand. After nitric acid oxidation, gold can be dissolved:

$$Au + HNO_3 + 4HCl \Longrightarrow HAuCl_4 + 2H_2O + NO\uparrow \quad (5.9)$$

Procedures

According to the experimental principle, some green solvents are reasonably designed and selected as ligands to achieve the effect of gold dissolution. For example, sodium chloride is used to play the role of ligands. After Cl^- as ligand forms a complex with gold, it can effectively reduce its electrode potential. Acetic acid is selected to provide protons, which can dissolve gold under the oxidation of nitric acid. After dissolution, use vitamin C (ascorbic acid) to reduce gold, so as to realize the recycling of solid waste in a safer and more environmentally protection way.

Experiment 5.8 Extraction, Content Analysis and Cosmetic Application of Natural Pigment Betacyanin

Objectives

(1) Learn about the extraction method of betacyanin.
(2) Learn about the properties and applications of betacyanin.
(3) Master the use of UV-vis spectrophotometers.
(4) Learn how to draw absorption curves and standard curves.

Principles

Betacyanin is a purplish-red substance combined by betalanic acid and glycosylated derivatives of cyclodopa or cyclodopa itself. The skeletal structure of its molecule is shown in Fig. 5.2. Glycosylation or acylation of R_1 or R_2 can produce betacyanin with a variety of structures. The phenolic hydroxyl and amine groups contained in the betacyanin molecule make it have a strong antioxidant capacity, and also promote the application of betacyanin in the treatment of oxidative stress-related diseases, such as cancer, inflammation, diabetes, hypertension, hyperlipidemia, obesity, Alzheimer's disease, etc. Therefore, it is valuable to apply the betacyanin in the fields of food, health products, medicine, and cosmetics. It is also of great significance to establish an analytical and detection method for betacyanin.

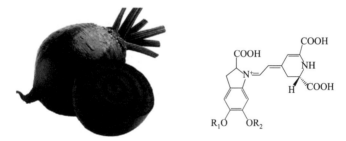

Fig. 5.2 The illustration of red beet and the structure of betacyanin molecule

The betacyanin molecule contains enriched hydrophilic functional groups such as carboxyl and hydroxyl groups, making it soluble in water. This property facilitates the development of clean and non-toxic extraction methods. The multi-ring structure contained in the molecular structure of betacyanin enables the obvious absorption in blue and green light region with maximum absorption peak in the range of 470-550 nm. According to the law of light complementarity, betacyanin shows intrinsic purplish-red in color. Taking advantage of its UV-vis absorption properties, content analysis method can be established based on Lambert-Beer's law.

Lambert-Beer's law describes the relationship between the absorbance and the content of substances, which finds that the absorbance is proportional to the concentration of the solution and proportional to the thickness of the solution layer. The formula $A = \varepsilon b c$ can express this principle, where ε is molar absorbance coefficient, b is the thickness of the solution layer, and c is the solution concentration of the substance. According to this law, betacyanin can be quantitatively analyzed. Usually, the standard curve method is often used to measure the content in real sample. The standard curve method is a data processing and analytical determination method based on univariate linear regression. The operation process is as follows. First, prepare a series of standard solutions with different concentrations and determine the absorbance. Then, find the mathematical relationship between absorbance (A) and solution concentration (c) according to the least squares method or using data processing software, which is generally expressed as $A=kc+b$. Afterwards, the absorbance of the real sample is determined, and the absorbance value is substituted into the above equation $A=kc+b$ to obtain the concentration.

Lipstick is one of the most used cosmetics in modern life. The history of humans applying red to their mouth dates back to the Neolithic period, five thousand years ago. The earliest lipstick found archaeologically was made from lead powder and red minerals more than three thousand years ago. Important Arts for Peoples's Walfare (Qi Min Yao Shu) recorded the ancient method of making lipstick in China, which is characterized by the obtaining of fragrant wine by soaking spices in warm wine, the addition of animal fat, cinnabar, and clear oil, cooling and solidification. The ancient people also used various plants and flowers instead of cinnabar to make lipstick. The development of the modern chemical industry provides various raw materials and methods to produce lipstick, promoting the popularity of lipstick. Recently, as the increases of people's demand for personalization of beauty products, ancient lipsticks and natural lipsticks have become new products of trendy makeup. Shenzhen and Enshi have also listed the production techniques of ancient lipsticks on the list of intangible cultural heritage. In this experiment, we improved the ancient method and used beet red pigment as the color source to make lipsticks, which is not only the inheritance of intangible cultural heritage techniques, but also the exploration of betacyanin in the application of beauty makeup.

 Materials

Instruments: wall-breaker, ultrasonic cleaner, centrifuge, balance, ultraviolet visible spectrophotometer, oven, 25 mL volumetric flask, 50 mL volumetric flask, 25 mL colorimetric tube.

Reagents: beet, distilled water, betacyanin standard, beeswax, glycerol, olive oil.

 Procedures

(1) Extraction of betacyanin   Break up fresh red beet into pieces in a wall-breaker. Transfer 10 g of the broken pieces in a 50 mL glass beaker, add 20 mL of distilled water, stir well and put it into the ultrasonic cleaner for 10 min at 360 W. Then, the extract and residue were poured into a 50 mL centrifugal tube, and centrifuged at 4,000 r/min for 5 min. Afterwards, the supernatant was collected in a beaker, and pipetted accurately 1 mL of the supernatant to 25 mL volumetric flasks, followed by the addition of distilled water till the concave solution surface reaches the volume mark.

Investigation on the experiment conditions:

① Extraction temperature: 20 ℃, 30 ℃, 40 ℃, 50 ℃;

② Extraction time: 20 min, 30 min, 40 min, 50 min;

③ pH value: 3, 4, 5, 6;
④ Mass ratio of the red beet and the water: 1∶2, 1∶3, 1∶4, 1∶5.

(2) Selection of the measurement wavelength

① Preparation of betacyanin stock solution. 0.5000 g of betacyanin was accurately weighed, dissolved in distilled water and transferred to a 50 mL volumetric flask, and distilled water was added to obtain a stock solution of betacyanin with a concentration of 10 mg/mL.

② Determination of optimum absorption wavelength. Transfer 1.0 mL of betacyanin stock solution into a 25 mL colorimetric cylinder, dilute it to 10 mL to obtain a solution with a concentration of 1 mg/mL. Then, 1 mL of the solution was transferred into a cuvette of 1 cm and the optimum absorption wavelength was measured in the range of 460-550 nm with a UV-visible spectrophotometer. Between 460-520 nm, it was measured every 10 nm, and between 520-550 nm, it was measured every 2 nm.

(3) Content analysis of betacyanin

① The standard curve. A series of standard solutions of different concentrations were obtained by adding 0.5 mL, 1.0 mL, 1.5 mL, 2.0 mL, 2.5 mL of the betacyanin stock solution (10 mg/mL) into five 25 mL colorimetric tubes and diluting the solutions to 10 mL with distilled water. The absorbance of each solution was measured with a 1 cm cuvette at the optimal absorption wavelength using a reagent blank as reference, and the standard curve was plotted.

② Determination of betacyanin in real samples. The extracted betacyanin solution from red beet was transferred into a cuvette, and the absorbance was measured at the optimal absorption wavelength. Then the content was calculated with the standard curve (expressed in mg/mL).

(4) Making handmade lipstick   0.50 g of betacyanin and 1 mL of glycerol were added in a 25 mL plastic beaker and stirred to make a homogeneous mixture. Then, 4.00 g beeswax and 10 mL of olive oil were added in a 100 mL plastic beaker and heated in boiled water bath till the beeswax dissolved and the mixture formed a yellow transparent solution. Subsequently, the beeswax solution was poured into the betacyanin solution while it is still hot. Stir the mixture till the solution become a paste with the decrease of solution temperature. After that, heat the mixture in water bath to make it a fluid state, and pour it into lipstick molds while it is still hot. Check and try to use the lipstick after it cooled down.

✻ Data Recording and Processing

(1) Absorption curve

Wavelength λ/nm										
Absorbance										
Wavelength λ/nm										
Absorbance										
Wavelength λ/nm										
Absorbance										

(2) Determination of betacyanin

Concentration/(mg/mL)	0.5	1.0	1.5	2.0	2.5	Unknown Sample
Absorbance(A)						

The standard curve was plotted using the absorbance value A as the vertical coordinate and the concentration value as the horizontal coordinate. Calculate the concentration of beet red pigment in the unknown sample.

Questions

(1) What is the effect of different extraction agents on the extraction of beet red pigment? What are the reasons for this effect?

(2) In this experiment only a crude extraction was used to obtain a product containing betacyanin. According to the references you have, how do you purify the crude product obtained in this experiment?

(3) How do different solution environments affect the absorbance of betacyanin? What is the reason for this?

(4) What is the basis for quantitative analysis by UV-vis spectrophotometry?

(5) When making lipstick by hand, olive oil, betacyanin, and beeswax are added. What is the purpose of each component?

Appendix

Appendix 1 Concentrations and Densities of Common Acid-Base Solutions

Reagent	Density/(g/mL)	$w/\%$	$c/(\text{mol/L})$
Hydrochloric acid	1.18-1.19	36-38	11.6-12.4
Nitric acid	1.39-1.40	65.0-68.0	14.4-15.2
Sulfuric acid	1.83-1.84	95-98	17.8-18.4
Phosphoric acid	1.69	85	14.6
Perchloric acid	1.68	70.0-72.0	11.7-12.0
Glacial acetic acid	1.05	99.8(G.R.) 99.0(A.R.,C.P.)	17.4
Hydrofluoric acid	1.13	40	22.5
Hydrobromic acid	1.49	47.0	8.6
Ammonia	0.88-0.90	25.0-28.0	13.3~14.8

Appendix 2 Common Primary Standard Substances and Their Drying Conditions and Applications

Primary standard substances		Composition after drying	Drying condition, $t/℃$	Calibration object
Name	Molecular formula			
Sodium bicarbonate anhydrous	$NaHCO_3$	Na_2CO_3	270-300	acid
Sodium carbonate	Na_2CO_3	Na_2CO_3	270-300	acid
Borax	$Na_2B_4O_7 \cdot 10H_2O$	$Na_2B_4O_7 \cdot 10H_2O$	Store in a desiccator containing NaCl and sucrose saturated solution	acid
Potassium bicarbonate	$KHCO_3$	K_2CO_3	270-300	acid
Oxalic acid	$H_2C_2O_4 \cdot 2H_2O$	$H_2C_2O_4 \cdot 2H_2O$	Air drying at room temperature	base or $KMnO_4$

(continued)

Primary standard substances		Composition after drying	Drying condition, t/℃	Calibration object
Name	Molecular formula			
Potassium hydrogen phthalate	$KHC_8H_4O_4$	$KHC_8H_4O_4$	105-110(1 h)	base
Potassium dichromate	$K_2Cr_2O_7$	$K_2Cr_2O_7$	120(1 h)	reducing agent
Potassium bromate	$KBrO_3$	$KBrO_3$	120(1-2 h)	reducing agent
Potassium iodate	KIO_3	KIO_3	110	reducing agent
Copper	Cu	Cu	Store in a dryer at room temperature	reducing agent
Arsenic trioxide	As_2O_3	As_2O_3	Store in a dryer at room temperature	oxidant
Calcium carbonate	$CaCO_3$	$CaCO_3$	110	EDTA
Sodium oxalate	$Na_2C_2O_4$	$Na_2C_2O_4$	110(2 h)	$KMnO_4$
Zinc	Zn	Zn	Store in a dryer at room temperature	EDTA
Zinc oxide	ZnO	ZnO	900-1000	EDTA
Sodium chloride	NaCl	NaCl	500-600	$AgNO_3$
Potassium chloride	KCl	KCl	500-600	$AgNO_3$
Silver nitrate	$AgNO_3$	$AgNO_3$	280-290	chloride
Aminosulfonic acid	$HOSO_2NH_2$	$HOSO_2NH_2$	Store in vacuum H_2SO_4 dryer for 48 h	base

Appendix 3 Preparation of Common Buffer

Buffer	pH	Preparation
Phosphate buffer	2.0	Liquid A: take 16.6 mL phosphoric acid, add water to 1 000 mL, and shake well. Liquid B: take 71.63 g disodium hydrogen phosphate and add water to 1000 mL. Mix 72.5 mL liquid A and 27.5 mL liquid B and shake well
Phosphate buffer	2.5	Take 100 g potassium dihydrogen phosphate, add 800 mL water, adjust the pH to 2.5 with hydrochloric acid, and dilute to 1 000 mL with water
Acetic acid-Lithium salt	3.0	Take 50 mL glacial acetic acid, mix it with 800 mL water, adjust the pH value to 3.0 with lithium hydroxide, and then dilute it to 1 000 mL with water
Phosphoric acid-Triethylamine	3.2	Take about 4 mL phosphoric acid and about 7 mL triethylamine, add 50% methanol to dilute to 1000 mL, and adjust the pH value to 3.2 with phosphoric acid
Sodium formate	3.3	Take 25 mL 2 mol/L formic acid solution, add 1 drop phenolphthalein indicator solution, neutralize with 2 mol/L sodium hydroxide solution, add 75 mL 2 mol/L formic acid solution, dilute with water to 200 mL, and adjust the pH value to 3.25-3.30

(continued)

Buffer	pH	Preparation
Acetate	3.5	Take 25 g ammonium acetate, dissolve it with 25 mL water, add 38 mL 7 mol/L hydrochloric acid solution, accurately adjust the pH value to 3.5 (indicated by potentiometric method) with 2 mol/L hydrochloric acid solution or 5 mol/L ammonia solution, and dilute it to 100 mL with water
Acetic acid-Sodium acetate	3.6	Take 5.1 g sodium acetate, add 20 mL glacial acetic acid, and then add water to dilute to 250 mL
Acetic acid-Sodium acetate	3.7	Take 20 g anhydrous sodium acetate, dissolve it with 300 mL water, add 1 mL bromophenol blue indicator solution and 60-80 mL of glacial acetic acid until the solution changes from blue to pure green, and then dilute it to 1000 mL with water
Ethanol-Ammonium acetate	3.7	Take 15.0 mL 5 mol/L acetic acid solution, add 60 mL ethanol and 20 mL water, adjust the pH value to 3.7 with 10 mol/L ammonium hydroxide solution, and dilute it to 1000 mL with water
Acetic acid-Sodium acetate	3.8	Take 13 mL 2 mol/L sodium acetate solution and 87 mL 2 mol/L acetic acid solution, add 0.5 mL copper sulfate solution (1 mg copper per mL), and then add water to dilute to 1000 mL
Citric acid-Disodium hydrogen phosphate	4.0	Liquid A: take 21 g citric acid or 19.2 g anhydrous citric acid, add water to dissolve it into 1000 mL, and store it in the refrigerator. Liquid B: take 71.63 g disodium hydrogen phosphate and add water to dissolve it into 1000 mL. Mix 61.45 mL liquid A and 38.55 mL liquid B, shake well
Acetic acid-Potassium acetate	4.3	Take 14 g potassium acetate, add 20.5 mL glacial acetic acid, and add water to dilute to 1000 mL
Acetic acid-Ammonium acetate	4.5	Take 7.7 g ammonium acetate, dissolve it with 50 mL water, add 6 mL glacial acetic acid, and dilute it to 1000 mL with water
Acetic acid-Sodium acetate	4.5	Take 18 g sodium acetate, add 9.8 mL glacial acetic acid, and add water to dilute to 1000 mL
Acetic acid-Sodium acetate	4.6	Take 5.4 g sodium acetate, dissolve it with 50 mL water, adjust the pH value to 4.6 with glacial acetic acid, and dilute it to 100 mL with water
Phosphate	5.0	Take a certain amount of 0.2 mol/L sodium dihydrogen phosphate solution and adjust the pH value to 5.0 with sodium hydroxide solution
Phthalate	5.6	Take 10 g potassium hydrogen phthalate, add 900 mL water, stir to dissolve, adjust the pH value to 5.6 with sodium hydroxide test solution (dilute with hydrochloric acid if necessary), and dilute to 1000 mL with water, mix well
Phosphate	5.8	Take 8.34 g potassium dihydrogen phosphate and 0.87 g dipotassium hydrogen phosphate, and add water to dissolve them into 1000 mL
Acetic acid-Ammonium acetate	6.0	Take 100 g ammonium acetate, add 300 mL water to dissolve it, add 7 mL glacial acetic acid and shake well
Acetic acid-Sodium acetate	6.0	Take 54.6 g sodium acetate, add 20 mL 1 mol/L acetic acid solution to dissolve it, and add water to dilute it to 500 mL
Citrate	6.2	Take 4.2 g citric acid, add 40 mL 1 mol/L 20% ethanol to prepare sodium hydroxide solution to dissolve it, and dilute it to 100 mL with 20% ethanol. Citrate buffer (pH 6.2): take 2.1% citric acid aqueous solution and adjust the pH value to 6.2 with 50% sodium hydroxide solution
Phosphate	6.5	Take 0.68 g potassium dihydrogen phosphate, add 15.2 mL 0.1 mol/L sodium hydroxide solution, and dilute it to 100 mL with water
Phosphate	6.6	Take 1.74 g sodium dihydrogen phosphate, 2.7 g disodium hydrogen phosphate and 1.7 g sodium chloride, and add water to dissolve it into 400 mL

(continued)

Buffer	pH	Preparation
Phosphate buffer (containing trypsin)	6.8	Take 6.8 g potassium dihydrogen phosphate, add 500 mL water to dissolve it, and adjust the pH value to 6.8 with 0.1 mol/L sodium hydroxide solution; in addition, take 10 g trypsin and add appropriate amount of water to dissolve it. After mixing the two solutions, add water to dilute to 1000 mL
Phosphate	6.8	Take 250 mL 0.2 mol/L potassium dihydrogen phosphate solution, add 118 mL 0.2 mol/L sodium hydroxide solution, dilute to 1000 mL with water, and shake well
Phosphate	7.0	Take 0.68 g potassium dihydrogen phosphate, add 29.1 mL 0.1 mol/L sodium hydroxide solution, and dilute to 100 mL with water
Phosphate	7.2	Take 50 mL 0.2 mol/L potassium dihydrogen phosphate solution and 35 mL 0.2 mol/L sodium hydroxide solution, and dilute to 200 mL with cool water after boiling, and shake well
Phosphate	7.3	Take 1.9734 g disodium hydrogen phosphate and 0.2245 g potassium dihydrogen phosphate, add water to 1000 mL, and adjust the pH value to 7.3
Barbital	7.4	Take 4.42 g barbital sodium, add water to dissolve and dilute to 400 mL, adjust the pH value to 7.4 with 2 mol/L hydrochloric acid solution, and filter it
Phosphate	7.4	Take 1.36 g potassium dihydrogen phosphate, add 79 mL 0.1 mol/L sodium hydroxide solution, and dilute to 200 mL with water
Phosphate	7.6	Take 27.22 g potassium dihydrogen phosphate, add water to dissolve it into 1000 mL, take 50 mL, add 42.4 mL 0.2 mol/L sodium hydroxide solution, and add water to dilute it to 200 mL
Barbital-Sodium chloride	7.8	Take 5.05 g sodium barbital, add 3.7 g sodium chloride and an appropriate amount of water to dissolve it; in addition, take 0.5 g gelatin with an appropriate amount of water, heat and dissolve it, and then incorporate it into the above solution. Then adjust the pH value to 7.8 with 0.2 mol/L hydrochloric acid solution, and then dilute it to 500 mL with water
Phosphate	7.8	Liquid A: take 35.9 g disodium hydrogen phosphate, dissolve it with water and dilute it to 500 mL. Liquid B: take 2.76 g sodium dihydrogen phosphate, dissolve it with water and dilute it to 100 mL. Mix 91.5 mL liquid A and 8.5 mL liquid B, and shake well
Phosphate	7.8-8.0	Take 5.59 g potassium dihydrogen phosphate and 0.41 g potassium dihydrogen phosphate, and add water to dissolve it into 1000 mL
Trimethylol aminomethane	8.0	Take 12.14 g trimethylolaminomethane, add 800 mL water, stir to dissolve, dilute to 1 000 mL, and adjust the pH value to 8.0 with 6 mol/L hydrochloric acid solution
Borax-Calcium chloride	8.0	Take 0.572 g borax and 2.94 g calcium chloride, dissolve them with 800 mL water, adjust the pH value to 8.0 with 2.5 mL 1 mol/L hydrochloric acid solution, and dilute them to 1 000 mL with water
Ammonia-Ammonium chloride	8.0	Take 1.07 g ammonium chloride, add water to dissolve it into 100 mL, and then add diluted ammonia solution (1→30) to adjust the pH value to 8.0
Trimethylol aminomethane	8.1	Take 0.294 g calcium chloride, add 40 mL 0.2 mol/L trimethylaminomethane solution to dissolve it, adjust the pH value to 8.1 with 1 mol/L hydrochloric acid solution, and dilute it to 100 mL with water
Barbital	8.6	Take 5.52 g barbital and 30.9 g sodium barbital, add water to dissolve it to 2000 mL

(continued)

Buffer	pH	Preparation
Trimethylol aminomethane	9.0	Take 6.06 g trimethylolaminomethane, add 3.65 g lysine salt, 5.8 g sodium chloride and 0.37 g disodium ethylenediamine tetraacetate, add water to dissolve it into 1000 mL, and adjust the pH value to 9.0
Boric acid-Potassium chloride	9.0	Take 3.09 g boric acid, add 500 mL 0.1 mol/L potassium chloride solution to dissolve it, and add 210 mL 0.1 mol/L sodium hydroxide solution
Ammonia-Ammonium chloride	10.0	Take 5.4 g ammonium chloride, dissolve it with 20 mL water, add 35 mL concentrated ammonia, and then dilute it with water to 100 mL
Borax-Sodium carbonate	10.8-11.2	Take 5.30 g anhydrous sodium carbonate and add water to dissolve it into 1000 mL; in addition, take 1.91 g borax and add water to dissolve it into 100 mL. Take 973 mL sodium carbonate solution and 27 mL borax solution before use, and mix well

Appendix 4 Common Indicators

Indicator	pH range of color change	Color change	Preparation
Cresol red (first color-changing range)	0.2-1.8	Red-Yellow	0.1% ethanol solution
Cresol red (second color-changing range)	7.2-8.8	Yellow-Purple red	0.1% ethanol solution
Thymol blue (first color-changing range)	1.2-2.8	Red-Yellow	0.1% ethanol solution, add 4.3mL 0.05 mol/L NaOH
Thymol blue (second color-changing range)	8.0-9.6	Yellow-Blue	0.1% ethanol solution, add 4.3 mL 0.05 mol/L NaOH
Methyl orange	3.0-4.4	Red-Orange	0.1% aqueous solution
Bromophenol blue	3.0-4.6	Yellow-Blue	0.1% ethanol solution, add 4.3 mL 0.05 mol/L NaOH
Congo red	3.0-5.2	Blue purple-Red	0.1% aqueous solution
Alizarin red S (first color-changing range)	3.7-5.2	Yellow-Purple	0.1% aqueous solution
Alizarin red S (second color-changing range)	10.0-12.0	Purple-Light yellow	0.1% aqueous solution
Methyl red	4.4-6.2	Red-Yellow	0.1% ethanol solution
Litmus	5.0-8.0	Red-Blue	0.1% ethanol solution
Bromothymol blue	7.2-8.8	Yellow-Purple red	0.1% ethanol solution
Phenothalin	8.2-10.0	Colourless-Purple red	0.1% ethanol solution
Tartar yellow	12.0-13.0	Yellow-Red	0.1% aqueous solution

Appendix 5 Dissociation Constants of Weak Acids and Bases

(1) Dissociation constants of inorganic acid in aqueous solution (25℃)

No.	Name	Chemical formula	K_a	pK_a
1	Metaaluminic acid	$HAlO_2$	6.3×10^{-13}	12.2
2	Arsenite	H_3AsO_3	6.0×10^{-10}	9.22
3	Arsenic acid	H_3AsO_4	$6.3 \times 10^{-3} (K_1)$ $1.05 \times 10^{-7} (K_2)$ $3.2 \times 10^{-12} (K_3)$	2.2 6.98 11.5
4	Boric acid	H_3BO_3	$5.8 \times 10^{-10} (K_1)$ $1.8 \times 10^{-13} (K_2)$ $1.6 \times 10^{-14} (K_3)$	9.24 12.74 13.8
5	Hypobromic acid	$HBrO$	2.4×10^{-9}	8.62
6	Hydrocyanic acid	HCN	6.2×10^{-10}	9.21
7	Carbonic acid	H_2CO_3	$4.2 \times 10^{-7} (K_1)$ $5.6 \times 10^{-11} (K_2)$	6.38 10.25
8	Hypochlorous acid	$HClO$	3.2×10^{-8}	7.5
9	Hydrofluoric acid	HF	6.61×10^{-4}	3.18
10	Germanic acid	H_2GeO_3	$1.7 \times 10^{-9} (K_1)$ $1.9 \times 10^{-13} (K_2)$	8.78 12.72
11	Periodate	HIO_4	2.8×10^{-2}	1.56
12	Nitrous acid	HNO_2	5.1×10^{-4}	3.29
13	Hypophosphoric acid	H_3PO_2	5.9×10^{-2}	1.23
14	Phosphite	H_3PO_3	$5.0 \times 10^{-2} (K_1)$ $2.5 \times 10^{-7} (K_2)$	1.3 6.6
15	Phosphoric acid	H_3PO_4	$7.52 \times 10^{-3} (K_1)$ $6.31 \times 10^{-8} (K_2)$ $4.4 \times 10^{-13} (K_3)$	2.12 7.2 12.36
16	Pyrophosphatic acid	$H_4P_2O_7$	$3.0 \times 10^{-2} (K_1)$ $4.4 \times 10^{-3} (K_2)$ $2.5 \times 10^{-7} (K_3)$ $5.6 \times 10^{-10} (K_4)$	1.52 2.36 6.6 9.25
17	Hydrosulphuric acid	H_2S	$1.3 \times 10^{-7} (K_1)$ $7.1 \times 10^{-15} (K_2)$	6.88 14.15
18	Sulphurous acid	H_2SO_3	$1.23 \times 10^{-2} (K_1)$ $6.6 \times 10^{-8} (K_2)$	1.91 7.18
19	Sulphuric acid	H_2SO_4	$1.0 \times 10^{3} (K_1)$ $1.02 \times 10^{-2} (K_2)$	−3 1.99
20	Thiosulfuric acid	$H_2S_2O_3$	$2.52 \times 10^{-1} (K_1)$ $1.9 \times 10^{-2} (K_2)$	0.6 1.72
21	Hydroselenic acid	H_2Se	$1.3 \times 10^{-4} (K_1)$ $1.0 \times 10^{-11} (K_2)$	3.89 11
22	Selenite	H_2SeO_3	$2.7 \times 10^{-3} (K_1)$ $2.5 \times 10^{-7} (K_2)$	2.57 6.6
23	Selenic acid	H_2SeO_4	$1 \times 10^{3} (K_1)$ $1.2 \times 10^{-2} (K_2)$	−3 1.92

No.	Name	Chemical formula	K_a	pK_a
24	Silicic acid	H_2SiO_3	$1.7 \times 10^{-10}\,(K_1)$ $1.6 \times 10^{-12}\,(K_2)$	9.77 11.8
25	Tellurite acid	H_2TeO_3	$2.7 \times 10^{-3}\,(K_1)$ $1.8 \times 10^{-8}\,(K_2)$	2.57 7.74

(2) Dissociation constants of organic acid in aqueous solution (25℃)

No.	Name	Chemical formula	K_a	pK_a
1	Formic acid	HCOOH	1.8×10^{-4}	3.75
2	Acetic acid	CH_3COOH	1.74×10^{-5}	4.76
3	Glycolic acid	$CH_2(OH)COOH$	1.48×10^{-4}	3.83
4	Oxalic acid	$(COOH)_2$	$5.4 \times 10^{-2}\,(K_1)$ $5.4 \times 10^{-5}\,(K_2)$	1.27 4.27
5	Glycine	$CH_2(NH_2)COOH$	1.7×10^{-10}	9.78
6	Monochloroacetic acid	$CH_2ClCOOH$	1.4×10^{-3}	2.86
7	Dichloroacetic acid	$CHCl_2COOH$	5.0×10^{-2}	1.3
8	Trichloroacetic acid	CCl_3COOH	2.0×10^{-1}	0.7
9	Propionic acid	CH_3CH_2COOH	1.35×10^{-5}	4.87
10	Acrylic acid	$CH_2=CHCOOH$	5.5×10^{-5}	4.26
11	Lactic acid	$CH_3CHOHCOOH$	1.4×10^{-4}	3.86
12	Malonic acid	$HOCOCH_2COOH$	$1.4 \times 10^{-3}\,(K_1)$ $2.2 \times 10^{-6}\,(K_2)$	2.85 5.66
13	2-Propargylic acid	$HC \equiv CCOOH$	1.29×10^{-2}	1.89
14	Glyceric acid	$HOCH_2CHOHCOOH$	2.29×10^{-4}	3.64
15	Pyruvic acid	$CH_3COCOOH$	3.2×10^{-3}	2.49
16	α-Propionic acid	CH_3CHNH_2COOH	1.35×10^{-10}	9.87
17	β-Propionic acid	$CH_2NH_2CH_2COOH$	4.4×10^{-11}	10.36
18	N-butyric acid	$CH_3(CH_2)_2COOH$	1.52×10^{-5}	4.82
19	Isobutyric acid	$(CH_3)_2CHCOOH$	1.41×10^{-5}	4.85
20	3-Butylenic acid	$CH_2=CHCH_2COOH$	2.1×10^{-5}	4.68
21	Isobutylene acid	$CH_2=C(CH_2)COOH$	2.2×10^{-5}	4.66
22	Fumaric acid	$HOCOCH=CHCOOH$	$9.3 \times 10^{-4}\,(K_1)$ $3.6 \times 10^{-5}\,(K_2)$	3.03 4.44
23	Maleic acid	$HOCOCH=CHCOOH$	$1.2 \times 10^{-2}\,(K_1)$ $5.9 \times 10^{-7}\,(K_2)$	1.92 6.23
24	Tartaric acid	$HOCOCH(OH)$ $CH(OH)COOH$	$1.04 \times 10^{-3}\,(K_1)$ $4.55 \times 10^{-5}\,(K_2)$	2.98 4.34
25	N-valeric acid	$CH_3(CH_2)_3COOH$	1.4×10^{-5}	4.86
26	Isovaleric acid	$(CH_3)_2CHCH_2COOH$	1.67×10^{-5}	4.78
27	2-Pentenoic acid	$CH_3CH_2CH=CHCOOH$	2.0×10^{-5}	4.7
28	3-Pentenoic acid	$CH_3CH=CHCH_2COOH$	3.0×10^{-5}	4.52
29	4-Pentenoic acid	$CH_2=CHCH_2CH_2COOH$	2.10×10^{-5}	4.677
30	Glutaric acid	$HOCO(CH_2)_3COOH$	$1.7 \times 10^{-4}\,(K_1)$ $8.3 \times 10^{-7}\,(K_2)$	3.77 6.08
31	Glutamic acid	$HOCOCH_2CH_2$ $CH(NH_2)COOH$	$7.4 \times 10^{-3}\,(K_1)$ $4.9 \times 10^{-5}\,(K_2)$ $4.4 \times 10^{-10}\,(K_3)$	2.13 4.31 9.358
32	N-hexanoic acid	$CH_3(CH_2)_4COOH$	1.39×10^{-5}	4.86
33	Isocaproic acid	$(CH_3)_2CH(CH_2)_3COOH$	1.43×10^{-5}	4.85

(continued)

No.	Name	Chemical formula	K_a	pK_a
34	(E)-2-Hexenoic acid	$H(CH_2)_3CH=CHCOOH$	1.8×10^{-5}	4.74
35	(E)-3-Hexenoic acid	$CH_3CH_2CH=CHCH_2COOH$	1.9×10^{-5}	4.72
36	Adipic acid	$HOCOCH_2CH_2CH_2CH_2COOH$	$3.8\times10^{-5}(K_1)$ $3.9\times10^{-6}(K_2)$	4.42 5.41
37	Citric acid	$HOCOCH_2C(OH)(COOH)$ CH_2COOH	$7.4\times10^{-4}(K_1)$ $1.7\times10^{-5}(K_2)$ $4.0\times10^{-7}(K_3)$	3.13 4.76 6.4
38	Phenol	C_6H_5OH	1.1×10^{-10}	9.96
39	Pyrocatechol	$o\text{-}C_6H_4(OH)_2$	$3.6\times10^{-10}(K_1)$ $1.6\times10^{-13}(K_2)$	9.45 12.8
40	Resorcinol	$m\text{-}C_6H_4(OH)_2$	$3.6\times10^{-10}(K_1)$ $8.71\times10^{-12}(K_2)$	9.3 11.06
41	Hydroquinone	$p\text{-}C_6H_4(OH)_2$	1.1×10^{-10}	9.96
42	2,4,6-Trinitrophenol	$2,4,6\text{-}(NO_2)_3C_6H_2OH$	5.1×10^{-1}	0.29
43	Gluconic acid	$CH_2OH(CHOH)_4COOH$	1.4×10^{-4}	3.86
44	Benzoic acid	C_6H_5COOH	6.3×10^{-5}	4.2
45	Salicylic acid	$C_6H_4(OH)COOH$	$1.05\times10^{-3}(K_1)$ $4.17\times10^{-13}(K_2)$	2.98 12.38
46	o-Nitrobenzoic acid	$o\text{-}NO_2C_6H_4COOH$	6.6×10^{-3}	2.18
47	m-Nitrobenzoic acid	$m\text{-}NO_2C_6H_4COOH$	3.5×10^{-4}	3.46
48	p-Nitrobenzoic acid	$p\text{-}NO_2C_6H_4COOH$	3.6×10^{-4}	3.44
49	Phthalic acid	$o\text{-}C_6H_4(COOH)_2$	$1.1\times10^{-3}(K_1)$ $4.0\times10^{-6}(K_2)$	2.96 5.4
50	Isophthalic acid	$m\text{-}C_6H_4(COOH)_2$	$2.4\times10^{-4}(K_1)$ $2.5\times10^{-5}(K_2)$	3.62 4.6
51	Terephthalic acid	$p\text{-}C_6H_4(COOH)_2$	$2.9\times10^{-4}(K_1)$ $3.5\times10^{-5}(K_2)$	3.54 4.46
52	1,3,5-Trimesic acid	$C_6H_3(COOH)_3$	$7.6\times10^{-3}(K_1)$ $7.9\times10^{-5}(K_2)$ $6.6\times10^{-6}(K_3)$	2.12 4.1 5.18
53	Phenyl hexacarboxylic acid	$C_6(COOH)_6$	$2.1\times10^{-1}(K_1)$ $6.2\times10^{-3}(K_2)$ $3.0\times10^{-4}(K_3)$ $8.1\times10^{-6}(K_4)$ $4.8\times10^{-7}(K_5)$ $3.2\times10^{-8}(K_6)$	0.68 2.21 3.52 5.09 6.32 7.49
54	Sebacic acid	$HOOC(CH_2)_8COOH$	$2.6\times10^{-5}(K_1)$ $2.6\times10^{-6}(K_2)$	4.59 5.59
55	Ethylene diamine tetraacetic acid(EDTA)	$CH_2-N(CH_2COOH)_2$ $\|$ $CH_2-N(CH_2COOH)_2$	$1.0\times10^{-2}(K_1)$ $2.14\times10^{-3}(K_2)$ $6.92\times10^{-7}(K_3)$ $5.5\times10^{-11}(K_4)$	2 2.67 6.16 10.26

(3) Dissociation constants of inorganic base in aqueous solution(25℃)

No.	Name	Chemical formula	K_b	pK_b
1	Aluminum hydroxide	$Al(OH)_3$	$1.38\times10^{-9}(K_3)$	8.86
2	Silver hydroxide	$AgOH$	1.10×10^{-4}	3.96
3	Calcium hydroxide	$Ca(OH)_2$	$3.72\times10^{-3}(K_1)$ $3.98\times10^{-2}(K_2)$	2.43 1.4

(continued)

No.	Name	Chemical formula	K_b	pK_b
4	Ammonia	$NH_3 + H_2O$	1.78×10^{-5}	4.75
5	Hydrazine	$N_2H_4 + H_2O$	$9.55 \times 10^{-7} (K_1)$ $1.26 \times 10^{-15} (K_2)$	6.02 14.9
6	Hydroxylamine	$NH_2OH + H_2O$	9.12×10^{-9}	8.04
7	Lead hydroxide	$Pb(OH)_2$	$9.55 \times 10^{-4} (K_1)$ $3.0 \times 10^{-8} (K_2)$	3.02 7.52
8	Zinc hydroxide	$Zn(OH)_2$	9.55×10^{-4}	3.02

(4) Dissociation constants of organic base in aqueous solution (25℃)

No.	Name	Chemical formula	K_b	pK_b
1	Methylamine	CH_3NH_2	4.17×10^{-4}	3.38
2	Urea	$CO(NH_2)_2$	1.5×10^{-14}	13.82
3	Ethylamine	$CH_3CH_2NH_2$	4.27×10^{-4}	3.37
4	Ethanolamine	$H_2N(CH_2)_2OH$	3.16×10^{-5}	4.5
5	Ethylenediamine	$H_2N(CH_2)_2NH_2$	$8.51 \times 10^{-5} (K_1)$ $7.08 \times 10^{-8} (K_2)$	4.07 7.15
6	Dimethylamine	$(CH_3)_2NH$	5.89×10^{-4}	3.23
7	Trimethylamine	$(CH_3)_3N$	6.31×10^{-5}	4.2
8	Triethylamine	$(C_2H_5)_3N$	5.25×10^{-4}	3.28
9	Propylamine	$C_3H_7NH_2$	3.70×10^{-4}	3.432
10	Isopropylamine	$i\text{-}C_3H_7NH_2$	4.37×10^{-4}	3.36
11	1,3-Diaminopropane	$NH_2(CH_2)_3NH_2$	$2.95 \times 10^{-4} (K_1)$ $3.09 \times 10^{-6} (K_2)$	3.53 5.51
12	1,2-Diaminopropane	$CH_3CH(NH_2)CH_2NH_2$	$5.25 \times 10^{-5} (K_1)$ $4.05 \times 10^{-8} (K_2)$	4.28 7.393
13	Tripropylamine	$(CH_3CH_2CH_2)_3N$	4.57×10^{-4}	3.34
14	Triethanolamine	$(HOCH_2CH_2)_3N$	5.75×10^{-7}	6.24
15	Butylamine	$C_4H_9NH_2$	4.37×10^{-4}	3.36
16	Isobutylamine	$C_4H_9NH_2$	2.57×10^{-4}	3.59
17	Tert-Butylamine	$C_4H_9NH_2$	4.84×10^{-4}	3.315
18	Hexamine	$H(CH_2)_6NH_2$	4.37×10^{-4}	3.36
19	Octylamine	$H(CH_2)_8NH_2$	4.47×10^{-4}	3.35
20	Aniline	$C_6H_5NH_2$	3.98×10^{-10}	9.4
21	Benzylamine	C_7H_9N	2.24×10^{-5}	4.65
22	Cyclohexylamine	$C_6H_{11}NH_2$	4.37×10^{-4}	3.36
23	Pyridin	C_5H_5N	1.48×10^{-9}	8.83
24	Hexamethylenetetramine	$(CH_2)_6N_4$	1.35×10^{-9}	8.87
25	2-Chlorophenol	C_6H_5ClO	3.55×10^{-6}	5.45
26	3-Chlorophenol	C_6H_5ClO	1.26×10^{-5}	4.9
27	4-Chlorophenol	C_6H_5ClO	2.69×10^{-5}	4.57
28	o-Aminophenol	$o\text{-}H_2NC_6H_4OH$	$5.2 \times 10^{-5} (K_1)$ $1.9 \times 10^{-5} (K_2)$	4.28 4.72
29	m-Aminophenol	$m\text{-}H_2NC_6H_4OH$	$7.4 \times 10^{-5} (K_1)$ $6.8 \times 10^{-5} (K_2)$	4.13 4.17
30	p-Aminophenol	$p\text{-}H_2NC_6H_4OH$	$2.0 \times 10^{-4} (K_1)$ $3.2 \times 10^{-6} (K_2)$	3.7 5.5
31	o-Toluidine	$o\text{-}CH_3C_6H_4NH_2$	2.82×10^{-10}	9.55
32	m-Toluidine	$m\text{-}CH_3C_6H_4NH_2$	5.13×10^{-10}	9.29
33	p-Toluidine	$p\text{-}CH_3C_6H_4NH_2$	1.20×10^{-9}	8.92
34	8-hydroxyquinoline (20℃)	$8\text{-}HOC_9H_6N$	6.5×10^{-5}	4.19

(continued)

No.	Name	Chemical formula	K_b	pK_b
35	Diphenylamine	$(C_6H_5)_2NH$	7.94×10^{-14}	13.1
36	Benzidine	$H_2NC_6H_4C_6H_4NH_2$	$5.01 \times 10^{-10} (K_1)$ $4.27 \times 10^{-11} (K_2)$	9.3 10.37

Appendix 6 Stability Constants of Coordination Compounds Formed by Metal Ions and EDTA (18-25℃, I = 0.1 mol/L)

Metal ions	lg$K$ EDTA	Metal ions	lg$K$ EDTA	Metal ions	lg$K$ EDTA
Ag^+	7.32	Fe^{3+}	25.1	Sc^{3+}	23.1
Al^{3+}	16.3	Ga^{3+}	20.3	Sn^{2+}	22.11
Ba^{2+}	7.86	Hg^{2+}	21.7	Sr^{2+}	8.73
Be^{2+}	9.2	In^{3+}	25.0	Th^{4+}	23.2
Bi^{3+}	27.94	Li^+	2.79	Ti^{2+}	17.3
Ca^{2+}	10.69	Mg^{2+}	8.7	Tl^{3+}	37.8
Cd^{2+}	16.46	Mn^{2+}	13.87	U^{4+}	25.8
Co^{2+}	16.31	Mo^{5+}	~28	V^{2+}	18.8
Co^{3+}	36	Na^+	1.66	Y^{3+}	18.09
Cr^{3+}	23.4	Ni^{2+}	18.62	Zn^{2+}	16.50
Cu^{2+}	18.80	Pb^{2+}	18.04	Zr^{4+}	29.5
Fe^{2+}	14.32	Pd^{2+}	18.5	REE	16-20

Appendix 7 Standard Atomic Weights (1995, IUPAC)

Name	Symbol	Atomic Weight	Name	Symbol	Atomic Weight	Name	Symbol	Atomic Weight
Sliver	Ag	107.8682	Erbium	Er	167.26	Lutetium	Lu	174.967
Aluminum	Al	26.98154	Europium	Eu	151.96	Magnesium	Mg	24.305
Argon	Ar	39.948	Fluorine	F	18.998403	Manganese	Mn	54.9380
Arsenic	As	74.9216	Iron	Fe	55.847	Molybdenum	Mo	95.94
Gold	Au	196.9655	Gallium	Ga	69.72	Nitrogen	N	14.0067
Boron	B	10.81	Gadolinium	Gd	157.25	Sodium	Na	22.98977
Barium	Ba	137.33	Germanium	Ge	72.59	Neodymium	Nd	144.24
Beryllium	Be	9.01218	Hydrogen	H	1.00794	Neon	Ne	20.179
Bismuth	Bi	208.9804	Helium	He	4.00260	Nickel	Ni	58.69
Bromine	Br	79.904	Hafnium	Hf	178.49	Neptunium	Np	237.0482
Carbon	C	12.011	Mercury	Hg	200.59	Oxygen	O	15.9994
Calcium	Ca	40.08	Holmium	Ho	164.9304	Osmium	Os	190.2
Cadmium	Cd	112.41	Iodine	I	126.9045	Phosphorus	P	30.97376
Cerium	Ce	140.12	Indium	In	114.82	Lead	Pb	207.2
Chlorine	Cl	35.453	Iridium	Ir	192.22	Palladium	Pd	106.42
Cobalt	Co	58.9332	Potassium	K	39.083	Praseodymium	Pr	140.9077
Cesium	Cs	132.9054	Krypton	Kr	83.80	Platinum	Pt	195.08
Copper	Cu	63.543	Lanthanum	La	138.9055	Radium	Ra	226.0254
Dysprosium	Dy	162.50	Lithium	Li	6.941	Rubidium	Rb	85.4678

(continued)

Name	Symbol	Atomic Weight	Name	Symbol	Atomic Weight	Name	Symbol	Atomic Weight
Rhenium	Re	186.207	Tin	Sn	118.69	Vanadium	V	50.9415
Rhodium	Rh	102.9055	Strontium	Sr	87.62	Tungsten	W	183.85
Ruthenium	Ru	101.07	Tantalum	Ta	180.9479	Xenon	Xe	131.29
Sulfur	S	32.06	Terbium	Tb	158.9254	Yttrium	Y	88.9059
Antimony	Sb	121.75	Tellurium	Te	127.60	Ytterbium	Yb	173.04
Scandium	Sc	44.9559	Thorium	Th	232.0381	Zinc	Zn	65.38
Selenium	Se	78.96	Thallium	Tl	204.383	Zirconium	Zr	91.22
Silicon	Si	28.0855	Thulium	Tm	168.9342			
Samarium	Sm	150.36	Uranium	U	238.0289			

第1章 分析化学实验基础知识

1.1 分析化学实验的任务和要求

分析化学是化学学科下的一个重要分支，是化学、医药、食品、环境等专业的基础必修课程。分析化学主要研究物质的组成、结构和含量等化学信息的分析方法及原理，具有很强的综合性、实践性和应用性。分析化学实验是分析化学课程教学的重要组成部分。通过本课程的学习，学生可以加深对分析化学理论知识的认识和理解；正确、熟练地掌握分析化学实验的基本操作技能；确立严格的"量""误差"和"有效数字"的概念，掌握实验数据记录及处理的方法；通过探索性实验，培养独立分析和解决实际问题的能力；培养严谨的科学作风和实事求是的科学态度。

为了完成上述任务，需要做到以下几点：

（1）实验前认真预习。结合实验教材和理论学习，明确实验目的、领会实验原理、熟悉实验内容和方法、清楚实验步骤、操作方法和注意事项。写出预习报告，列出数据记录表格。

（2）准时进入实验室，认真听讲。实验过程中，保证操作严格规范，仔细观察实验现象并如实记录实验数据。学生应有专门的实验记录本（可与预习报告共用），绝不允许将数据记在单页纸或小纸片上。文字记录应整洁清晰，切忌篡改数据、弄虚作假。

（3）实验过程中保证实验台面整洁和良好的实验秩序。药品及仪器使用后及时放回原处，保持实验室卫生。

1.2 实验室安全知识

化学实验室中，经常使用大量易破损的玻璃仪器、水、电、气以及有腐蚀性、易燃、易爆的化学试剂等。因此，为了确保实验人员的人身安全和实验的顺利进行，必须严格遵守实验室安全操作规范。

（1）进入实验室，必须穿实验服。

（2）实验室内禁止饮食、吸烟。一切化学药品严禁入口，取用药品要选用药匙等专用器具，不能用手直接拿取，实验结束后及时洗手。

（3）注意用电安全，不要用湿手接触电源，实验结束后及时切断电源。

（4）能产生有毒的、有刺激性气体的实验必须在通风橱内进行。嗅闻气体时，应用手将气体少量扇向自己，不要直接俯向容器嗅闻。

（5）使用浓酸、浓碱及其他强腐蚀性试剂时，切勿溅在皮肤和衣服上，更要戴好护目镜，防止溅入眼睛里。

(6) 使用易燃的有机溶剂（例如乙醇、苯、丙酮、乙醚等）时要远离火源和热源；低沸点的有机试剂不能在明火上直接加热，而应水浴加热。

(7) 使用有毒试剂（例如氰化物、砷化物、汞盐、铅盐等）时，严防接触伤口或进入口内，其废液不能随意倒入下水道，应倒入指定的回收瓶统一回收处理。

(8) 严禁任意混合各种化学试剂，以免发生意外事故。

(9) 实验中如发生烫伤、割伤等应及时处理，严重者应立即送医院治疗。

(10) 实验室若发生火灾，应立即切断电源和气源，并根据起火原因采取针对性灭火措施。

1.3 定量分析实验基础知识

1.3.1 分析实验用水

1.3.1.1 分析化学实验用水规格

实验室用水根据分析任务和要求的不同，需选用不同规格的纯水。国家标准规定了我国分析实验室用水的技术指标、制备方法及检验方法。表 1.1 为分析实验室用水的级别和主要指标。

表 1.1 分析实验室用水级别和主要指标（引自 GB/T 6682—2008）

名称	一级	二级	三级
pH 范围(25℃)	—	—	5.0～7.5
电导率(25℃)/(mS/m)	≤0.01	≤0.10	≤0.50
可氧化物质含量(以 O 计)/(mg/L)	—	≤0.08	≤0.40
吸光度(254nm,1cm 光程)	≤0.001	≤0.01	—
蒸发残渣(105℃±2℃)/(mg/L)	—	≤1.0	≤2.0
可溶性硅(以 SiO_2 计)/(mg/L)	≤0.01	≤0.02	—

注：1. 由于一级水、二级水的纯度下，难以测定其真实的 pH，因此对一级水、二级水的 pH 范围不作规定。

2. 由于在一级水的纯度下，难以测定可氧化物质和蒸发残渣，对其限量不作规定。可用其他条件和制备方法保证一级水的质量。

1.3.1.2 纯水的制备

分析化学实验对水的要求较高，用于溶解和稀释的水都必须经过纯化。根据具体实验要求不同，对水的纯度要求也不同。纯水的制备方法常用的有蒸馏法、离子交换法、电渗析法等。

(1) 蒸馏法 蒸馏法利用水与杂质的沸点不同而将水与杂质分离。蒸馏法操作简单、成本低廉，能除去水中不挥发的杂质，但不能除去易溶于水的气体及某些低沸点易挥发物。因此，一次蒸馏水只能用于定性分析或一般工业分析。

(2) 离子交换法 化学实验室广泛采用离子交换树脂来分离出水中的杂质离子，这种方法叫离子交换法，制得的纯水称为去离子水。这种方法的优点是成本低、制备的水量大、去除杂质能力强，但不能除去有机物等非电解质杂质，并会有微量树脂溶在水中。

(3) 电渗析法 电渗析法是在离子交换法的基础上发展起来的一种方法。它是在外加电场的作用下，利用阴阳离子交换膜对水中阴、阳离子的选择性透过，使杂质离子从水中分离出来的方法。该方法不能除去非离子型杂质，只适用于要求不高的实验。

1.3.1.3 纯水的检验

纯水的检验方法一般分为物理方法和化学方法两种。根据分析实验室的要求，检验纯水通常有以下几种项目。

(1) pH 由于空气中的 CO_2 可溶于水，故纯水的 pH 一般在 6.0 左右。

(2) 电导率 25℃时，纯水的电阻率为 $(1\sim10)\times10^6\Omega\cdot cm$，高纯水的电阻率大于 $10\times10^6\Omega\cdot cm$。

(3) Ca^{2+} 和 Mg^{2+} 取适量水样，加入 $NH_3\cdot H_2O\text{-}NH_4Cl$ 缓冲溶液，再加入 0.2%铬黑T指示剂1滴，不显红色。

(4) Cl^- 取适量水样，加入 HNO_3 酸化，再加入 1% $AgNO_3$ 溶液2滴，摇匀，无浑浊产生。

1.3.2 化学试剂

1.3.2.1 化学试剂的规格

试剂的纯度对实验结果的准确度影响很大，不同实验对试剂纯度要求也不相同。因此，必须了解化学试剂的分类标准。化学试剂以所含杂质多少分为若干个等级，其规格和使用范围如表 1.2 所示。

表 1.2 化学试剂的规格

等级	名称	符号	标签颜色	适用范围
一级试剂	优级纯	G.R.	绿色	精密分析实验
二级试剂	分析纯	A.R.	红色	一般分析实验
三级试剂	化学纯	C.P.	蓝色	一般定性实验
四级试剂	实验试剂	L.R.	棕色	一般化学制备实验
生化试剂	生物试剂	B.R.	黄色	生物化学实验

除此之外，还有一些特殊用途的高纯试剂，如基准试剂、色谱试剂、光谱试剂等。其中，基准试剂又称标准试剂，纯度相当于或高于一级试剂，可作为滴定分析中的基准物质，用于直接配制标准溶液；光谱纯试剂是以光谱分析时出现的干扰谱线的数目和强度来衡量的；色谱纯试剂是在最高灵敏度（10^{-10}g）条件下无杂质峰表示的。

应根据实验要求分别选用不同规格的试剂。通常情况下，分析实验中所用的一般试剂可选用 A.R. 级（分析纯）试剂，并用蒸馏水或去离子水配制。

1.3.2.2 化学试剂的存放

化学试剂应保存在通风良好、干净、干燥的房间，远离火源并防止水分、灰尘和其他物质污染。

(1) 固体试剂应保存在广口瓶内，液体试剂应盛在细口瓶或滴瓶中。

(2) 见光易分解的试剂（如硝酸银、过氧化氢、高锰酸钾等）应放在棕色瓶内并置于暗处。

(3) 容易侵蚀玻璃的试剂（如氢氟酸、含氟盐、氢氧化钠等）应保存在塑料瓶中。

(4) 吸水性强的试剂（如无水碳酸钠、氢氧化钠等）试剂瓶口应严格密封。

(5) 相互易发生作用的试剂应分开保存。易燃易爆的试剂应分开存放在阴凉通风、不受阳光直射的地方。

（6）剧毒试剂（如氰化物、三氧化二砷等）必须设有专人保管，取用时严格记录。

（7）每个试剂瓶都要贴有标签，标明试剂的名称、浓度、纯度以及配制的日期等。

1.3.2.3 化学试剂的使用

化学试剂的选用原则是在满足实验要求的前提下，合理选择相应级别的试剂。选择试剂的级别应就低而不就高，既不能越级造成浪费，也要保证分析结果的准确度。在取用试剂时，应注意以下几点：

（1）任何试剂不能直接用手拿取。固体试剂应用干净的药匙取用，药匙不可混用。取用药品时，取下的瓶盖应倒放在桌面上，并在取用后立即盖回，防止污染和变质。

（2）在滴瓶中取用液体试剂时，滴管不可触碰到所用的容器器壁，以免污染原试剂。在细口瓶取用液体试剂时，先将瓶塞倒放在桌面上，倾倒时使标签一面朝向手心，以免瓶口残留的少量液体流下而腐蚀标签。

（3）酌量取用试剂，试剂一经取出不得倒回，以免污染原瓶中的试剂。

1.3.3 溶液的浓度及其配制方法

浓度是表示在一定量的溶液或溶剂中所含溶质的量，溶液浓度常有以下几种表示方式。

（1）百分浓度

① 质量-质量百分浓度（m/M,%）是指每100g溶液中所含溶质的质量（g），即质量-质量百分浓度（m/M,%）=溶质质量（g）/[溶质质量（g）+溶剂质量（g）]×100%。市售的酸、碱常用此法表示，例如37%的盐酸溶液即100g的盐酸溶液中含有37g的纯HCl和63g的水。

② 质量-体积百分浓度（m/V,%）是指每100mL溶液中所含溶质的质量（g），即质量-体积百分浓度（m/V,%）=溶质质量(g)/溶液体积(mL)×100%。例如1%硝酸银溶液是指1g硝酸银溶于适量水中，再以水稀释至100mL。

（2）体积比浓度（V/V）是指液体用水稀释或与其他液体相互混合时所制得的浓度。体积比浓度(V/V)=溶质体积(mL)/溶剂体积(一般指水，mL)。例如1:2盐酸是由1体积浓盐酸与2体积水混合而成的。

（3）体积百分比浓度（V/V,%）是指每100mL溶液中所含溶质的体积（mL）。例如95%乙醇溶液，即100mL溶液中含有95mL乙醇和5mL水。

（4）物质的量浓度（也称为摩尔浓度，mol/L）是指1L溶液中含有溶质的物质的量（mol）。

$$物质的量浓度 = \frac{溶质的物质的量}{溶液体积} = \frac{\frac{溶质质量}{溶质摩尔质量}}{溶液体积}$$

1.4 定量分析实验仪器及基本操作

1.4.1 分析天平

1.4.1.1 分析天平的使用方法

（1）称量前先将分析天平罩取下叠好。检查分析天平水平仪，若水平仪的气泡不在圆圈中心，应调节水平调节脚，使气泡回到中心。

(2) 检查天平门是否关闭。打开天平开关,按下"Tare"键,显示为"0.0000g"。

(3) 按所需方法进行称量。称量时从侧门取放物质,待显示的数稳定后读数,读数时应关闭侧门以免空气流动引起读数波动。

(4) 称量结束后,关闭并清理天平,罩上分析天平罩。

1.4.1.2 称量方法

根据不同的称量对象和称量要求,采用相应的称量方法。常用的称量方法有以下三种。

(1) **直接称量法** 直接称量法适用于称量洁净干燥的器皿、块状或棒状的金属等。

方法:按"Tare"键显示"0.0000g"后,将被称物置于称量盘上,待天平平衡后,所得读数即为被称物的质量。

(2) **固定质量称量法** 固定质量称量法用于称取某一固定质量的试剂。该法适用于称量不易吸水、在空气中稳定、无腐蚀性的粉末状或颗粒状物质。

方法:将干燥的器皿放在天平称量盘上,待天平平衡后按"Tare"键显示"0.0000g"。向器皿中逐步加入试样至所需质量。

(3) **差减法** 差减法用于称取一定质量范围内的试样。该法适用于称量易吸湿、易氧化或易与 CO_2 反应的试样。

方法:将适量试样装入干燥洁净的称量瓶内,盖上瓶盖。用干净的纸条套住称量瓶[图 1.1(a)],将称量瓶置于分析天平称量盘上,准确称取试样与称量瓶的总质量,记为 m_1。取出称量瓶,置于盛放试样的容器上方,用纸片包住瓶盖,打开瓶盖,并将称量瓶倾斜,用瓶盖轻敲瓶口上方,使试样慢慢落入容器内,如图 1.1(b)所示。当倾倒出的试样接近所需质量时,用瓶盖轻敲瓶口上部,使粘在瓶口的试样落下,同时缓缓将称量瓶竖起,并盖好瓶盖。将称量瓶置于天平的称量盘上,

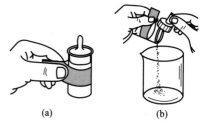

图 1.1 差减法称量

记下读数 m_2。m_1 与 m_2 的差值即为试样的质量。如果一次倒入试样质量超过所需范围,必须弃去重称,试样切勿放回称量瓶。

按上述方法可连续称取多份试样。

第一份试样质量(g)= $m_1 - m_2$

第二份试样质量(g)= $m_2 - m_3$

第三份试样质量(g)= $m_3 - m_4$

1.4.2 容量玻璃器皿

1.4.2.1 容量瓶及使用

容量瓶是用于准确测量盛装溶液体积的量入式量器。外形是细颈梨形的平底玻璃瓶,带有磨口玻璃塞或塑料塞,颈上有刻度线。容量瓶的容量定义为:在指定温度下(一般为 20℃),液体的弯月面下缘与刻度线相切时,溶液的体积与瓶上标注的体积相等,以 mL 计。容量瓶主要用于配制准确浓度的标准溶液或将溶液稀释至准确体积的稀溶液。

(1) **容量瓶检漏** 容量瓶在使用前应先对其进行检漏。其操作方法是:加水至容量瓶标线附近,盖好瓶塞,一手食指顶住瓶塞,其余手指拿住瓶颈标线以上部分,另一只手五指托住瓶底边缘,将容量瓶倒置 2min,如不漏水,将容量瓶直立并将瓶塞转动 180°后,再倒置 2min,如仍不漏水,即可

容量瓶的试漏

使用。

(2) 溶液的配制

① 溶解样品：准确称取固体物质于体积合适的烧杯中，加入少量纯水，搅拌使其完全溶解。

② 溶液转移：将溶液定量转移至容量瓶中，方法是一手拿玻璃棒使其悬空伸入容量瓶瓶口，玻璃棒下端靠在瓶颈内壁低于刻线处，上端不碰瓶口；另一手拿烧杯，烧杯嘴边缘紧贴玻璃棒中下部，慢慢倾斜烧杯，使溶液沿玻璃棒流入容量瓶，如图1.2(a)所示。待溶液流尽后，慢慢将烧杯和玻璃棒直立，同时将烧杯沿玻璃棒缓缓上提，使附在玻璃棒与烧杯嘴之间的液滴流回到烧杯中，并将玻璃棒放回烧杯中。用纯水冲洗玻璃棒和烧杯内壁3～4次，按上述方法将洗涤液全部转移至容量瓶。

③ 定容：加水至容量瓶体积的2/3处，拿起容量瓶按水平方向旋转几周，使溶液初步混匀。继续加水至标线1cm处，放置1～2min，使附着在瓶颈内壁的溶液流下。用洗瓶或滴管加水至弯月面下缘与标线相切［图1.2(b)］，盖紧瓶塞。

④ 摇匀：一手食指顶住瓶塞，另一只手指尖托住瓶底边缘，倒转容量瓶，使瓶内气泡上升到顶部，轻轻振摇，再倒转回来，重复操作多次使溶液充分混匀，如图1.2(c)。

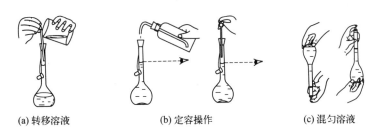

(a) 转移溶液　　　　　(b) 定容操作　　　　　(c) 混匀溶液

图1.2　溶液的配制

(3) 溶液的稀释　用移液管移取一定体积溶液于容量瓶中，按上述方法定容、摇匀。

注意事项：

(1) 不能在容量瓶中进行溶质的溶解。

(2) 容量瓶只可用于配制溶液，不能储存溶液。配制好的溶液应转移到干燥、洁净的磨口试剂瓶中保存。

(3) 热溶液应冷却至室温后再进行转移，否则会造成体积误差。

(4) 容量瓶不能加热或在烘箱中烘干。

(5) 需避光的溶液应以棕色容量瓶配制。

(6) 容量瓶使用完毕后应立即用水冲洗干净，若长期不用，磨口处应洗净擦干并衬有纸片。

(7) 容量瓶与瓶塞配套使用，不能随意调换。使用容量瓶时，应用细绳或橡皮筋将瓶塞系在瓶颈上，以防瓶塞被玷污或混用。

1.4.2.2　移液管及使用

移液管是准确移取一定体积液体的量出式量器，包括单标线移液管和分刻度移液管。单标线移液管是一根细长而中间有膨大部分的玻璃管，管颈上部刻有环形标线，膨大部分标有指定温度下的容积［图1.3(a)］。在标明温度下，移取溶液的弯月面下缘与标线相切时，使溶液自由流出，则流出溶液的体积与管上标明的体积相同。分刻度移液管是一种具有均匀刻度的玻璃管［图1.3(b)］，可用于准确移取标示范围内任意体积的液体，其准确度比移液管稍差。

(1) 单标线移液管的使用

① 润洗：用洗液、自来水、蒸馏水洗至内壁不挂水珠为止。移取液体前，应先用滤纸将管尖口内外的水吸干，并用待移取溶液润洗3次，以除去管内残留的水分，润洗溶液由管尖口放出。润洗的具体操作是，倒少许待取溶液于干燥洁净的烧杯中，用洗耳球将待测液吸至移液管1/4处左右，迅速移去洗耳球的同时用右手食指按住管口。将移液管取出，左手扶住管的下端，慢慢松开右手食指，一边转动移液管，一边降低上管口，将管平置，使溶液接触到标线上面的部分和全管内壁。最后将管直立，使溶液从管的下端放出并弃去。

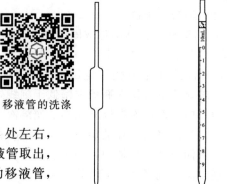

(a) 单标线移液管　　(b) 分刻度移液管

图1.3　移液管

② 吸取溶液：吸取液体时，右手拇指和中指拿住移液管管颈标线上方，使移液管尖端伸入待取溶液的液面下约1~2cm，太深会使外壁粘附过多溶液，影响量取溶液体积的准确性，太浅会由于液面下降而吸入空气。左手拿洗耳球，先将球内空气压出，接着将洗耳球尖头紧接在移液管上管口上，慢慢松开洗耳球，溶液逐渐吸入管内，此时移液管尖口应随容器中溶液液面的下降而下降，如图1.4(a)所示。当管中液面上升至标线以上5mm左右时，迅速移去洗耳球并用右手食指按住上管口，并将移液管提离液面。微微倾斜容器，使移液管尖紧贴容器内壁，微微松动食指并用拇指和中指轻轻捻转移液管，使液面平稳下降，同时平视刻度，直至溶液的弯月面下缘与标线相切时，立即用食指压紧上管口使溶液不再流出。

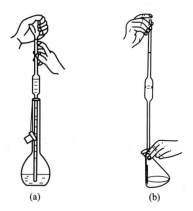

图1.4　移液管的使用

③ 放液：将管下部粘附的少量溶液用滤纸擦干，将移液管移至盛接溶液的容器中，使管垂直，盛接容器倾斜约30°，管尖紧靠容器内壁。松开右手食指使溶液自由沿壁流下，如图1.4(b)所示。待溶液全部放出后，等待15s，取出移液管。对于管尖端的残留溶液，如移液管上未标有"吹"字的，切勿将残留溶液吹出，因为在生产移液管时，已经考虑了末端残留溶液的体积。

(2) 分刻度移液管的使用：使用分刻度移液管吸取溶液的方法大致与上述移液管操作相同。需要注意的是，每次吸取溶液时液面应调至最高刻线，然后使管内液面平稳下降，直至所需体积后按紧食指，移去移液管。

注意事项：

(1) 移液管不能在烘箱中加热烘干，也不能移取过热或过冷的溶液。

(2) 移液管在使用后应洗净放在移液管架上。

(3) 移液管和容量瓶一般配套使用，因而使用前应做相对容积的校准。

(4) 实验过程中应使用同一支移液管，减小实验误差。

液体的移取

1.4.2.3 滴定管及使用

滴定管上刻有精密的刻度，是滴定时用来准确测量流出溶液体积的量出式量器。常量分析中所用的滴定管标称容量为25mL和50mL，此外还有标称容量为10mL、5mL、2mL、1mL的半微量或微量滴定管。其中，最常使用的标称容量为50mL的滴定管最小刻度为

0.1mL，读数时应精确到 0.01mL。

根据盛装溶液的性质不同，滴定管分为酸式滴定管和碱式滴定管（图 1.5）。酸式滴定管下端带有玻璃旋塞，适用于盛装酸性溶液和氧化性溶液，不能盛装碱性溶液，因碱性溶液会使玻璃旋塞与旋塞槽黏合，以致难以转动。碱式滴定管下端连接一段橡胶管，橡胶管内有一大小适中的玻璃珠以控制溶液的流出速度。碱式滴定管主要用来盛装碱性溶液，具有氧化性的溶液或其他能与橡胶管反应的溶液，如 $KMnO_4$、I_2、$AgNO_3$ 等，不能装在碱式滴定管中。另有通用型滴定管，外观与酸式滴定管一样，其下端旋塞材质为聚四氟乙烯，其不受溶液酸碱性的限制，可用来盛放各种溶液滴定。

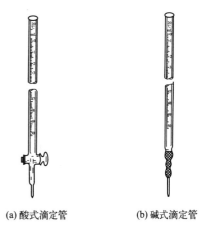

(a) 酸式滴定管　　(b) 碱式滴定管

图 1.5　滴定管

（1）滴定管使用前的准备　酸式滴定管在使用前应检查旋塞转动是否灵活，是否漏水。将旋塞关闭，滴定管内注满水后，垂直固定在滴定管架上放置 2min，观察滴定管口及旋塞两端是否有水渗出；将旋塞转动 180°，再放置 2min，观察是否有水渗出。若不渗水即可使用，否则应将旋塞取出，重新涂抹凡士林并检漏后方可使用。

涂凡士林的具体操作如下：将旋塞取出，用滤纸将旋塞和旋塞槽擦干。用手指蘸少量凡士林均匀地涂一薄层于旋塞两头（图 1.6 中 1 和 2 处），并将旋塞插入旋塞槽内。向同一方向转动旋塞，使旋塞内油膜均匀透明，旋塞转动灵活。涂好后，用橡胶圈套好旋塞。

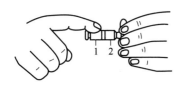

图 1.6　涂凡士林的方法

碱式滴定管要求玻璃珠和橡胶管大小合适，使用前检查滴定管是否漏水，液滴是否能够灵活控制。如不符合要求，应重新更换。

旋塞为聚四氟乙烯材质的滴定管，在使用前检查旋塞是否灵活转动，是否漏水。由于聚四氟乙烯旋塞有弹性，可通过调节旋塞尾部的螺帽来调节旋塞与旋塞槽的紧密度，因此此类滴定管无须涂抹凡士林。

（2）溶液的装入

① 润洗：在装入溶液前，用待装溶液对滴定管润洗 2～3 次，每次 5～10mL 以除去管内残留的水分，确保滴定溶液浓度不变。润洗时，关闭旋塞，两手平端滴定管，慢慢转动，使溶液布满滴定管。打开旋塞，使少量溶液从滴定管下端流出，剩余溶液从管口倒出。装入溶液时应直接从试剂瓶倒入滴定管，不得借助其他容器以免溶液的浓度改变或造成污染。

滴定管的洗涤

② 排气：装好溶液后确保滴定管下端无气泡，否则在滴定过程中气泡溢出会影响溶液体积的准确性。酸式滴定管排气时迅速转动旋塞，使溶液急速流出的同时将气泡赶出。对于碱式滴定管可把橡胶管向上弯曲，玻璃尖嘴斜向上方，捏挤玻璃球，使溶液从尖嘴喷出，即可排出气泡（图 1.7）。最后调整液面至 0.00mL 刻度处或稍下一点的位置，记下初始读数。

图 1.7　碱式滴定管排气泡

（3）滴定管的操作　滴定时，应将滴定管固定在滴定管架上，保持滴定管垂直。滴定最好在锥形瓶中进行，必要时也可以在烧杯、碘量瓶等中进行。

① 酸式滴定管的操作：使用酸式滴定管时，用左手控制滴定管旋塞部分，拇指在前，食指和中指在后，轻轻转动旋塞，使溶液逐滴流出，如图 1.8(a) 所示。转动旋塞时要注意手心空握，以免旋塞松动或被顶出，造成漏液。

② 碱式滴定管的操作：使用碱式滴定管时，左手无名指和小指夹住出口玻璃管，拇指在前，食指在后，捏住橡胶管内玻璃球稍上处，使橡胶管与玻璃球之间形成一条缝隙，溶液即可流出 [图 1.8(b)]。操作时，不要用力捏挤玻璃球，不能使玻璃球上下移动；不能捏挤玻璃球下部的橡胶管，以免空气进入形成气泡。停止滴定时，应先松开拇指和食指，然后再松开无名指和小指。

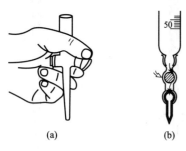

滴定操作

图 1.8　滴定管滴定操作

③ 滴定操作方法：滴定时，右手拇指、食指和中指握住锥形瓶，滴定管尖端伸入瓶口下 1~2cm。边滴加溶液边摇动锥形瓶，摇动时应向同一方向做圆周运动，使瓶内溶液混合均匀、反应快速进行完全 [如图 1.9(a)]。滴定时左手不能离开旋塞让溶液自行流下，锥形瓶也不能离开滴定管尖端。滴定时要注意锥形瓶内溶液颜色的变化，临近终点时，滴定速度要减慢，加一滴，摇几下，最后每次加入半滴，直至溶液出现明显的颜色变化，到达终点为止。加半滴溶液的方法如下：微微转动旋塞或轻轻捏挤玻璃球，使溶液悬于管尖端上，形成半滴，将锥形瓶内壁与管口相接触，使液滴沾落，再用洗瓶以少量纯水将附于瓶壁上的溶液冲下。

在烧杯中进行滴定时，滴定管尖端伸入烧杯内 1~2cm，不要靠壁，左手控制滴加溶液，右手持玻璃棒在烧杯中不断搅动溶液。注意玻璃棒要做圆周运动，但不要接触烧杯壁和杯底，如图 1.9(b) 所示。

在碘量瓶等具塞锥形瓶中滴定时，瓶塞要夹在右手的中指与无名指之间，不能放在其他地方，以免玷污，如图 1.9(c) 所示。

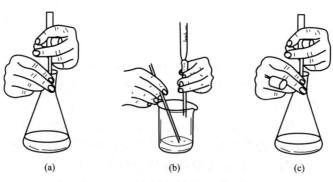

图 1.9　滴定操作方法

（4）滴定管的读数　滴定管读数不准确通常是引起滴定分析误差的主要来源之一。因此滴定管的读数应遵循以下规则：

① 装入溶液或滴定结束后，静置 1~2min，待附着在内壁上的溶液流下后再读数。每次读数前检查管壁是否挂有液珠，下管口是否有气泡，管尖是否挂液滴。

② 读数时应将滴定管从滴定管架上取下，用右手拇指和食指捏住滴定管上端无刻度处，使滴定管保持自然垂直向下再进行读数。

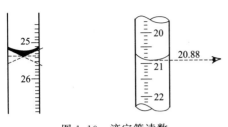

图 1.10 滴定管读数

③ 由于水溶液浸润玻璃,在附着力和内聚力的作用下,滴定管内的液面呈现弯月形。对于无色溶液和浅色溶液,读数时视线应与弯月面下缘的最低点在同一水平线上,如图 1.10 所示。对于深色溶液,如 $KMnO_4$、I_2 溶液,由于其弯月面不够清晰,读数时视线应与液面的上边缘在同一水平。

④ 每次滴定前应将液面调节在 0.00mL 或稍下一点的位置,这样可固定在某一段体积范围内滴定,以减少体积误差。

⑤ 读数要求读到小数点后第二位,即准确到 ±0.01mL。

1.4.3 pH 计

pH 计也称酸度计,是用来测定溶液 pH 的精密仪器。pH 计由电极和电位计两部分组成,其中电极又分为指示电极、参比电极和复合电极。

1.4.3.1 基本电极

(1) 指示电极 玻璃电极是测量 pH 的指示电极,其结构如图 1.11 所示。该电极下端的玻璃球泡是用特殊玻璃吹制而成的 pH 敏感膜,能对氢离子活度有选择性响应;电极内部装有特定的内参比溶液,溶液中插入一支 Ag-AgCl 内参比电极。玻璃电极的电极电位随溶液 pH 的变化而改变。

(2) 参比电极 通常用饱和甘汞电极作为参比电极,其结构见图 1.12。饱和甘汞电极由汞、Hg_2Cl_2 和饱和 KCl 溶液组成。饱和甘汞电极电位在一定温度下恒定,不随溶液 pH 的变化而变化。当指示电极、参比电极与试液组成工作电池时,电位计在零电流条件下测量其电动势,即可测定溶液的 pH。

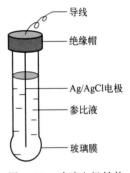

图 1.11 玻璃电极结构

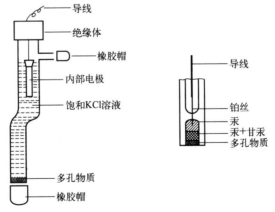

图 1.12 饱和甘汞电极结构

(3) 复合电极 目前使用较多的是复合电极,它是由指示电极与参比电极组合而成的,其结构如图 1.13 所示。复合电极使用方便,且玻璃球膜被有效地保护,不易损坏。

1.4.3.2 使用方法

(1) pH 电极的准备 将复合电极下端的电极保护瓶拔下,拉下电极上端的橡胶套,露出上端小孔,用蒸馏水清洗电极并用滤纸吸干残留水分。

(2) 标定　仪器使用前要标定,通常使用二点标定法标定电极斜率。标定缓冲溶液一般第一次使用 pH=6.86 的溶液,第二次用接近被测溶液 pH 的缓冲溶液,如被测溶液为酸性时,应选 pH=4.00 的缓冲溶液;被测溶液为碱性时则选 pH=9.18 的缓冲溶液。

(3) pH 的测量　经过标定的仪器,即可用来测量被测溶液。每次测量前用蒸馏水清洗电极头部,接着用滤纸轻轻吸干电极上残余的水分或用被测溶液清洗。测量时为保证精度,应使电极头球泡完全浸入到溶液中,电极离容器 1~2cm,同时溶液应保持均匀流动且无气泡,待读数稳定后即可读取数据。

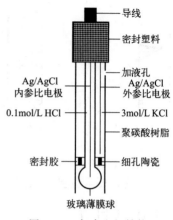

图 1.13　复合电极结构

(4) 电极电位值(mV)的测量　将仪器的显示调整为"mV"挡,其他步骤同上,待读数稳定后即可读取。

1.5　实验数据的记录、处理和实验报告

定量分析的任务是测定试样中某种成分的含量,因此对实验原始数据进行规范记录和正确处理,合理报告分析结果对实验结果的准确性至关重要。

1.5.1　实验数据的记录

(1) 数据应记录在专门的实验记录本或预习报告上,绝不允许将数据记在单页纸或小纸片上。

(2) 实验过程中的各种测量数据和有关现象,应及时、准确、清晰地记录下来。切忌篡改数据、弄虚作假。

(3) 记录实验过程中的数据时,应注意其有效数字的位数。例如,使用台秤称重时,要求记录到 0.01g;使用分析天平称重时,记录至 0.0001g;滴定管和移液管的体积要记录至 0.01mL。

1.5.2　实验数据的处理

定量分析实验通常需平行测定 3~5 次,用平均值来表示测定结果,用相对平均偏差(或标准偏差)来衡量分析结果的精密度。为了简单、清晰、正确地处理实验数据,通常采用表格的形式进行记录和处理。

例如,采用邻苯二甲酸氢钾(KHP)基准物质标定 NaOH 溶液浓度的数据记录及处理表如下:

实验序号	1	2	3
m_{KHP}/g			
V_{NaOH}(初始)/mL			
V_{NaOH}(终点)/mL			
V_{NaOH}(消耗)/mL			
$c_{NaOH}/(mol/L)$			
$\bar{c}_{NaOH}/(mol/L)$			
$|d_i|$			
$\bar{d}_r/\%$			

1.5.3 实验报告

实验报告是实验的总结，是使学生对实验从直观感性认识上升到理性认识的过程，是培养学生归纳总结、分析问题能力和书写能力的有效方式。实验结束后，学生应根据实验记录进行整理总结，完成实验报告的书写。实验报告应包括实验题目、实验目的、实验原理、实验步骤、实验数据的记录和处理、实验结果和思考讨论等几部分。以下是实验报告格式模板，供参考。

实验报告格式模板

实验日期：_____　　实验地点：_____
姓　　名：_____　　学　　号：_____

实验名称：食用醋中总酸度的测定

实验目的

(1) 熟练掌握滴定管、容量瓶、移液管的使用方法和滴定操作；
(2) 掌握氢氧化钠标准溶液的配制和标定方法；
(3) 了解强碱滴定弱酸的反应原理及指示剂的选择；
(4) 学会食醋中总酸度的测定方法。

实验原理

食醋中的主要成分是醋酸（有机弱酸，$K_a=1.8\times10^{-5}$）。此外，还含有少量的其他弱酸，如乳酸等，与 NaOH 反应产物为 NaAc：

$$HAc+NaOH\longrightarrow NaAc+H_2O$$

化学计量点时 pH≈8.7，滴定突跃在碱性范围内（如 0.1mol/L NaOH 滴定 0.1mol/L HAc 的突跃范围为 pH 7.74～9.70）。若使用在酸性范围内变色的指示剂，如甲基橙，将引起很大的滴定误差（该反应化学计量点时溶液呈弱碱性，酸性范围内变色的指示剂变色时，溶液呈弱酸性，则滴定不完全）。因此，在此应选择在碱性范围内变色的指示剂酚酞（8.0～9.6）。（指示剂的选择主要以滴定突跃范围为依据，指示剂的变色范围应全部或一部分在滴定突跃范围内，则终点误差小于 0.1%）。

因此，可选用酚酞作指示剂，利用 NaOH 标准溶液测定 HAc 含量。食醋中总酸度用 HAc 的含量来表示。

实验用品

仪器：锥形瓶、量筒、分析天平（万分之一）、碱式滴定管、移液管、容量瓶。
试剂：邻苯二甲酸氢钾（G.R.）、NaOH 固体（A.R.）、酚酞指示剂。

实验流程

(1) 0.1mol/L NaOH 溶液的配制和标定

配制：台秤迅速称取 NaOH 1.00g→用去离子水稀释至 250mL。

标定：分析天平称邻苯二甲酸氢钾 0.4～0.6g（三份）→250mL 锥形瓶→加 20～30mL 去离子水、2～3 滴酚酞→用 NaOH 溶液滴定→溶液由无色变成微红色 30s 不褪→记下 V_{NaOH}。

（2）食用醋总酸度的测定

稀释：移液管准确移取 25.00mL 食用醋→定容摇匀。

滴定：移液管移取试液 25.00mL 三份→250mL 锥形瓶→2～3 滴酚酞→用 NaOH 溶液滴定→溶液由无色变成微红色 30s 不褪→记下 V_{NaOH}。

数据记录与处理

（1）0.1mol/L NaOH 溶液的标定

实验序号	1	2	3
m_{KHP}/g			
$V_{\text{NaOH}}(\text{初始})/\text{mL}$			
$V_{\text{NaOH}}(\text{终点})/\text{mL}$			
$V_{\text{NaOH}}(\text{消耗})/\text{mL}$			
$c_{\text{NaOH}}/(\text{mol/L})$			
$\bar{c}_{\text{NaOH}}/(\text{mol/L})$			
$|d_i|$			
$\bar{d}_r/\%$			

（2）食用醋总酸度的测定

实验序号	1	2	3
$V_{\text{醋}}/\text{mL}$			
$V_{\text{NaOH}}(\text{初始})/\text{mL}$			
$V_{\text{NaOH}}(\text{终点})/\text{mL}$			
$V_{\text{NaOH}}(\text{消耗})/\text{mL}$			
$c_{\text{HAc}}/(\text{g/L})$			
$\bar{c}_{\text{HAc}}/(\text{g/L})$			
$|d_i|$			
$\bar{d}_r/\%$			

注意事项

（1）食醋中醋酸的浓度较大，故必须稀释后再进行滴定。

（2）测定醋酸含量时，所用的蒸馏水不能含有二氧化碳，否则会溶于水中生成碳酸，将同时被滴定。

思考题

（1）滴定醋酸时为什么要用酚酞作指示剂？

（2）该方法的测定原理是什么？

（3）酚酞指示剂使溶剂变红后，在空气中放置一段时间又变为无色，原因是什么？

（4）标准溶液的浓度应保留几位有效数字？

第 2 章
常用样品制备及分离技术

2.1 样品的制备

样品的制备通常是指样品的采集和分解。

(1) 样品的采集是指从大批物料中采取少量样本作为原始试样，所采试样应具有高度的代表性，采取的试样的组成能代表全部物料的平均组成。对于固体样品（土壤、沉积物、金属等）具有不均匀性，液体样品的化学组成容易发生变化等问题，要具体样品具体考虑，保证取样的准确性和代表性。

(2) 样品分解常用的方法有：溶解法、熔融法、灰化法和消化法。样品分解时，要注意对被测组分的保护，并结合干扰组分的分离，简单、快速地进行测定。

此部分内容在分析化学教材中有详细讲解，本书不再赘述。

2.2 样品的分离

2.2.1 分离方法的选择

在一个分离体系中，通常必须设计不同的相，物质在不同相之间转移才能使不同物质在空间上分离开，而多数分离过程选择两相体系。寻找适当的两相体系，使各种被分离组分在两相间的作用势能之差增大，从而使它们选择性地分配于不同的相中。

分离方法的种类很多，分类方法也很多，从不同的角度可将分离方法分成若干各有特色的类型。下面简单介绍在分析化学实验中常用到的几种分离方法。

2.2.2 沉淀分离

沉淀是一种经典的分离方法，其基本原理是在沉淀过程中，两种或两种以上的物质在固、液两相的分配比方面有一定的差距，利用这种差距进行多次沉淀，从而达到分离的目的。

提高沉淀分离效果的一些措施：(1) 利用有机沉淀剂（成盐、螯合、吸附）；(2) 利用络合掩蔽提高沉淀分离效果；(3) 利用选择性氧化还原；(4) 利用共沉淀载体分离或富集痕量组分；(5) 均相沉淀。

沉淀分离后，往往需要配合离心或过滤进行分离。

2.2.3 蒸馏

将液体加热至沸腾，使液体变为气体，然后再将蒸气冷凝为液体，这两个过程的联合操作称为蒸馏（图 2-1）。蒸馏是分离和纯化液体有机混合物的重要方法之一，其中减压蒸馏是用真空泵将蒸馏系统抽到一定的真空度进行蒸馏的过程。蒸馏可以将易挥发和不易挥发的物质分离开，也可以将沸点不同的液体混合物分离开，主要适用于以下几个方面：

① 分离沸点有显著区别（相差 30℃以上）的液体混合物；
② 常量法测定沸点及判断液体的纯度；
③ 除去液体中所含的不挥发性物质；
④ 回收溶剂或因浓缩液体的需要而蒸出部分的溶剂。

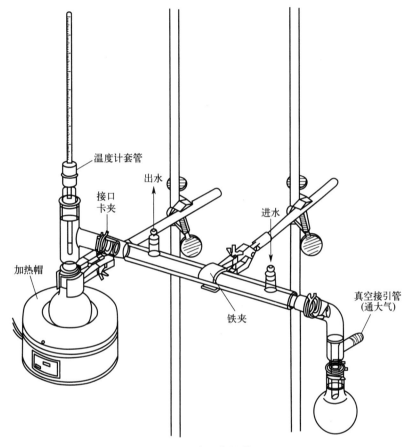

图 2.1　常用蒸馏装置

2.2.4　萃取分离

广义的萃取分离法是将样品相中的目标化合物选择性地转移到另一相中或选择性地保留在原来的相中（即转移非目标化合物），从而使目标化合物与原来的复杂基体相互分离的方法。萃取分离法大体上分为液相萃取、固相萃取和超临界流体萃取等。

溶剂萃取是利用不同物质在互不相溶的两相（水相和有机相）间的分配系数的差异，使水相样品中目标物质与基体物质相互分离的方法。溶剂萃取既可用于有机物的分离，也可用于无机物的分离。溶剂萃取至今仍是实验室和工厂中相当常用的分离技术，其优势主要表现在仪器设备简单、操作方便、分离选择性比较高和应用范围广等方面。萃取分离过程中，样品以溶液状态被转移，其溶解过程与分子间的作用力直接相关，而分子间的作用力大小与分子极性有关。一般顺序：非极性物质＜极性物质＜氢键物质＜离子型物质。

用于提取所需要物质的溶剂叫萃取剂。加入萃取剂的目的是使那些亲水性溶质通过与萃取剂反应生成各种类型的疏水性化合物后而易于进入有机相。特别是在金属离子的溶剂萃取中，萃取剂的种类相当丰富。对萃取剂有以下基本要求：（1）具有至少一个萃取功能基团；（2）具有足够的疏水性；（3）良好的选择性；（4）有较高的萃取容量。萃取操作的基本仪器有分液漏斗、索氏提取器（图2.2）等。

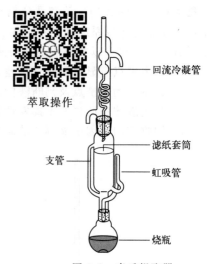

图2.2 索氏提取器

溶剂微萃取也称液相微萃取（LPME），是1996年才发展起来的新的溶剂萃取技术。它结合了液-液萃取和固相微萃取的优点，只需极少量的有机溶剂，装置简单，操作方便，成本低。LPME技术适合萃取在水溶液中溶解度小、含有酸性或碱性官能团的痕量目标物。LPME技术还可以方便地与后续分析仪器连接，实现在线样品前处理。例如LPME后续气相色谱分析时，可采用微量进样针直接进样，克服了固相微萃取（SPME）解吸速度慢、涂层降解、记忆效应大的缺点。与不使用溶剂的SPME相比，LPME的主要缺点是在进行色谱分析时有溶剂峰，有时会掩盖目标组分的色谱峰。采用分散液-液微萃取分离富集水样中拟除虫菊酯类农药残留的一个实例是：取5.0mL经0.45μm微孔滤膜真空过滤后的水样于10mL具塞尖底离心试管中，加入10μL萃取溶剂氯苯和1.0mL分散剂丙酮，轻轻振荡1min，即形成一个水/丙酮/氯苯的乳浊液体系。氯苯均匀地分散在水相中，室温放置2min，以5000r/min离心5min，分散在水相中的萃取溶剂氯苯沉积到试管底部，用微量进样器吸取1μL萃取溶剂直接进样做气相色谱分析。

当通过将萃取剂固定在其他支撑载体上，使溶剂萃取中的溶剂损失和相分离难的问题得到了解决。然而，固定化溶剂的稳定性不好、支撑材料的耐溶剂能力不够等问题的出现，使微胶囊技术被用到了溶剂萃取中，产生了溶剂微胶囊技术，即在微胶囊形成过程中将用于萃取的溶剂包覆于微胶囊的空腔中。溶剂微胶囊的制备方法有很多，一般包括溶剂分散和溶剂包覆两个步骤。由于具有萃取树脂的优点，溶剂微胶囊避免了传统溶剂萃取的乳化和分相问题，在萃取剂包覆量和防止萃取剂流失方面具有明显优势。

2.2.5 重结晶

将晶体置于溶剂中加热溶解，经冷却重新析出的过程即是重结晶，它是纯化固体有机化合物最常用的方法之一。固体物质的溶解度一般随温度升高而增大。把待提纯晶体溶解在热溶剂中制成饱和溶液，趁热过滤，溶解度小的杂质留在滤纸上被除去；冷却时由于溶解度降低变成过饱和溶液形成晶体析出，经减压抽滤，晶体留在滤纸上，溶解度大的杂质留在溶剂中与晶体分离。

重结晶操作重要的是选择溶剂。合适的溶剂必须符合下列条件：
① 与重结晶物质不发生反应；
② 高温时溶解度大，低温时溶解度小；
③ 杂质溶解度或者很大或者很小；
④ 与重结晶物质易分离。

另外还要考虑无毒、不易燃等因素。重结晶操作的一般过程如下：

(1) 溶解　将含有杂质的固体粗产物置于锥形瓶中，加入少量溶剂，安装回流冷凝管，加热至沸腾，滴加溶剂直至固体全部溶解，继续加入大约20%溶剂，以补充溶剂的挥发和防止热过滤时晶体析出。

(2) 脱色　如果粗产物含有有色杂质，可以将上述溶液稍微冷却后加入大约5%的活性炭，再次加热至沸腾，有色杂质即可被吸附在活性炭上。

不可将活性炭加入沸腾的溶液中，防止暴沸；活性炭也不要加入过多，因为它在吸附杂质的同时也会吸附产物，造成产物的损失。

(3) 热过滤　趁热将溶液过滤以除去不溶性杂质。热过滤的要点是系统要"热"，操作要"快"，因而在进行热过滤前应做好充分的准备工作，使操作步骤周详、紧凑，尽量减少产物损失。

(4) 冷却结晶　将热过滤的溶液冷却使晶体析出，从而与溶液中的部分杂质分离。快速冷却得到的晶体较小，自然冷却得到的晶体较大。如果晶体不易析出，可以用玻璃棒摩擦容器的内壁引发结晶或加入晶种。

(5) 抽滤和洗涤　将晶体与溶液（一般称为母液）分离，并以合适的溶剂洗涤。

(6) 干燥。

2.2.6 色谱法

色谱法是一种物理化学分离和分析方法，它基于物质溶解度、蒸气压、吸附能力、立体结构或离子交换等物理化学性质的微小差异，使其在流动相和固定相之间的分配系数不同。而当两相做相对运动时，组分在两相间进行连续多次分配，从而达到彼此分离。

薄层色谱法（TLC）是将合适的固定相均匀涂布于平面载体上，点样，然后以合适的溶剂展开，达到分离、鉴定和定量的目的。薄层色谱法设备简单、操作方便、分离快速、灵敏度及分辨率高。

薄层色谱可以用来分离混合物，鉴定和精制化合物，跟踪化学反应进程及作为探索柱色谱条件的先导。它分离速度快，分离效率高，需要样品少；特别适用于挥发性小、高温易发生变化、不宜用气相色谱分析的化合物。

(1) 薄层板的制备　通常采用湿法制板，有平铺法或涂铺法两种方法。平铺法是在待铺玻璃板两边用玻璃作边框，将调好的吸附剂倒在玻璃板上面刮平、去掉边框即成。涂铺法是用涂铺器进行铺层的方法。湿法制板中常用的黏合剂有煅石膏（10%～15%）和羧甲基纤维素钠（CMCNa）水溶液（0.2%～1%），可选择加入其中的一种或两种。铺成的薄层板在室温下自然干燥后，可以直接使用。如果吸附力太弱，可在105～120℃下活化后使用。

(2) 常用的薄层色谱展开剂类型　①电子授受体溶剂苯、甲苯、乙酸乙酯、丙酮等；②质子给予体溶剂异丙醇、正丁醇、甲醇及无水乙醇等；③强质子给予体溶剂氯仿、冰醋酸、甲酸及水等；④质子受体溶剂三乙胺、乙醚等；⑤偶极作用溶剂二氯甲烷；⑥惰性溶剂（非极性溶剂）环己烷、正己烷等。在选择混合溶剂时，可在保持溶剂总极性不变的前提下，通过改变其中的某一组分获得最佳选择性。

(3) 上样与展开　样品浓度以样品能均匀分布于吸附剂表面而不析出沉淀为宜，通常为5%～10%。一般采用带状点样，样品带应尽可能窄，以获得更好的分离效果。另外，也可采用多次展开可以提高PTLC的分离效果，即在PTLC一次展开结束后，先将板干燥，再放入容器内展开。根据色带的R_f值，可多次重复上述操作。

(4) 样品带显色与收集　有颜色的化合物，可直接观察斑点；能产生荧光的物质可在紫外光下观察。

其中，常用的通用型薄层色谱显色剂的配制及显色方法如下：
① 硫酸：常用硫酸乙醇（1∶1）溶液喷后于110℃烤15min，不同有机化合物显不同颜色。
② 0.05%高锰酸钾溶液：易还原化合物在淡红背景上显黄色。
③ 酸性重铬酸钾试剂：5%重铬酸钾浓硫酸溶液，必要时于150℃烤薄层。
④ 5%磷钼酸乙醇溶液喷后于120℃烘烤，还原性化合物显蓝色，再用氨气熏，则背景变为无色。

在确定谱带位置后，用铅笔或针尖连接两边，勾画出薄层上的谱带分布情况，再用刮刀或与真空收集器相连的管形刮离器将该谱带从板上刮下，最后以极性尽可能低的溶剂将化合物从吸附剂中洗脱下来（通常1g吸附剂约使用5mL溶剂）。

柱色谱又叫柱层析，是用色谱柱分离化合物的方法。将吸附剂装在色谱柱内制成色谱柱，液体从柱顶加入，其中的组分被吸附剂吸附，加入溶剂（洗脱剂）洗脱，吸附力强的组分移动慢，吸附力弱的组分移动快，从而实现分离。

常规柱色谱法是靠重力驱动流动相流经固定相的一种分离方法，分离时可将样品溶解在少量初始洗脱溶剂中，加到固定相顶端。当待分离样品在洗脱剂中溶解度不佳时，可采用固体上样法，即先将样品溶在一定溶剂中，然后加入2~5倍量的固定相（或硅藻土），将该混合物在低温下用旋转蒸发仪蒸干或自然晾干，然后把所得的粉末加到色谱柱的上部。在洗脱之前，可在样品上覆盖一层沙子或玻璃珠，以防样品界面被破坏。常规柱色谱法通常用于粗提物的制备或 Rf 值差别很大的混合物的分离。采用梯度洗脱可以提高常规柱色谱法的分辨能力。常用溶剂极性大小次序为：石油醚＜二硫化碳＜四氯化碳＜三氯乙烯＜苯＜二氯甲烷＜氯仿＜乙醚＜乙酸乙酯＜乙酸甲酯＜丙酮＜正丙醇＜甲醇＜水。

2.2.7　离子交换分离

离子交换分离是利用离子交换剂与溶液中欲分离的离子之间发生交换反应而实现分离的方法，是在固液两相中进行的单元操作。

离子交换过程的特点：（1）选择性高，主要去除离子化的物质，并进行等物质的量交换；（2）去除效率高；（3）适用性强，应用范围广（无机、有机及高纯物的制备）；（4）交换剂可反复使用；（5）操作简便，分离容易。

离子交换反应一般为可逆反应，进行的方向主要取决于树脂相和溶液相中各种离子的相对浓度、即离子浓度差作为推动力，推动着交换反应的进行。当反应进行到一定程度时，就达到了离子交换平衡状态。此时，每一种物质在树脂相中的化学势与在溶液相中的化学势相等。

影响交换速率的因素有很多，如：（1）树脂的粒度：粒度小且均匀的树脂交换速率高；（2）树脂的交联度：交联度越大，树脂的溶胀性越差，影响了离子在树脂内部的扩散；（3）温度：提高温度利于交换；（4）溶液浓度：提高浓度有利于增加膜扩散速率；（5）搅拌强度：适当提高搅拌速度可提高扩散速率；（6）待交换离子的性质：价态和半径的大小。

2.2.8　吸附分离

吸附是利用吸附剂对液体或气体中某一组分具有选择性吸附的能力，使其富集在吸附剂表面的过程。具有一定吸附能力的固体材料成为吸附剂，被吸附的物质成为吸附质。

影响吸附效果的主要因素：

（1）吸附剂的性质　一般来说，比表面积越大，孔隙度越高，吸附容量就越大；颗粒度越小，吸附速度就越快；孔径适当，有利于吸附物向空隙中扩散。要吸附相对分子质量大的物质时，就应该选择孔径大的吸附剂；要吸附相对分子质量小的物质，则需选择比表面积大及孔径较小的吸附剂；极性化合物需选择极性吸附剂；非极性化合物应选择非极性吸附剂。

（2）吸附质的性质　能使表面张力降低的物质，易为表面所吸附，也就是说，固体的表面张力越小，液体被固体吸附得越多；溶质从较易溶解的溶剂中被吸附时，吸附量较少。

（3）温度　吸附一般是放热的，所以只要达到了吸附平衡，升高温度会使吸附量降低。但在低温时，有些吸附过程往往在短时间内达不到平衡，而升高温度会使吸附速度加快，出现吸附量增加。

（4）溶液 pH 值　溶液的 pH 值会影响吸附剂或吸附质的解离情况，进而影响吸附量。一般在等电点附近吸附量最大。各种溶质吸附的最佳 pH 值需通过实验确定。例如，有机酸在酸性下、胺类在碱性下较易为非极性吸附剂所吸附。

（5）吸附质浓度与吸附剂用量　在吸附达到平衡时，吸附质的浓度称为平衡浓度。普遍的规律是：吸附质的平衡浓度愈大，吸附量也愈大。如用活性炭脱色和去热原时，为了避免对有效成分的吸附，往往将料液适当稀释后进行。

2.2.9　膜分离

膜分离兼有分离、浓缩、纯化和精制的功能，按照其开发的年代先后有微孔过滤、透析、电渗析、反渗透、超滤、气体分离和纳滤。

与传统分离技术相比，具有以下特点：

（1）分离效率较高　在按物质颗粒大小分离的领域，以重力为基础的分离技术最小极限是微米，而膜分离可以分离的颗粒大小为纳米级。与扩散过程相比，在蒸馏过程中物质的相对挥发度的比值大都小于10，难分离的混合物有时刚刚大于1，而膜分离的分离系数则要大得多。

（2）分离过程的能耗较低　大多数膜分离过程都不发生相变化，而相变化的潜热很大。另外，很多膜分离过程是在室温附近进行的，被分离物料加热或冷却的消耗很小。

（3）多数膜分离过程的工作温度在室温附近，特别适合热敏物质的处理。

但是，膜分离技术也存在一些不足之处，如膜的强度较差，使用寿命不长，易于被玷污而影响分离效率等。

2.2.10　电化学分离法

根据原子或分子的电性质以及离子的带电性质和行为进行化学分离的方法称为电化学分离法。除电解分离法外，近年来又创立了一些新的电化学分离分析技术，如：电泳分离法、化学修饰电极分离富集法、介质交换伏安法就是近些年来发展起来的高选择性、高灵敏度的分离分析法。它们在富集分离痕量物质、消除性质相近物质的干扰方面，扮演着重要的角色。

与其他化学分离方法相比，电化学分离法的特点是：操作简单，往往可以同时进行多种试样的分离；除了需要消耗一定的电能外，化学试剂的消耗量小，放射性污物也比较少；除自发电沉积和电渗析外，其他电化学方法的分离速度都比较快。尤其是近年来高压电泳的发展，即使对于比较复杂的样品也能进行快速而有效的分离。

例如，电泳是指带电颗粒在电场的作用下向着与其电性相反的方向迁移的现象。电泳技术利用待分离样品中各种分子带电性质以及分子本身大小、形状的差异，使带电分子在电场中产生不同的迁移速度，从而对样品进行分离。

第3章 定量分析基础实验

实验3.1 分析天平的称量练习

实验目的

(1) 了解分析天平的构造和称量原理；
(2) 掌握分析天平的称量程序和使用规则；
(3) 掌握用差减法称量样品的操作。

实验原理

见 1.4.1。

实验用品

仪器：分析天平（万分之一）、台秤（百分之一）、称量瓶。
试剂：石英砂。

实验步骤

(1) 固定称量法（每份称取石英砂 0.5000g） 取一个干净的表面皿，放入电子天平内，待显示准确质量后，按 TARE（清零键），直接用药匙向表面皿中慢慢加样，至天平屏幕上显示 0.5000g。

(2) 差减称量法 称取 0.3~0.4g 石英砂 3 份，所有操作均需垫滤纸条，用来隔离手指和玻璃仪器。

① 取三个干净的小烧杯，分别在分析天平上准确称取质量。
② 取一个干净的称量瓶，加入约 1g 石英砂粉末，在分析天平上准确称取质量。
③ 取出称量瓶，用盖轻轻敲击称量瓶，转移石英砂 0.3~0.4g 于小烧杯中，并准确称出称量瓶和剩余石英砂的质量，计算出称量瓶减少的质量。
④ 准确称出小烧杯加试样的质量，计算出小烧杯增加的质量。
⑤ 以同样方法转移试样 0.3~0.4g 于另外两个小烧杯，再准确称取并计算相应的质量。

数据记录与处理

石英砂的差减法称量练习（表 3.1）

表 3.1 石英砂的差减称量

实验序号	1	2	3
空烧杯的质量/g			

实验序号	1	2	3
称量瓶+样品的质量/g			
倒出样品后称量瓶的质量/g			
倒入样品后烧杯的质量/g			
称量瓶减少的质量/g			
烧杯增加的质量/g			

思考题

(1) 试述电子天平的称量方法。

(2) 分析天平的灵敏度越高,称量的准确度是否就越高?

(3) 差减法称量过程中能不能用小药勺取样?

(4) 在差减法称出样品的过程中,若称量瓶内的试样吸湿,对称量会造成什么误差?若试样倾入坩埚后再吸湿,对称量是否有影响?

电子天平的使用

(5) 用万分之一分析天平准确称取 1g $K_2Cr_2O_7$ 试样,可记几位有效数字?

拓展阅读

一份珍藏的试卷

在上海交通大学至今保存着钱学森先生求学期间的一张试卷。在一次考试中,他获得了满分,可对于一向严谨的他来说,却发现有一处漏掉了一个字母 s,于是他将这份试卷的分数改成了 96 分,并上报给了老师。从这件事中我们看到了钱学森先生严谨的态度、对自己的高要求。钱学森先生是一代又一代学子学习的榜样。

实验 3.2 滴定分析的基本操作练习——酸碱滴定

实验目的

(1) 掌握滴定分析常用仪器的洗涤和使用方法;

(2) 掌握酸碱标准溶液的配制方法、酸碱溶液相互滴定比较;

(3) 熟悉甲基橙、酚酞指示剂的使用和终点的正确判断,初步掌握酸碱指示剂的选择方法。

实验原理

滴定是常用的测定溶液浓度的方法。将标准溶液(已知其准确浓度的溶液)由滴定管加入到待测溶液中去(也可反过来滴加),直至反应达到终点,这种操作称为滴定。

滴定反应的终点是借助指示剂的颜色变化来确定的,强酸滴定强碱一般用甲基橙作指示剂,而强碱滴定强酸一般用酚酞作指示剂。

利用酸碱中和反应的中和滴定很容易测得酸碱的物质的量浓度,根据化学方程式来计算。例如,$NaOH + HCl == NaCl + H_2O$。

实验用品

仪器:分析天平、量筒、烧杯、试剂瓶、酸式滴定管、碱式滴定管、锥形瓶、移液管。

试剂:NaOH (s)、盐酸(A.R.)、酚酞指示剂(0.2%乙醇溶液)、甲基橙指示剂(0.2%)。

实验步骤

(1) 0.1mol/L NaOH 溶液的配制　用天平迅速称取 1g NaOH 固体于 100mL 小烧杯

中,加约 50mL 无 CO_2 的去离子水溶解,然后转移至试剂瓶中,用去离子水稀释至 250mL,摇匀后,用橡胶塞塞紧。贴好标签,写好试剂名称、浓度(空一格,待填写准确浓度)。

(2) 0.1mol/L HCl 溶液的配制　用洁净量筒量取浓 HCl 约 9.0mL(预习中应计算)倒入 250mL 试剂瓶中用去离子水稀释至 250mL,盖上玻璃塞,充分摇匀。贴好标签,备用。

(3) 酸碱溶液的相互滴定操作练习

① 酸式和碱式滴定管的准备。准备好酸式和碱式滴定管各一支。分别用 5～10mL HCl 和 NaOH 溶液润洗酸式和碱式滴定管 2～3 次。再分别装入 HCl 和 NaOH 溶液,排除气泡,调节液面至零刻度或稍下一点的位置,静置 1min 后,记下初读数。

② 以酚酞作指示剂用 NaOH 溶液滴定 HCl。从酸式滴定管中放出 25.00mL HCl 于锥形瓶中,加入 1～2 滴酚酞,在不断摇动下,用 NaOH 溶液滴定,注意控制滴定速度,当滴加的 NaOH 落点处周围红色褪去较慢时,表明临近终点,用洗瓶洗涤锥形瓶内壁,控制 NaOH 溶液一滴一滴地或半滴半滴地滴出,至溶液呈微红色,且半分钟不褪色即为终点,记下读数。又由酸式滴定管放入 1～2mL HCl,再用 NaOH 溶液滴定至终点。如此反复练习滴定、终点判断及读数若干次。

③ 以甲基橙作指示剂用 HCl 溶液滴定 NaOH。从碱式滴定管中放出 25.00mL NaOH 于锥形瓶中,加入 1～2 滴甲基橙,在不断摇动下,用 HCl 溶液滴定至溶液由黄色恰呈橙色为终点。再由碱式滴定管中放入 1～2mL NaOH,继续用 HCl 溶液滴定至终点,如此反复练习滴定、终点判断及读数若干次。

(4) HCl 和 NaOH 溶液体积比 V_{HCl}/V_{NaOH} 的测定　从酸式滴定管以 10mL/min 的流速放出 20mL HCl 于锥形瓶中,加 1～2 滴酚酞,用 NaOH 溶液滴定至溶液呈微红色,且半分钟不褪色即为终点。读取并准确记录 HCl 和 NaOH 的体积,平行测定三次。计算 V_{HCl}/V_{NaOH},要求相对平均偏差不大于 0.3%。

体积比的测定也可以采用甲基橙作指示剂,以 HCl 溶液滴定 NaOH,平行测定三次。如果时间允许,这两个相互滴定均可进行,将所得结果进行比较,并加以讨论。

数据记录与处理

HCl 和 NaOH 溶液体积比 V_{HCl}/V_{NaOH} 的测定(表 3.2)

表 3.2　HCl 和 NaOH 溶液体积比 V_{HCl}/V_{NaOH} 的测定

实验序号	1	2	3
V_{HCl}(初始)/mL			
V_{HCl}(终点)/mL			
V_{HCl}(消耗)/mL			
V_{NaOH}(初始)/mL			
V_{NaOH}(终点)/mL			
V_{NaOH}(消耗)/mL			
V_{HCl}/V_{NaOH}			
V_{HCl}/V_{NaOH} 的平均值			
$|d_i|$			
$d_r/\%$			

思考题

(1) 在实验中,用酚酞作指示剂时,为什么要求 NaOH 溶液滴定至溶液呈微红色,且半分钟不褪色即为终点?

(2) 下列操作是否准确:

① 每次洗涤后的废液从移液管的上口倒出。

氢氧化钠溶液的配制

② 为了加速溶液的流出，用洗耳球把移液管内溶液吹出。
③ 吸取溶液时，移液管末端伸入溶液太多；转移溶液时，任其自然流下。
④ 烧杯只用自来水冲洗干净。
⑤ 滴定过程中活塞漏水。
⑥ 滴定管下端气泡未赶尽。
⑦ 滴定过程中，向烧杯内加少量蒸馏水。
⑧ 滴定管内壁挂有液滴。
（3）NaOH 和 HCl 能否直接配制准确浓度的溶液？为什么？
（4）怎样合理选择指示剂？

实验 3.3　容量仪器的校准

实验目的

（1）了解容量仪器校准的意义和方法；
（2）初步掌握移液管的校准和容量瓶与移液管间相对校准的操作；
（3）掌握滴定管、容量瓶、移液管的使用方法；
（4）进一步熟悉分析天平的称量操作。

实验原理

滴定管、移液管和容量瓶是分析实验中常用的玻璃量器，都具有刻度和标称容量。量器产品都允许有一定的容量误差。在准确度要求较高的分析测试中，对自己使用的一套量器进行校准是完全必要的。

校准的方法有称量法和相对校准法。称量法的原理是：准确称量被校量器中量入和量出的纯水的质量 m，再根据纯水的密度 ρ 计算出被校量器的实际容量。由于玻璃的热胀冷缩，所以在不同温度下，量器的容积也不同。因此，规定使用玻璃量器的标准温度为 20℃。各种量器上标出的刻度和容量，均为在标准温度 20℃ 下量器的标称容量。但是，在实际校准工作中，容器中水的质量是在室温下和空气中称量的。因此，必须考虑如下三方面的影响：

（1）空气浮力使质量改变。
（2）水的密度随温度而改变。
（3）玻璃容器本身的容积随温度而改变。

考虑了上述的影响，可得出 20℃ 容量为 1L 的玻璃容器，在不同温度时所盛水的质量，不同温度下的纯水密度见表 3.3。据此计算量器的校正值十分方便。

表 3.3　不同温度下的纯水密度（ρ_w）

t/℃	ρ_w/(g/mL)	t/℃	ρ_w/(g/mL)	t/℃	ρ_w/(g/mL)
8	0.9886	15	0.9979	22	0.9968
9	0.9985	16	0.9978	23	0.9966
10	0.9984	17	0.9976	24	0.9963
11	0.9983	18	0.9975	25	0.9961
12	0.9982	19	0.9973	26	0.9959
13	0.9981	20	0.9972	27	0.9956
14	0.9980	21	0.9970	28	0.9954

不同温度下1L水（20℃）的质量（在空气中用黄铜砝码称量）。

如某只25mL移液管在25℃放出的纯水的质量为24.921g，密度为0.99617g/mL，计算该移液管在20℃时的实际容积。$V_{20}=24.921/0.99617=25.02(\text{mL})$。

则这只移液管的校正值为25.02mL－25.00mL＝0.02(mL)。

需要特别指出的是，校准不当和使用不当都是产生容积误差的主要原因，其误差甚至可能超过允许或量器本身的误差。因此，在校准时务必正确、仔细地进行操作，尽量减小校准误差。凡是使用校准值的，其允许次数不应少于两次，且两次校准数据的偏差应不超过该量器允许的1/4，并取其平均值作为校准值。

有时，只要求两种容器之间有一定的比例关系，而不需要知道它们各自的准确体积，这时可用容量相对校准法。经常配套使用的移液管和容量瓶，采用相对校准法更为重要。例如，用25mL移液管取蒸馏水于干净且倒立晾干的100mL容量瓶中，到第4次重复操作后，观察瓶颈处水的弯月面下缘是否刚好与刻线上缘相切，若不切，应重新做一记号为标线，以后此移液管和容量瓶配套使用时就用此次校准的标线。

 实验用品

仪器：分析天平、滴定管（50mL）、容量瓶（250mL）、移液管（25mL）、锥形瓶（50mL）、温度计。

试剂：蒸馏水。

 实验步骤

(1) 滴定管的绝对校准（称量法）

① 将已洗净且外表干燥的带磨口玻璃塞的锥形瓶放在分析天平上称量，得空瓶质量m_0，记录至0.001g。

② 将已洗净的滴定管盛满纯水，调至0.00mL刻度处，从滴定管中放出一定体积（记为V_1）的纯水于已称量的锥形瓶中，并塞紧塞子，称出"瓶+水"的质量m_1，两次质量之差（m_1-m_0）即为放出之水的质量$m_水$。用同法称量滴定管从0mL到10.00mL、0mL到15.00mL、0mL到20.00mL、0mL到25.00mL、0mL到30.00mL等刻度间的$m_水$，在实验水温下，用每次$m_水$除以水的密度，即可得到滴定管各部分的实际容量V_2（表3.4）。

重复校准一次，两次相应区间的水质量相差应小于0.02g，求出平均值，并计算校准值$\Delta V=(V_2-V_1)$。以V_1为横坐标，ΔV为纵坐标，绘制滴定管校准曲线。移液管和容量瓶也可用称量法进行校准。

(2) 移液管和容量瓶的相对校准　用25mL移液管取纯水于干净且晾干的250mL容量瓶中，重复操作10次，观察液面的弯月面下缘是否恰好与标线上缘相切，如不相切，则用胶布在瓶颈上另作标记，以后实验中，此移液管和容量瓶配套使用时，应以新标记为准。

 数据记录与处理

表3.4　滴定管的校准

滴定管放出水的容积 V_1/mL	空瓶质量 m_0/g	瓶+水质量 m_1/g	水的质量($m_水$) m_1-m_0/g	真实容量 V_2/mL	校正值 $\Delta V=V_2-V_1$
0.00～10.00					
0.00～15.00					
0.00～20.00					

滴定管放出水的容积 V_1/mL	空瓶质量 m_0/g	瓶+水质量 m_1/g	水的质量($m_水$) m_1-m_0/g	真实容量 V_2/mL	校正值 $\Delta V=V_2-V_1$
0.00~25.00					
0.00~30.00					

注：1. 拿取锥形瓶时，不能直接用手拿，要用纸条（三层以上）套取。

2. 锥形瓶磨口部位不要沾到水。

3. 测量实验水温时，须将温度计插入水中后才读数，读数时温度计球部位仍浸在水中。

思考题

(1) 校正滴定管时，为何锥形瓶和水的质量只需称到 0.001g？

(2) 容量瓶校准时为什么要晾干？在用容量瓶配制标准溶液时是否也要晾干？

(3) 分段校准滴定管时，为什么每次都要从 0.00mL 开始？

(4) 容量仪器校正的主要影响因素有哪些？

(5) 容量仪器为什么要校准？容量分析中如何应用校准值？

拓展阅读

卢嘉锡与小数点

著名化学家卢嘉锡在一次测验中因为小数点算错一位而被区嘉炜教授扣掉了 3/4 的分数，这使他很不服气，然而区教授却严肃地说，"假如设计一座桥梁，小数点错一位可就要出大问题、犯大错误，今天我扣你 3/4 的分数，就是扣你把小数点放错了地方。"在理解了教授的一片苦心之后，卢嘉锡把区教授的教诲牢记在心，此后不管干什么，都十分严谨。

实验 3.4　酸碱溶液的配制与标定

实验目的

(1) 掌握酸、碱标准溶液的配制和标定方法；

(2) 练习移液管的正确使用；

(3) 进一步熟练掌握滴定操作；

(4) 准确判断酸碱指示剂颜色的变化。

实验原理

标准溶液是指已知准确浓度的溶液，其配制方法通常有两种：直接法和标定法。

(1) 直接法　准确称取一定质量的物质，溶解后定量转移到容量瓶中，并稀释至刻度，摇匀。根据称取物质的质量和容量瓶的体积即可算出该标准溶液的准确浓度。适用此方法配制标准溶液的物质必须是基准物质。

(2) 标定法　大多数物质的标准溶液不宜用直接法配制，可选用标定法：即先近似配成所需浓度的溶液，再用基准物质或已知准确浓度的标准溶液标定其准确浓度。HCl 和 NaOH 标准溶液在酸碱滴定中最常用，但由于浓盐酸易挥发，NaOH 固体易吸收空气中 CO_2 和水蒸气，故只能选用标定法来配制。其浓度一般在 0.01~1mol/L 之间，通常配制 0.1mol/L 的溶液。

(3) 标定碱的基准物质　常用标定碱标准溶液的基准物质有邻苯二甲酸氢钾、草酸等。

① 邻苯二甲酸氢钾。它易制得纯品，在空气中不吸水，容易保存，摩尔质量较大，是一种较好的基准物质，标定反应如下：

$$\text{邻-COOH/COOK} + \text{NaOH} \longrightarrow \text{邻-COONa/COOK} + H_2O \tag{3.1}$$

化学计量点时，溶液呈弱碱性（pH=9.20），可选用酚酞作指示剂。

邻苯二甲酸氢钾通常在 105～110℃下干燥 2h，干燥温度过高，则脱水成为邻苯二甲酸酐。

② 草酸（$H_2C_2O_4 \cdot 2H_2O$）。它在相对湿度为 5%～95%时不会风化失水，故将其保存在磨口玻璃瓶中即可，草酸固体状态比较稳定，但溶液状态的稳定性较差，空气能使草酸溶液慢慢氧化，光和 Mn^{2+} 能催化其氧化。因此，草酸溶液应置于暗处存放。

标定反应如下：

$$2NaOH + H_2C_2O_4 =\!=\!= Na_2C_2O_4 + 2H_2O \tag{3.2}$$

反应产物为 $Na_2C_2O_4$，在水溶液中显碱性，可选用酚酞作指示剂。

(4) 标定酸的基准物质　常用于标定酸的基准物质有无水碳酸钠和硼砂。其浓度还可通过与已知准确浓度的 NaOH 标准溶液进行标定。

① 无水碳酸钠。它易吸收空气中的水分，先将其置于 270～300℃干燥 1h，然后保存于干燥器中备用。标定反应如下：

$$Na_2CO_3 + 2HCl =\!=\!= 2NaCl + H_2O + CO_2 \uparrow \tag{3.3}$$

到达计量点时，为 H_2CO_3 饱和溶液，pH 为 3.9，以甲基橙作指示剂应滴至溶液呈橙色为终点。为使 H_2CO_3 的饱和部分不断地分解逸出，临近终点时应将溶液剧烈摇动或加热。

② 硼砂（$Na_2B_4O_7 \cdot 10H_2O$）。它易于制得纯品，吸湿性小，摩尔质量大但由于含有结晶水，当空气中相对湿度小于 39%时，有明显的风化而失水的现象，常保存在相对湿度为 60%的恒温器（下置饱和的蔗糖溶液）中。其标定反应为：

$$Na_2B_4O_7 + 2HCl + 5H_2O =\!=\!= 2NaCl + 4H_3BO_3 \tag{3.4}$$

产物为 H_3BO_3，其水溶液 pH 约为 5.1，可用甲基红作指示剂。

③ 与已知准确浓度的 NaOH 标准溶液进行比较标定。0.1mol/L HCl 和 0.1mol/L NaOH 溶液的比较标定是强酸强碱的滴定，化学计量点时 pH=7.00，滴定突跃范围比较大（pH=4.30～9.70）。因此，凡是变色范围全部或部分落在突跃范围内的指示剂，如甲基橙、甲基红、酚酞、甲基红-溴甲酚绿混合指示剂，都可用来指示终点。比较滴定中可以用酸溶液滴定碱溶液，也可用碱溶液滴定酸溶液。若用 HCl 溶液滴定 NaOH 溶液，选用甲基橙为指示剂。

实验用品

仪器：分析天平、量筒、滴定管、锥形瓶、烧杯。

试剂：盐酸（0.1mol/L）、NaOH 溶液（0.1mol/L）、酚酞指示剂（0.2%乙醇溶液）、甲基橙指示剂（0.1%）、邻苯二甲酸氢钾（基准试剂）、无水碳酸钠（基准试剂）。

实验步骤

(1) 0.1mol/L HCl 溶液浓度的标定

① 方法一：以无水碳酸钠为基准物质

用称量瓶在分析天平上准确称量（　）至（　）g 无水 Na_2CO_3（摩尔质量 105.99g/mol）三份，分别置于 250mL 锥形瓶中（称量前将锥形瓶编号）。用量筒加入 20～30mL 水溶解，加入 1～2 滴甲基橙指示剂，用预标定的 HCl 溶液滴定至溶液由黄色变橙色即为终点。在接近终点时，为除去溶液中溶解的 CO_2，应剧烈振动，或加热驱赶 CO_2，冷却后再滴定至终点。根据基准物质 Na_2CO_3 质量和消耗的 HCl 体积，计算 HCl 溶液的浓度 c_{HCl} 和相对平均偏差 $\bar{d}_r$。

② 方法二：以硼砂为基准物质

准确称取（　）至（　）g $Na_2B_4O_7·10H_2O$（摩尔质量 381.37g/mol）三份，分别置于 250mL 锥形瓶中，加入 20~30mL 水溶解，加入 1~2 滴甲基橙指示剂，用预标定的 HCl 溶液滴定至溶液由黄色变橙色即为终点。根据硼砂的质量和消耗的 HCl 体积，计算 HCl 溶液的浓度 c_{HCl} 和相对平均偏差。

（2）0.1mol/L NaOH 溶液浓度的标定

① 方法一：以邻苯二甲酸氢钾为基准物质

准确称取（　）至（　）g $KHC_8H_4O_4$（摩尔质量 204.23g/mol）三份，分别置于 250mL 锥形瓶中，加入 20~30mL 水溶解，加入 2~3 滴酚酞指示剂，用预标定的 NaOH 溶液滴定至溶液由无色变成微红色在 30s 不褪色即为终点。根据称取的基准物质的质量和消耗 NaOH 体积，计算 NaOH 溶液的浓度 c_{NaOH} 和相对平均偏差 $\bar{d}_r$。

② 方法二：以草酸为基准物质

准确称取（　）至（　）g $H_2C_2O_4·2H_2O$（摩尔质量 126.07g/mol）三份，分别置于 250mL 锥形瓶中，加入 20~30mL 水溶解，加入 2~3 滴酚酞指示剂，用预标定的 NaOH 溶液滴定至溶液由无色变微红色在 30s 不褪色即为终点。根据称取的基准物质的质量和消耗的 NaOH 体积，计算 NaOH 溶液的浓度 c_{NaOH} 和相对平均偏差 $\bar{d}_r$。

数据记录与处理

（1）0.1mol/L HCl 溶液的标定（表 3.5）

表 3.5　0.1mol/L HCl 溶液的标定

实验序号	1	2	3
m(基准物质)/g			
V_{HCl}(初始)/mL			
V_{HCl}(终点)/mL			
V_{HCl}(消耗)/mL			
c_{HCl}/(mol/L)			
$\bar{c}_{HCl}$/(mol/L)			
$|d_i|$			
$\bar{d}_r/\%$			

（2）0.1mol/L NaOH 溶液的标定（表 3.6）

表 3.6　0.1mol/L NaOH 溶液的标定

实验序号	1	2	3
m(基准物质)/g			
V_{NaOH}(初始)/mL			
V_{NaOH}(终点)/mL			
V_{NaOH}(消耗)/mL			
c_{NaOH}/(mol/L)			
$\bar{c}_{NaOH}$/(mol/L)			
$|d_i|$			
$\bar{d}_r/\%$			

思考题

（1）如何计算称取基准物邻苯二甲酸氢钾或 Na_2CO_3 的质量范围？称得太多或太少对标定有何影响？

（2）溶解基准物质时加入 20～30mL 水，是用量筒量取，还是用移液管移取？为什么？

（3）如果基准物未烘干，将使标准溶液浓度的标定结果偏高还是偏低？

（4）用 NaOH 标准溶液标定 HCl 溶液浓度时，以酚酞作指示剂，若 NaOH 溶液因贮存不当吸收了 CO_2，问对测定结果有何影响？

（5）HCl 和 NaOH 溶液能直接配制准确浓度吗？为什么？

实验 3.5　食用醋中总酸度的测定

实验目的

（1）熟练掌握滴定管、容量瓶、移液管的使用方法和滴定操作。
（2）掌握氢氧化钠标准溶液的配制和标定方法。
（3）了解强碱滴定弱酸的反应原理及指示剂的选择。
（4）学会食醋中总酸度的测定方法。

实验原理

食醋中的主要成分是醋酸（有机弱酸，$K_a = 1.8 \times 10^{-5}$）。此外，还含有少量的其他弱酸，如乳酸等，与 NaOH 反应产物为 NaAc：

$$HAc + NaOH = NaAc + H_2O \tag{3.5}$$

化学计量点时 pH≈8.7，滴定突跃在碱性范围内（如 0.1mol/L NaOH 滴定 0.1mol/L HAc 突跃范围为 pH：7.74～9.70）。若使用在酸性范围内变色的指示剂，如甲基橙，将引起很大的滴定误差（该反应化学计量点时溶液呈弱碱性，酸性范围内变色的指示剂变色时，溶液呈弱酸性，则滴定不完全）。因此，在此应选择在碱性范围内变色的指示剂酚酞（8.0～9.6）。（指示剂的选择主要以滴定突跃范围为依据，指示剂的变色范围应全部或一部分在滴定突跃范围内，则终点误差小于 0.1%）。

因此，可选用酚酞作指示剂，利用 NaOH 标准溶液测定 HAc 含量。食醋中总酸度用 HAc 的含量来表示。

实验用品

仪器：锥形瓶、量筒、分析天平（万分之一）、碱式滴定管、移液管、容量瓶。
试剂：邻苯二甲酸氢钾（G.R.）、NaOH 固体（A.R.）、酚酞指示剂。

实验步骤

（1）0.1mol/L NaOH 溶液的配制与标定（参照实验 3.4）。

（2）用标定好的 NaOH 溶液润洗碱式滴定管，然后装入 NaOH 溶液。准确吸取食用醋试样 10.00mL 置于 250mL 容量瓶中，用蒸馏水稀释至刻度，摇匀。

（3）用移液管吸取 25.00mL 上述稀释后的试液于 250mL 锥形瓶中。加入 25mL 新煮沸并冷却的蒸馏水，加入 2 滴酚酞指示剂。用上述已标定的 NaOH 标准溶液滴至溶液呈微红色且 30s 不褪色即为终点，平行测定 3 次。根据 NaOH 标准溶液的用量，计算食用醋的总酸度。

（4）用甲基橙作指示剂，用上法滴定。计算结果，比较两种指示剂结果之间的差别。

注：① 食醋中醋酸的浓度较大，且颜色较深，故必须稀释后再进行滴定。

② 测定醋酸含量时，所用的蒸馏水不能含有二氧化碳，否则会溶于水中生成碳酸，将同时被滴定。

数据记录与处理

（1）0.1mol/L NaOH 的标定（表 3.7）

表 3.7 0.1mol/L NaOH 溶液的标定

实验序号	1	2	3
$m_{邻苯二甲酸氢钾}$/g			
V_{NaOH}(初始)/mL			
V_{NaOH}(终点)/mL			
V_{NaOH}(消耗)/mL			
c_{NaOH}/(mol/L)			
$\bar{c}_{NaOH}$/(mol/L)			
$|d_i|$			
$\bar{d}_r$/%			

（2）食用醋中总酸度的测定（表 3.8）

表 3.8 食用醋中总酸度的测定

实验序号	1	2	3
$V_{醋}$/mL			
V_{NaOH}(初始)/mL			
V_{NaOH}(终点)/mL			
V_{NaOH}(消耗)/mL			
c_{HAc}/(g/L)			
$\bar{c}_{HAc}$/(g/L)			
$|d_i|$			
$\bar{d}_r$/%			

思考题

（1）滴定醋酸时为什么要用酚酞作指示剂？

（2）该方法的测定原理是什么？

（3）酚酞指示剂使溶剂变红后，在空气中放置一段时间又变为无色，原因是什么？

（4）标准溶液的浓度应保留几位有效数字？

 拓展阅读

古代诗词与醋文化

"醋"文化在我国由来已久，众多优秀传统文学作品中可见对醋的描写，承载了中华民族世代相传的底蕴与智慧。宋代吴自牧在《梦粱录》中提到："盖人家每日不可阙者，柴米油盐酱醋茶"。自宋朝之后，醋逐渐成为日常生活的必需品，被视为"早晨起来七件事"（元·武汉臣·《玉壶春》）之一，如"芜荑酱醋吃煮葵"（唐·颜真卿·《七言滑语联句》）、"持螯更喜桂阴凉，泼醋擂姜兴欲狂"（清·曹雪芹·《螃蟹咏》）。

实验 3.6　铵盐中氮含量的分析——甲醛法

实验目的

（1）掌握用甲醛法测定铵盐中氮的原理和方法；
（2）熟练滴定操作和滴定终点的判断。

实验原理

铵盐是常见的无机化肥，是强酸弱碱盐，可用酸碱滴定法测定其含量，但由于 NH_4^+ 的酸性太弱（$K_a=5.6\times10^{-10}$），直接用 NaOH 标准溶液滴定有困难。生产和实验室中广泛采用甲醛法测定铵盐中的含氮量。

甲醛法是基于甲醛与一定量铵盐作用，生成相当量的酸（H^+）和六亚甲基四铵盐（$K_a=7.1\times10^{-6}$），反应如下：

$$4NH_4^+ + 6HCHO = (CH_2)_6N_4H^+ + 6H_2O + 3H^+ \tag{3.6}$$

由反应式可知，4mol NH_4^+ 与甲醛作用定量地生成 3mol 的 H^+ 和 1mol 的质子化的 $(CH_2)_6N_4H^+$（$K_a=7.1\times10^{-6}$），即 1mol 的 NH_4^+ 转换为较强的 1mol 酸，并与 1mol 的 NaOH 完全反应。

由于在化学计量点时，溶液中存在的六亚甲基四胺是一种很弱的碱（$K_a=1.5\times10^{-9}$），溶液的 pH 约为 8.7，故选用酚酞作指示剂。铵盐与甲醛的反应在室温下进行较慢，加入甲醛后，需放置几分钟，使反应完全。甲醛中常含有少量甲酸，使用前必须先以酚酞为指示剂，用 NaOH 溶液中和，否则会使测定结果偏高。若铵盐中含有游离酸，需中和除去，即以甲基红为指示剂，用 NaOH 溶液滴定至橙色，加以扣除。

实验用品

仪器：滴定管、容量瓶、移液管、锥形瓶、称量瓶、量筒、分析天平。

试剂：0.1mol/L NaOH，0.1%甲基红乙醇溶液，0.2%酚酞乙醇溶液，20%甲醛溶液：甲醛（40%）与水等体积混合，$(NH_4)_2SO_4$ 试样。

实验步骤

（1）NaOH 溶液浓度的标定（同实验 3.5，用邻苯二甲酸氢钾标定）。

（2）甲醛溶液的处理　甲醛中常含有微量甲酸是由于甲醛受空气氧化所致，应除去，否则产生正误差。处理方法如下：取原装甲醛（40%）的上层清液于烧杯中，用水稀释一倍，加入 1~2 滴 0.2%酚酞指示剂，用 0.1mol/L NaOH 溶液中和至甲醛溶液呈淡红色。

（3）试样中含氮量的测定　准确称取（　）至（　）g $(NH_4)_2SO_4$ 于烧杯中，用适量蒸馏水溶解，然后定量移至 250mL 容量瓶中，最后用蒸馏水稀释至刻度，摇匀。用移液管移取试液 25mL 于锥形瓶中，加 1~2 滴甲基红指示剂，溶液呈红色。用 0.1mol/L NaOH 溶液中和至红色转为橙色，然后加入 8mL 已中和的 1:1 甲醛溶液，再加入 1~2 滴酚酞指示剂摇匀，静置 1min 后，用 0.1mol/L NaOH 标准溶液滴定至溶液淡红色持续半分钟不褪，即为终点。记录结果，平行测定 3 次。根据 NaOH 标准溶液的浓度和滴定消耗的体积，计算试样中氮的含量。

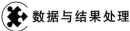 **数据与结果处理**

（1）0.1mol/L NaOH 溶液的标定（表 3.9）

表 3.9　0.1mol/L NaOH 溶液的标定

实验序号	1	2	3
$m_{邻苯二甲酸氢钾}$/g			
V_{NaOH}（初始）/mL			
V_{NaOH}（终点）/mL			
V_{NaOH}（消耗）/mL			
c_{NaOH}/(mol/L)			
$\bar{c}_{NaOH}$/(mol/L)			
$\lvert d_i \rvert$			
$\bar{d}_r$/%			

（2）铵盐中氮含量的测定（表 3.10）

表 3.10　铵盐中氮含量的测定

实验序号	1	2	3
$V_{铵盐}$/mL			
V_{NaOH}（初始）/mL			
V_{NaOH}（终点）/mL			
V_{NaOH}（消耗）/mL			
X_N/%			
$\bar{X}_N$/%			
$\lvert d_i \rvert$			
$\bar{d}_r$/%			

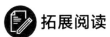 **思考题**

（1）铵盐中氮的测定为何不采用 NaOH 直接滴定法？

（2）为什么中和甲醛试剂中的甲酸以酚酞作指示剂；而中和铵盐试样中的游离酸则以甲基红作指示剂？

（3）NH_4HCO_3 中含氮量的测定，能否用甲醛法？

（4）$(NH_4)_2SO_4$ 试样溶于水后，能否用 NaOH 溶液直接测定氮含量？为什么？

（5）尿素 $CO(NH_2)_2$ 中含氮量的测定，先加 H_2SO_4 加热消化，全部变为 $(NH_4)_2SO_4$ 后，按甲醛法同样测定，试写出含氮量的计算公式。

拓展阅读

"三聚氰胺"毒奶粉事件

奶粉中蛋白质含量是其质量好坏的评价标准，氮含量可间接指示蛋白质含量。三聚氰胺中氮含量比较高，把它加入奶粉中，可提高检测结果中牛奶含氮量。目前我国已制定三聚氰胺含量检测的国家标准，严格管控其在食品中的含量。

"三聚氰胺"事件其实是一个伦理道德缺失的问题，作为分析检测人员我们应该具有高度的责任感和使命感，并树立正确的职业精神和职业道德。

实验 3.7 混合碱的分析——双指示剂法

实验目的

(1) 了解酸碱滴定法的应用；
(2) 掌握双指示剂法测定混合碱的原理和组成成分的判别及计算方法。

实验原理

混合碱是 Na_2CO_3 与 NaOH 或 Na_2CO_3 与 $NaHCO_3$ 的混合物。欲测定同一份试样中各组分的含量，可用 HCl 标准溶液滴定，选用两种不同指示剂分别指示第一、第二化学计量点的到达。根据到达两个化学计量点时消耗的 HCl 标准溶液的体积，便可判别试样的组成及计算各组分含量。

在混合碱试样中加入酚酞指示剂，此时，溶液呈红色，用 HCl 标准溶液滴定到溶液由红色恰好变为浅粉色（参考老师提供的对照色瓶），则试液中所含 NaOH 完全被中和，Na_2CO_3 则被中和到 $NaHCO_3$，若溶液中含 $NaHCO_3$，则未被滴定，滴定用去的 HCl 标准溶液的体积为 V_1（mL），反应如下：

$$NaOH + HCl = NaCl + H_2O \tag{3.7}$$
$$Na_2CO_3 + HCl = NaCl + NaHCO_3 \tag{3.8}$$

再加入甲基橙指示剂，继续用 HCl 标准溶液滴定到样品由黄色变为橙色。此时，样品试液中的 $NaHCO_3$（或者是 Na_2CO_3 第一步被中和生成的，或者是最初的试样中含有的）被中和反应生成 CO_2 和 H_2O。此时，又消耗的 HCl 标准溶液（即第一计量点到第二计量点消耗）的体积为 V_2（mL）。反应如下：

$$NaHCO_3 + HCl = NaCl + CO_2 \uparrow + H_2O \tag{3.9}$$

当 $V_1 > V_2$ 时，试样为 Na_2CO_3 与 NaOH 的混合物，中和 Na_2CO_3 所需 HCl 是分两批加入的，两次用量应该相等，即滴定 Na_2CO_3 所消耗的 HCl 的体积为 $2V_2$，而中和 NaOH 所消耗的 HCl 的体积为 (V_1-V_2)，计算 NaOH 和 Na_2CO_3 的含量公式应为：

$$c_{NaOH} = \frac{(V_1-V_2)c_{HCl}M_{NaOH}}{1000m_s} \times 100\% \tag{3.10}$$

$$c_{Na_2CO_3} = \frac{V_2 c_{HCl} M_{Na_2CO_3}}{1000m_s} \times 100\% \tag{3.11}$$

当 $V_1 < V_2$ 时，试样为 Na_2CO_3 与 $NaHCO_3$ 的混合物，此时，V_1 为第一步中和 Na_2CO_3 时所消耗的 HCl 的体积，故两步中和 Na_2CO_3 所消耗的 HCl 的总体积为 $2V_1$，中和 $NaHCO_3$ 消耗的 HCl 的体积为 (V_2-V_1)，计算 $NaHCO_3$ 和 Na_2CO_3 含量的公式为：

$$c_{NaHCO_3} = \frac{(V_2-V_1) \times c_{HCl} M_{NaHCO_3}}{1000m_s} \times 100\% \tag{3.12}$$

$$c_{Na_2CO_3} = \frac{V_1 c_{HCl} M_{Na_2CO_3}}{1000m_s} \times 100\% \tag{3.13}$$

式中，m_s 为混合碱试样质量（g）。

实验用品

仪器：分析天平、滴定管、锥形瓶、烧杯、称量瓶。
试剂：0.1mol/L HCl 标准溶液、0.2%酚酞乙醇溶液、0.2%甲基橙水溶液、NaOH 和

Na₂CO₃ 或 Na₂CO₃ 与 NaHCO₃ 混合试样、基准物质 Na₂CO₃、混合碱试样；pH=8.3 的参比溶液：0.05mol/L 的 $Na_2B_4O_7$ 溶液和 0.1mol/L HCl 标准溶液，以 6∶4 比例配成缓冲溶液，加入酚酞指示剂，置于磨口瓶中，溶液为浅红色。

实验步骤

（1）0.1mol/L HCl 标准溶液（基准 Na₂CO₃ 标定，参照实验 3.4）。

（2）混合碱的测定（双指示剂法）　准确称取 2.0000g 试样于 250mL 烧杯中，加水使之溶解后，定量转入 250mL 容量瓶中，用水稀释至刻度，充分摇匀。

平行移取 3 份上述溶液于锥形瓶中，每份 25.00mL。加入 3 滴酚酞指示剂，用 HCl 溶液滴定至溶液呈浅粉色。近终点时，以参比为对照，缓慢滴加 HCl 溶液，每加一滴，均需充分摇动，慢慢滴到溶液颜色与参比溶液颜色一样为止。记录所消耗 HCl 标准液的体积 V_1（mL）。

再在上述溶液中加入 2 滴甲基橙指示剂，继续用 HCl 溶液滴定至溶液由黄色变为橙色。接近终点时应剧烈摇动试液，以免形成 CO_2 过饱和溶液而使终点提前。记录消耗 HCl 体积 V_2（mL）。

平行测定 3 次，按消耗 HCl 体积 V_1 和 V_2 判断该试样的组分，计算 Na₂CO₃ 和 NaHCO₃ 或 Na₂CO₃ 与 NaOH 的含量（g/L），并计算以 Na₂O（g/L）表示的试样总碱量。

数据记录与处理

（1）0.1mol/L HCl 溶液的标定（表 3.11）

表 3.11　0.1mol/L HCl 溶液的标定

实验序号	1	2	3
$m_{基准物质}$/g			
V_{HCl}(初始)/mL			
V_{HCl}(终点)/mL			
V_{HCl}(消耗)/mL			
c_{HCl}/(mol/L)			
$\bar{c}_{HCl}$/(mol/L)			
$|d_i|$			
$\bar{d}_r$/%			

（2）混合碱的分析（表 3.12）

表 3.12　混合碱的分析

实验序号	1	2	3
$m_{混合碱}$/g			
V_1(初始)/mL			
V_1(终点)/mL			
V_1(消耗)/mL			
V_2(初始)/mL			
V_2(终点)/mL			
V_2(消耗)/mL			
判断混合碱组成	A　　　B		
w_A/%			
$\bar{w}_A$/%			
$\bar{d}_r$/%			
w_B/%			

续表

实验序号	1	2	3
$\bar{w}_B/\%$			
$\bar{d}_r/\%$			
$w_{Na_2O}/\%$			
$\bar{w}_{Na_2O}/\%$			
$\bar{d}_r/\%$			

思考题

（1）在同一份溶液中，采用双指示剂法测定混合碱，试判断下列五种情况下，混合碱的成分是什么？

①$V_1=0$；②$V_2=0$；③$V_1>V_2$；④$V_1<V_2$；⑤$V_1=V_2$。

（2）计算两个化学计量点的pH值。

（3）在滴定混合碱中，第一化学计量点之前，若滴定速度过快，摇动锥形瓶不够，会对测定造成什么影响？为什么？

（4）简述双指示剂法的优缺点。

拓展阅读

侯氏制碱法

侯德榜国外求学八年，始终不忘青年时代"工业救国"的初心，获得博士学位后毅然回国。他应爱国实业家范旭东的邀请，出任永利碱业公司的总工程师，开启了他半个多世纪的科学救国和实业救国的人生历程。侯德榜经历了由不合格的红色纯碱到白色纯碱、再到战争期间的工厂被迫搬迁、原料改变导致工艺成本过高、国外的技术封锁等困难。他毫不气馁，不断摸索，终于成功开发了"联合制碱法"新工艺，即"侯氏制碱法"，可使盐的利用率提高到96%，同时将污染环境的废物转化为化肥——氯化铵，开创了世界制碱工业的新纪元。

实验3.8　EDTA溶液的配制与标定

实验目的

（1）掌握配位滴定的原理及特点；

（2）掌握标定EDTA的基本原理及方法；

（3）熟悉金属指示剂的使用及终点判断。

实验原理

乙二胺四乙酸（简称EDTA）难溶于水，其标准溶液常用其二钠盐（$Na_2H_2Y \cdot 2H_2O$，分子量$M=392.28$g/mol）采用间接法配制。标定EDTA溶液的基准物质有Zn、ZnO、$CaCO_3$、Cu、$MgSO_4 \cdot 7H_2O$、Hg、Ni、Pb等。

用于测定Pb^{2+}（或Zn^{2+}）、Bi^{3+}含量的EDTA溶液可用ZnO或金属Zn作基准物进行标定。以二甲酚橙为指示剂，在pH=5～6的溶液中，二甲酚橙指示剂（XO）本身显黄色，而与Zn^{2+}的络合物显紫红色。EDTA能与Zn^{2+}形成更稳定的络合物，当使用EDTA溶液滴定至近终点时，EDTA会把与二甲酚橙络合的Zn^{2+}置换出来，而使二甲酚橙游离。因此，溶液由紫红色变为黄色。其变色原理可表达如下：

$$XO(黄色) + Zn^{2+} \rightleftharpoons ZnXO(紫红色) \tag{3.14}$$

$$ZnXO(紫红色) + EDTA \rightleftharpoons Zn\text{-}EDTA(无色) + XO(黄色) \tag{3.15}$$

EDTA 溶液若用于测定石灰石或白云石中的 CaO、MgO 的含量及测定水的硬度，最好选用 $CaCO_3$ 为基准物质进行标定。这样，基准物质和被测组分含有相同的成分，使得滴定条件一致，可以减小系统误差。将 $CaCO_3$ 用 HCl 溶解后，制成 Ca 标准溶液，调节酸度至 $pH \geqslant 12$，使用钙指示剂，用 EDTA 滴至溶液由紫红色变为纯蓝色。

实验用品

仪器：分析天平、滴定管、锥形瓶、容量瓶、烧杯、称量瓶。

试剂：乙二胺四乙酸二钠（$Na_2H_2Y \cdot 2H_2O$）、金属 Zn（A.R.）、二甲酚橙指示剂（0.2%水溶液）、1:1 盐酸、1:1 氨水、基准试剂 $CaCO_3$、铬黑 T 指示剂（1%乙醇溶液）；

氨性缓冲溶液（pH=10）：将 20g NH_4Cl 溶于少量水中，加入 100mL 氨水，用水稀释至 1L；

K-B 指示剂：0.2g 酸性铬兰 K 和 0.4g 萘酚绿 B 混溶于 100mL 水中；

20%六亚甲基四胺（pH=5.5）：将 20g 试剂溶于水，加 4mL 浓 HCl，稀释至 100mL。

实验步骤

（1）0.02mol/L EDTA 标准溶液的配制　称取 2g 乙二胺四乙酸二钠盐于 250mL 烧杯中，加入 10mL 水，加热溶解后，稀释至 250mL，存于试剂瓶中或聚乙烯塑料瓶中。

（2）标准锌溶液的配制　准确称取（　）至（　）g 纯锌于 250mL 烧杯中，盖上表面皿。从烧杯嘴中滴加 5~10mL 1:1 HCl，放置至 Zn 全部溶解后，定量转移到 250mL 容量瓶中，用水稀释至刻度，摇匀，计算其准确浓度 $c_{Zn^{2+}}$。

（3）$CaCO_3$ 标准溶液的配制　准确称取 120℃ 干燥过的 $CaCO_3$（　）至（　）g 于小烧杯中，加少量水润湿盖上表面皿，从烧杯嘴滴加 1:1 的 HCl 10mL，待 $CaCO_3$ 完全溶解后，定量转移至 250mL 容量瓶中，稀释定容，摇匀，计算其准确浓度 $c_{Ca^{2+}}$。

（4）EDTA 溶液浓度的标定

① 以锌标准溶液标定 EDTA 溶液

a. 以铬黑 T 为指示剂。用移液管吸取锌标准溶液 25.00mL 于 250mL 锥形瓶中，滴加 1:1 氨水至开始出现白色沉淀，加 10mL 氨性缓冲溶液（pH=10），加水 20mL，加 2~3 滴铬黑 T 指示剂，用 EDTA 标准溶液滴定至溶液由紫红色变为纯蓝色，即达终点。平行测 3 份，其滴定体积之差不超过 0.04mL。根据消耗的 EDTA 标准溶液的体积，计算其浓度 c_{EDTA}。

b. 以二甲酚橙为指示剂。用移液管吸取锌标准溶液 25.00mL 于 250mL 锥形瓶中，加 0.5%二甲酚橙指示剂 2~3 滴，然后滴加 20%六亚甲基四胺至溶液呈稳定的紫红色，再多加 5mL。用 EDTA 标准溶液滴定至溶液由紫红色变为亮黄色为终点，平行测定 3 次，其滴定体积之差不超过 0.04mL。按 EDTA 溶液消耗的体积，算出其浓度 c_{EDTA}。

② 以 $CaCO_3$ 标准溶液标定 EDTA 溶液。准确移取 25.00mL $CaCO_3$ 标准溶液于 250mL 锥形瓶中，加入 10mL 氨性缓冲溶液（pH=10），加入 3 滴 K-B 指示剂，用 EDTA 标准溶液滴定至溶液由紫红色变为蓝色即为终点。平行测定 3 次，其滴定体积之差不超过 0.04mL。根据消耗的 EDTA 标准溶液体积，计算其浓度 c_{EDTA}。

数据记录与处理

（1）0.02mol/L EDTA 溶液的标定（基准物质 $CaCO_3$）（表 3.13）

表 3.13　0.02mol/L EDTA 溶液的标定（基准物质 CaCO₃）

实验序号	1	2	3
m_{CaCO_3}/g			
V_{EDTA}（初始）/mL			
V_{EDTA}（终点）/mL			
V_{EDTA}（消耗）/mL			
$c_{EDTA}/(mol/L)$			
$\bar{c}_{EDTA}/(mol/L)$			
$|d_i|$			
$\bar{d}_r/\%$			

（2）0.02mol/L EDTA 溶液的标定（基准物质 Zn 粉）（表 3.14）

表 3.14　0.02mol/L EDTA 溶液的标定（基准物质 Zn 粉）

实验序号	1	2	3
m_{Zn}/g			
V_{EDTA}（初始）/mL			
V_{EDTA}（终点）/mL			
V_{EDTA}（消耗）/mL			
$c_{EDTA}/(mol/L)$			
$\bar{c}_{EDTA}/(mol/L)$			
$|d_i|$			
$\bar{d}_r/\%$			

思考题

（1）EDTA 标准溶液和 Zn 标准溶液的配制方法有何不同？

（2）配制 Zn 标准溶液时应注意些什么？

（3）用 Zn 作基准物，二甲酚橙作指示剂，标定 EDTA 溶液浓度，溶液的酸度应控制在什么范围？如何控制？如果溶液酸性较强，怎么办？

（4）用 CaCO₃ 为基准物，以钙指示剂指示终点标定 EDTA 时，应控制溶液的酸度为多大？为什么？如何控制？

（5）配位滴定中为什么要加入缓冲溶液？

实验 3.9　自来水硬度的测定

实验目的

（1）掌握 EDTA 法测定水硬度的原理和方法；

（2）学习水硬度的表示方法及常用硬度的计算方法。

实验原理

水的硬度主要由于水中含有钙盐和镁盐，其他金属离子如铁、铝、锰、锌也形成硬度，但一般含量甚少，测定工业用水总硬度时可忽略不计。

硬度有暂时硬度和永久硬度之分。以碳酸氢盐形式存在的钙盐/镁盐，加热即成碳酸盐沉淀而失去硬度，这类盐所形成的硬度叫作暂时硬度，反应如下：

$$Ca(HCO_3)_2 \xrightarrow{\triangle} CaCO_3 \downarrow + CO_2 \uparrow + H_2O \qquad (3.16)$$

$$Mg(HCO_3)_2 \xrightarrow{\triangle} \underset{\underset{\displaystyle \longrightarrow Mg(OH)_2 \downarrow + CO_2 \uparrow}{+ H_2O\Big\downarrow}}{MgCO_3(不完全沉淀) + CO_2 \uparrow + H_2O} \qquad (3.17)$$

水中含有钙、镁的硫酸盐、氯化物、硝酸盐等形式的硬度称为永久硬度，它们在加热时亦不沉淀（但在锅炉运转温度下，溶解度变小析出称为锅垢）。暂硬和永硬的总和称为"总硬"，由镁离子形成的硬度称为"镁硬"，由钙离子形成的硬度称为"钙硬"。测定水的总硬度常采用配位滴定法，在pH=10的NH_3-NH_4Cl缓冲溶液中，以铬黑T（EBT）或K-B指示剂，用EDTA标准溶液滴定至溶液由紫红色变为纯蓝色即为终点。

干扰离子的掩蔽：若水样中存在Fe^{3+}、Al^{3+}等微量杂质，可用三乙醇胺进行掩蔽，Cu^{2+}、Pb^{2+}、Zn^{2+}等重金属离子可用Na_2S或KCN掩蔽。

分别测钙、镁硬度：可控制pH介于12～13之间（此时，氢氧化镁沉淀），选用钙指示剂进行测定。镁硬度可由总硬度减去钙硬度求出。

水的硬度有许多表示方法。我国常用度（°）来表示。1°表示10万份水中含有一份CaO，也就是1L水中含有10mg CaO，即1°=10mg/L。

实验用品

仪器：分析天平、滴定管、移液管、锥形瓶、称量瓶、烧杯。

试剂：0.02mol/L的EDTA溶液、盐酸（1∶1）溶液、三乙醇胺（1∶1）溶液；NH_3-NH_4Cl缓冲溶液（pH=10）、铬黑T指示剂（0.5%）、$CaCO_3$（基准物质）、K-B指示剂。

实验步骤

(1) 0.02mol/L EDTA标准溶液的配制和标定（参照实验3.9）。

(2) 自来水总硬度的测定　移取水样100.00mL于250mL锥形瓶中，加入1～2滴1∶1 HCl微沸数分钟以除去CO_2，冷却后，加入3mL 1∶1三乙醇胺（若水样中含有重金属离子，则加入1mL 2% Na_2S溶液掩蔽），10mL氨缓冲溶液，2～3滴K-B指示剂，EDTA标准溶液滴定至溶液由紫红色变为纯蓝色，即为终点。注意接近终点时应慢滴多摇。平行测定3次，计算水的总硬度，以度（°）表示分析结果。

注：

① 铬黑T与Mg^{2+}显色灵敏度高，与Ca^{2+}显色灵敏度低，当水样中Ca^{2+}含量高而Mg^{2+}很低时，得到不敏锐的终点，可采用K-B混合指示剂。

② 水样中含铁量超过10mg/mL时用三乙醇胺掩蔽有困难，需用蒸馏水将水样稀释到Fe^{3+}不超过10mg/mL即可。

数据记录与处理

(1) 0.02mol/L EDTA溶液的标定（基准物质$CaCO_3$）（表3.15）

表3.15　0.02mol/L EDTA溶液的标定（基准物质$CaCO_3$）

实验序号	1	2	3
m_{CaCO_3}/g			
V_{EDTA}(初始)/mL			

续表

实验序号	1	2	3
V_{EDTA}(终点)/mL			
V_{EDTA}(消耗)/mL			
c_{EDTA}/(mol/L)			
$\bar{c}_{EDTA}$/(mol/L)			
$|d_i|$			
$\bar{d}_r$/%			

(2) 自来水总硬度的测定（表 3.16）

表 3.16　自来水总硬度的测定

实验序号	1	2	3
$V_水$/mL			
V_{EDTA}(初始)/mL			
V_{EDTA}(终点)/mL			
V_{EDTA}(消耗)/mL			
$H/°$			
$\bar{H}/°$			
$|d_i|$			
$\bar{d}_r$/%			

思考题

(1) 配制 $CaCO_3$ 溶液和 EDTA 溶液时，各采用何种天平称量？为什么？

(2) 铬黑 T 指示剂是怎样指示滴定终点的？

(3) 以 HCl 溶液溶解 $CaCO_3$ 基准物质时，操作中应注意些什么？

(4) 配位滴定中为什么要加入缓冲溶液？

(5) 用 EDTA 法测定水的硬度时，哪些离子的存在有干扰？如何消除？

(6) 配位滴定与酸碱滴定法相比，有哪些不同点？操作中应注意哪些问题？

拓展阅读

美国拉夫运河污染事件

拉夫运河位于纽约州，是为修建水电站而挖成的一条运河，于 20 世纪 40 年代干涸废弃。1942 年被美国一家电化学公司购买当作垃圾仓库来倾倒大量工业废弃物，并持续长达 11 年之久。1953 年，这条充满各种有毒废弃物的运河被该公司填埋覆盖，转赠给当地的教育机构，盖起了大量的住宅和学校。不幸的是，从 1977 年开始，这里的居民不断罹患各种怪病，孕妇流产、儿童夭折、婴儿畸形、癫痫等也频频发生，可见，环境污染会给人类带来无尽的伤害，环境保护的意义不仅在当下，更是在子孙后代。在揭开运河污染真相的道路上，人们所表现出来的坚持不懈的精神尤为珍贵，在一个资产上亿元的跨国公司和一个不给予回应的政府面前，拉夫运河的居民取得了胜利。

实验 3.10　工业硫酸铝中铝含量的分析

实验目的

(1) 了解返滴定法和置换滴定法的基本原理；

(2) 掌握工业硫酸铝中铝的测定原理和方法。

实验原理

工业硫酸铝作为一种无机原料被广泛应用于食品，制药，造纸，水处理等诸多领域。由于 Al^{3+} 易形成一系列多核羟基络合物，这些多核羟基络合物与 EDTA 络合缓慢，故通常采用返滴定法测定铝。先加入定量且过量的 EDTA 标准溶液，在 pH≈3.5 时煮沸几分钟，使 Al^{3+} 与 EDTA 配位完全，继而在 pH 为 5～6 时，以二甲酚橙为指示剂，用 Zn^{2+} 盐标准溶液返滴定过量的 EDTA 而得到铝的含量。

但是，返滴定法测定铝含量缺乏选择性，所有能与 EDTA 形成稳定络合物的离子都干扰。对于像合金、硅酸盐、水泥和炉渣等复杂试样中的铝，往往采用置换滴定法以提高选择性。即在用 Zn^{2+} 返滴过量的 EDTA 后，加入过量的 NH_4F，加热至沸，使 AlY^- 与 F^- 之间发生置换反应，释放出与 Al^{3+} 的物质的量相等的 H_2Y^{2-}（EDTA），再用 Zn^{2+} 盐标准溶液滴定释放出来的 EDTA 而得到铝的含量。

用置换滴定法测定铝，若试样中含 Ti^{4+}、Zr^{4+}、Sn^{4+} 等离子时，亦会发生与 Al^{3+} 相同的置换反应而干扰 Al^{3+} 的测定。这时，就要采用掩蔽的方法，把上述干扰离子掩蔽掉。例如，用苦杏仁酸掩蔽 Ti^{4+} 等。铝合金所含杂质主要有 Si、Mg、Cu、Mn、Fe、Zn，个别还含 Ti、Ni、Ca 等，通常用 HNO_3-HCl 混合酸溶解，亦可在银坩埚或塑料烧杯中以 $NaOH$-H_2O_2 分解后再用 HNO_3 酸化。

实验用品

仪器：分析天平、滴定管、移液管、锥形瓶、称量瓶、烧杯。

试剂：20% NaOH 溶液、1∶1 HCl 溶液、1∶3 HCl 溶液、0.02mol/L EDTA 溶液、0.2% 二甲酚橙溶液、1∶1 氨水、20% 六亚甲基四胺溶液、0.02mol/L Zn^{2+} 标准溶液、20% NH_4F 溶液（贮于塑料瓶中）。

实验步骤

(1) 0.02mol/L EDTA 溶液的标定（同实验 3.8，用金属锌标定）。

(2) 样品处理　准确称取 0.10～0.11g 工业硫酸铝于 50mL 烧杯中，加入 10mL NaOH，在沸水浴中使其完全溶解，稍冷后，加 1∶1 HCl 溶液至有絮状沉淀产生，再多加 10mL 1∶1 HCl 溶液，定量转移试液于 250mL 容量瓶中，加水至刻度，摇匀。

(3) 样品中铝含量的测定

① 返滴定法。准确移取上述试液 25.00mL 于 250mL 锥形瓶中，加 30.00mL EDTA，两滴二甲酚橙，此时，溶液为黄色，加氨水至溶液呈紫红色，再加 1∶3 HCl 溶液，使溶液呈现黄色。煮沸 3min，冷却。加 20mL 六亚甲基四胺，此时，溶液应为黄色，如果溶液呈红色，还需滴加 1∶3 HCl 溶液，使其变黄。用 Zn^{2+} 标准溶液滴定，当溶液由黄色恰好转变为紫红色时即为终点。

② 置换滴定法。先按照①的实验步骤操作，再向溶液中加 10mL NH_4F，加热至微沸，用水冷却，再补加 2 滴二甲酚橙，此时，溶液为黄色，若为红色，应滴加 1∶3 HCl 溶液使其变为黄色。再用 Zn^{2+} 标准溶液滴定，当溶液由黄色恰好转变为紫红色时即为终点，根据这次 Zn^{2+} 标准溶液所耗体积计算 Al 的质量分数。

数据记录与处理

(1) 0.02mol/L EDTA 溶液的标定（基准物质 Zn）（表 3.17）

表 3.17　0.02mol/L EDTA 溶液的标定（基准物质 Zn）

实验序号	1	2	3
m_{Zn}/g			
V_{EDTA}（初始）/mL			
V_{EDTA}（终点）/mL			
V_{EDTA}（消耗）/mL			
c_{EDTA}/(mol/L)			
$\bar{c}_{EDTA}$/(mol/L)			
$|d_i|$			
$\bar{d}_r$/%			

（2）样品中铝含量的测定（表 3.18）

表 3.18　工业硫酸铝中铝含量的测定

实验序号	1	2	3
$m_{铝样}$/g			
V_{EDTA}/mL			
$V_{Zn^{2+}}$（初始）/mL			
$V_{Zn^{2+}}$（终点）/mL			
$V_{Zn^{2+}}$（消耗）/mL			
w_{Al}/%			
$\bar{w}_{Al}$/%			
$|d_i|$			
$\bar{d}_r$/%			

思考题

（1）为什么测定简单试样中的 Al^{3+} 用返滴定法即可，而测定复杂试样中的 Al^{3+} 则需采用置换滴定法？

（2）用返滴定法测定简单试样中的 Al^{3+} 时，所加入过量 EDTA 溶液的浓度是否必须准确？为什么？

（3）本实验中使用的 EDTA 溶液要不要标定？

（4）为什么加入过量的 EDTA，第一次用 Zn^{2+} 标准溶液滴定时，可以不计所消耗的体积？但此时是否需准确滴定溶液由黄色变为紫红色？为什么？

实验 3.11　锌铋混合溶液中 Zn^{2+}、Bi^{3+} 含量的连续测定

实验目的

（1）掌握通过控制溶液酸度进行多种离子连续配位滴定的原理和方法。
（2）熟悉二甲酚橙指示剂的应用。

实验原理

如果要在同一溶液中分别测定 M、N 两种离子，需满足以下条件：$\lg(c_M K'_{MY}) \geqslant 5$、$\lg(c_N K'_{NY}) \geqslant 5$、$\Delta\lg(cK') \geqslant 5 (\Delta pM' = 0.2, E_t \leqslant 0.3\%)$。这时，测定 M 的适宜酸度范围是：

最高酸度：$\alpha_{Y(H)} \approx \alpha_{Y(N)} = 1 + K_{NY}[N]$ 对应的酸度； (3.18)

最低酸度：M 水解对应的 pH 值，即 $[OH^-] = \sqrt[n]{\dfrac{K_{sp}}{c_M}}$ (3.19)

Zn^{2+}、Bi^{3+} 均能与 EDTA 形成稳定的配合物，其 lgK 分别为 16.50 和 27.94，由于 $\Delta lg(cK) > 5$，故可控制溶液不同的酸度分别测定它们的含量。首先调节溶液的 pH=1，以二甲酚橙（XO）为指示剂，此时 Zn^{2+} 与 XO 不能形成稳定的配合物，不会对 Bi^{3+} 的分析造成干扰。Bi^{3+} 与 XO 形成紫红色配合物，Bi^{3+}-XO 的条件稳定常数为 9.6，可准确滴定。之后再向这份溶液中加入六亚甲基四胺调节 pH 至 5～6，此时 Zn^{2+} 与 XO 形成紫红色配合物，继续用 EDTA 标准溶液滴定至溶液由紫红色变为亮黄色，即为 Zn^{2+} 的终点。

pH=1 时的反应为：

滴定前：$Bi^{3+} + H_3In^{4-}$（黄色）$\longrightarrow BiH_3In^-$（紫红）

滴定开始至计量点前：$Bi^{3+} + H_2Y^{2-} \longrightarrow BiY^- + 2H^+$

计量点：$H_2Y^{2-} + BiH_3In^-$（紫红）$\longrightarrow BiY^- + H_3In^{4-}$（黄色）$+ 2H^+$

pH=5～6 时反应为：

滴定前：$Pb^{2+} + H_3In^{4-}$（黄色）$\longrightarrow PbH_3In^{2-}$（紫红）

滴定开始至计量点前：$Pb^{2+} + H_2Y^{2-} \longrightarrow PbY^{2-} + 2H^+$

计量点：$H_2Y^{2-} + PbH_3In^{2-}$（紫红）$\longrightarrow PbY^{2-} + H_3In^{4-}$（黄色）$+ 2H^+$

  实验用品

仪器：分析天平、滴定管、移液管、锥形瓶、称量瓶、烧杯。

试剂：0.02mol/L EDTA、基准 ZnO、0.2%二甲酚橙溶液、1∶1 HCl、20%六亚甲基四胺溶液、1∶1 氨水、锌铋混合液。

 实验步骤

(1) 0.02mol/L EDTA 标准溶液的配制　称取 2 g $Na_2H_2Y \cdot 2H_2O$ 于 100mL 去离子水中，加热溶解，稀释至 250mL，摇匀（长期放置应置于硬质玻璃瓶或聚乙烯瓶中）。

(2) 0.02mol/L EDTA 标准溶液的标定　准确称取氧化锌（　）至（　）g 于 250mL 烧杯中，盖上表面皿。从烧杯嘴中滴加 5～10mL 1∶1 HCl，放置至 ZnO 全部溶解后，定量转移到 250mL 容量瓶中，用水稀释至刻度，摇匀。计算其准确浓度 $c_{Zn}/(mol/L)$。用移液管吸取锌标准溶液 25.00mL 于 250mL 锥形瓶中，加 0.2%二甲酚橙指示剂 2～3 滴，然后滴加 20%六亚甲基四胺至溶液呈稳定的红紫色，再多加 5mL。用 EDTA 标准溶液滴定至溶液由红紫色变为亮黄色为终点，平行测定三次，其滴定体积之差不超过 0.04mL。按 EDTA 溶液消耗的体积，算出其浓度 $c_{EDTA}/(mol/L)$。

(3) 混合溶液的测定　准确移取锌铋混合液 25.00mL，加入 1～2 滴 0.2%二甲酚橙指示剂，用 EDTA 标准溶液滴定至溶液由紫红色变为亮黄色，即为滴定 Bi^{3+} 的终点。平行测定 3 次，计算混合液中 Bi^{3+} 的含量（g/L）。

在测完 Bi^{3+} 的溶液中再加 2～3 滴二甲酚橙指示剂，逐滴加入 1∶1 氨水（只需几滴，用试纸检测 pH 为 5），使溶液呈橙色，再滴加 20%六亚甲基四胺至溶液呈稳定的紫红色，并过量 5mL，用标准 EDTA 溶液滴定至溶液呈亮黄色，即为滴定的终点。平行测定 3 次，计算混合液中 Zn^{2+} 的含量（g/L）。

注：

① pH=1 时，$Bi(NO_3)_3$ 沉淀不会析出，二甲酚橙也不与 Zn^{2+} 配位；如果酸度太高，

二甲酚橙不与 Bi^{3+} 配位，溶液呈黄色。

② Bi^{3+} 与 EDTA 反应速度较慢，故滴定速度不宜过快，且要剧烈摇动。

数据记录与处理

（1）0.02mol/L EDTA 溶液浓度的标定（表 3.19）

表 3.19　0.02mol/L EDTA 溶液的标定（基准物质 ZnO）

实验序号	1	2	3
m_{Zn}/g			
V_{EDTA}（初始）/mL			
V_{EDTA}（终点）/mL			
V_{EDTA}（消耗）/mL			
c_{EDTA}/(mol/L)			
$\bar{c}_{EDTA}$/(mol/L)			
$|d_i|$			
$\bar{d}_r$/%			

（2）混合溶液的连续滴定（表 3.20）

表 3.20　混合溶液的连续滴定

实验序号	1	2	3
pH=1.0 测 Bi^{3+}			
V_1（初始）/mL			
V_1（终点）/mL			
V_1（消耗）/mL			
$w_{Bi^{3+}}$/%			
$\bar{w}_{Bi^{3+}}$/%			
$|d_i|$			
$\bar{d}_r$/%			
pH=5.0 测 Zn^{2+}			
V_2（初始）/mL			
V_2（终点）/mL			
V_2（消耗）/mL			
$w_{Zn^{2+}}$/%			
$\bar{w}_{Zn^{2+}}$/%			
$|d_i|$			
$\bar{d}_r$/%			

思考题

（1）能否在同一份试液中先滴定 Zn^{2+}，后滴定 Bi^{3+}？

（2）如果试液中含有 Fe^{3+}，一般加入抗坏血酸掩蔽，可用三乙醇胺掩蔽吗？

（3）在 pH≈1 的条件下用 EDTA 标准溶液测定 Bi^{3+}，共存的 Zn^{2+} 为何不干扰？

（4）能否取等量混合试液两份，一份控制 pH=1.0 滴定 Bi^{3+}，另一份控制 pH 为 5～6

锌铋连续滴定

滴定 Zn^{2+}、Bi^{3+} 总量？为什么？

实验 3.12 高锰酸钾标准溶液的配制与标定

实验目的

（1）正确配制一个相对稳定的 $KMnO_4$ 标准溶液；
（2）掌握用 $Na_2C_2O_4$ 标定 $KMnO_4$ 溶液的原理和方法。

实验原理

高锰酸钾是一种强氧化剂，它的氧化能力和还原产物与溶液酸度有很大关系。
在强酸性溶液中：
$$MnO_4^- + 5e + 8H^+ \rightleftharpoons Mn^{2+} + 4H_2O \quad E^\ominus = 1.51V \quad (3.20)$$
在弱酸性、中性、弱碱性溶液中：
$$MnO_4^- + 3e + 2H_2O \rightleftharpoons MnO_2 + 4OH^- \quad E^\ominus = 0.59V \quad (3.21)$$
在强碱性溶液中：
$$MnO_4^- + e \rightleftharpoons MnO_4^{2-} \quad E^\ominus = 0.564V \quad (3.22)$$

应用高锰酸钾法时，可根据被测物质的性质采用不同的方法。

市售 $KMnO_4$ 试剂纯度一般为 99%～99.5%，其中常含有少量的 MnO_2 和其他杂质。蒸馏水中常含有微量还原性的有机物质，它们可与 MnO_4^- 反应而析出 $MnO(OH)_2$ 沉淀。这些生成物以及光、热、酸、碱等外界条件的改变均会促使 $KMnO_4$ 的进一步分解。因此，$KMnO_4$ 标准溶液不能通过直接称量法获得。

为了获得相对稳定的 $KMnO_4$ 溶液，必须按照实验步骤中所述的方法配制。标定 $KMnO_4$ 溶液的基准物很多，如 $Na_2C_2O_4$、$H_2C_2O_4 \cdot 2H_2O$、As_2O_3、$(NH_4)_2Fe(SO_4)_2$ 和纯铁丝等，其中，$Na_2C_2O_4$ 较为常用。

在 H_2SO_4 溶液中，MnO_4^- 与 $C_2O_4^{2-}$ 的反应如下：
$$2MnO_4^- + 5C_2O_4^{2-} + 16H^+ \rightleftharpoons 2Mn^{2+} + 10CO_2\uparrow + 8H_2O \quad (3.23)$$

为了使这个反应能够定量地、较快地进行，应注意下列滴定条件：

（1）温度　在室温下此反应速率缓慢，常将溶液加热至 70～80℃，并趁热滴定，滴定完毕时的温度不应低于 60℃。但温度也不易过高，若高于 90℃，则 $H_2C_2O_4$ 部分分解，导致标定结果偏高。
$$H_2C_2O_4 \rightleftharpoons CO_2\uparrow + CO\uparrow + H_2O \quad (3.24)$$

（2）酸度　酸度过低，MnO_4^- 会部分被还原成 MnO_2；酸度过高，会促使 $H_2C_2O_4$ 分解，一般滴定开始时的酸度应控制在 0.5～1mol/L。为防止诱导氧化 Cl^- 的反应发生，应当在 H_2SO_4 介质中进行。

（3）滴定速度　开始滴定时，MnO_4^- 与 $C_2O_4^{2-}$ 的反应速度很慢，滴入的 $KMnO_4$ 褪色较慢。因此，滴定开始阶段滴定速度不宜太快，否则滴入的 $KMnO_4$ 溶液来不及与 $C_2O_4^{2-}$ 反应，就在热的酸性溶液中发生分解，导致标定结果偏低。
$$4MnO_4^- + 12H^+ \rightleftharpoons 4Mn^{2+} + 5O_2 + 6H_2O \quad (3.25)$$

（4）催化剂　滴定开始时，滴入的几滴 $KMnO_4$ 溶液褪色较慢，但随着这几滴 $KMnO_4$ 和 $Na_2C_2O_4$ 作用完毕后，滴定产物 Mn^{2+} 的生成，反应速率逐渐加快。若滴定前加入几滴 $MnSO_4$ 溶液，则在滴定一开始，反应速度就很快，可见 Mn^{2+} 在此反应中起着催化剂的作用。

(5) 指示剂 因为 $KMnO_4$ 本身具有颜色，溶液中稍有过量的 MnO_4^-，即可显示出粉红色。一般不必另外加入指示剂，$KMnO_4$ 可作为自身指示剂。

(6) 滴定终点 用 $KMnO_4$ 溶液滴定至终点后，溶液中出现的粉红色不能持久，这是因为空气中的还原性的气体和灰尘都能使 MnO_4^- 缓慢还原，故溶液的粉红色逐渐消失。所以，滴定时溶液中出现的粉红色如在 $0.5\sim1\mathrm{min}$ 内不褪色，就可认为已经达到滴定终点。

标定好的 $KMnO_4$ 溶液在放置一段时间后，若发现有 $MnO(OH)_2$ 沉淀析出，应重新过滤并标定。

实验用品

仪器：微孔玻璃漏斗、酸式滴定管、锥形瓶、称量瓶、量筒、烧杯等。

试剂：$KMnO_4$(A.R.)、$3\mathrm{mol/L}\ H_2SO_4$ 溶液、$Na_2C_2O_4$(105℃干燥 2 h 备用)。

实验步骤

(1) $KMnO_4$ 溶液的配制 称取稍多于理论计算用量的 $KMnO_4$ 固体约（　　）g，溶于 500mL 水中，盖上表面皿，加热至沸，并保持微沸状态约 1 h，随时补充蒸发掉的水分，冷却，将溶液在室温条件下静置 2～3 天后，用微孔玻璃漏斗过滤。滤液储存于棕色试剂瓶中，置于暗处保存，备用。

(2) 0.02mol/L $KMnO_4$ 溶液的标定（微量滴定） 准确称取（　　）至（　　）g 基准物质 $Na_2C_2O_4$，用去离子水溶解后转入 25mL 容量瓶定容。移取 5.0mL 草酸钠标准溶液 3 份于 25mL 锥形瓶中，加入 4mL 水稀释，加入 3mL 3mol/L H_2SO_4，在水浴中加热至约 70～80℃（溶液开始冒蒸汽），趁热用 0.02mol/L $KMnO_4$ 溶液滴定。第一滴 $KMnO_4$ 加入后褪色很慢，需剧烈振动，待溶液中产生了 Mn^{2+} 后，滴定速度可加快，直到溶液呈现微红色，并持续 1min 内不褪色即为终点，记录滴定所消耗的 $KMnO_4$ 溶液的体积。

数据记录与处理

0.02mol/L $KMnO_4$ 溶液的标定（表 3.21）

表 3.21 $KMnO_4$ 溶液的标定

实验序号	1	2	3
$m_{Na_2C_2O_4}/\mathrm{g}$			
V_{KMnO_4}(初始)/mL			
V_{KMnO_4}(终点)/mL			
V_{KMnO_4}(消耗)/mL			
$c_{KMnO_4}/(\mathrm{mol/L})$			
$\bar{c}_{KMnO_4}/(\mathrm{mol/L})$			
$\|d_i\|$			
$\bar{d}_r/\%$			

思考题

(1) 在配制 $KMnO_4$ 溶液时应注意哪些问题？为什么？

(2) 配制 $KMnO_4$ 溶液时，过滤后的滤器上残留的产物是什么？应选用什么物质清洗干净？

(3) $KMnO_4$ 溶液的配制过程中要用微孔玻璃漏斗过滤，能否用定量滤纸过滤？为什么？

(4) 配制 $KMnO_4$ 溶液时，必须将 $KMnO_4$ 溶液煮沸 1~2 h，其目的是什么？

(5) 用 $Na_2C_2O_4$ 标定 $KMnO_4$ 溶液的反应条件是什么？

(6) 为什么用 3mol/L H_2SO_4 控制溶液的酸度？

实验 3.13 双氧水中过氧化氢含量的测定

实验目的

(1) 掌握标定 $KMnO_4$ 溶液的方法；

(2) 掌握用 $KMnO_4$ 法直接测定 H_2O_2 含量的原理和方法。

实验原理

双氧水是医药卫生行业广泛使用的消毒剂，其主要成分 H_2O_2 分子中有一个过氧键—O—O—。在酸性溶液中，它是一个强氧化剂，但遇到氧化性更强的 $KMnO_4$ 时，H_2O_2 表现出还原性。测定过氧化氢的含量，就是利用在稀硫酸溶液中，室温条件下 $KMnO_4$ 氧化 H_2O_2，其反应式为：

$$5H_2O_2 + 2MnO_4^- + 6H^+ = 2Mn^{2+} + 5O_2\uparrow + 8H_2O \tag{3.26}$$

开始时反应速率很慢，滴入第一滴溶液不容易褪色，待 Mn^{2+} 生成后（Mn^{2+} 起自动催化作用）反应速率加快。当达到化学计量点时，微过量的 $KMnO_4$ 使溶液呈微红色，滴定结束。

实验用品

仪器：酸式滴定管、容量瓶、移液管、锥形瓶、称量瓶、量筒等。

试剂：$Na_2C_2O_4$（基准物质）、3mol/L H_2SO_4 溶液、0.02mol/L $KMnO_4$ 溶液、3% H_2O_2 试液（市售 H_2O_2 稀释后备用）。

实验步骤

(1) 配制 0.02mol/L $KMnO_4$ 溶液 250mL（提前一周完成，见实验 3.12）。

(2) 用基准 $Na_2C_2O_4$ 标定 $KMnO_4$ 溶液的浓度（见实验 3.12）。

(3) 双氧水中 H_2O_2 含量的测定（微量滴定） 用移液管准确移取 H_2O_2 试液 1.00mL 置于 25mL 容量瓶中，加水稀释至刻度，充分摇匀。

移取 5.00mL 上述试液 3 份，分别置于 25mL 锥形瓶中，加 4mL 水、2mL 3mol/L H_2SO_4 溶液，用 $KMnO_4$ 标准溶液滴定至溶液呈微红色且 30 s 内不褪色即为终点。

数据记录与处理

(1) 0.02mol/L $KMnO_4$ 溶液的标定（表 3.22）

表 3.22　$KMnO_4$ 溶液的标定

实验序号	1	2	3
$m_{Na_2C_2O_4}/g$			
V_{KMnO_4}（初始）/mL			
V_{KMnO_4}（终点）/mL			
V_{KMnO_4}（消耗）/mL			
$c_{KMnO_4}/(mol/L)$			
$\bar{c}_{KMnO_4}/(mol/L)$			
$\lvert d_i \rvert$			
$\bar{d}_r/\%$			

（2）H_2O_2 含量的测定（表 3.23）

表 3.23　H_2O_2 含量的测定

实验序号	1	2	3
$V_{H_2O_2}/mL$			
V_{KMnO_4}（初始）/mL			
V_{KMnO_4}（终点）/mL			
V_{KMnO_4}（消耗）/mL			
$\rho_{H_2O_2}/(g/100mL)$			
$\bar{\rho}_{H_2O_2}/(g/100mL)$			
$\lvert d_i \rvert$			
$\bar{d}_r/\%$			

 思考题

（1）对本实验来讲，$KMnO_4$ 滴定法的优缺点是什么？

（2）用 $KMnO_4$ 法测定双氧水中 H_2O_2 的含量，为什么要在酸性条件下进行？能否用 HNO_3、HCl 或 HAc 来调节溶液的酸度？为什么？

（3）试述铈量法和碘量法测定 H_2O_2 含量的原理。

拓展阅读

用 $KMnO_4$ 滴定法测定 H_2O_2 含量

在生物化学中常用 $KMnO_4$ 滴定法测定过氧化氢酶的活性。在血液中加入一定量的过氧化氢，由于过氧化氢酶能使过氧化氢分解，作用完后，在酸性条件下，用 $KMnO_4$ 滴定剩余的 H_2O_2，就可以了解过氧化氢酶的活性。H_2O_2 也是常用的工业漂白剂，在测定工业用 H_2O_2 时 $KMnO_4$ 法测定则不适用，因为产品中常加有少量乙酰苯胺等有机物作稳定剂，滴定时也要消耗 $KMnO_4$ 溶液，从而引起方法误差。遇此情况，应采用碘量法或铈量法进行滴定。碘量法是利用 H_2O_2 和 KI 作用，析出 I_2，然后用 $Na_2S_2O_3$ 溶液滴定。

双氧水过氧化氢含量测定实验——微量滴定

实验 3.14 铁矿石中铁含量的测定

实验目的

(1) 学习矿石试样的酸溶法。
(2) 掌握无汞法测定铁的原理及方法。
(3) 掌握氧化还原指示剂的变色原理。
(4) 了解重铬酸钾废液的处理方法，增强环保意识。

实验原理

重铬酸钾是一种常用的氧化剂，在酸性溶液中常被还原剂还原为 Cr^{3+}，半反应式为：

$$Cr_2O_7^{2-} + 14H^+ + 6e \Longrightarrow 2Cr^{3+} + 7H_2O \tag{3.27}$$

重铬酸钾的标准电极电位是 1.33V，但在酸性溶液中被还原时，其条件电位 $E^{\ominus\prime}$ 较标准电位 $E^{\ominus}$ 小。例如：在 3mol/L HCl 溶液中，$E^{\ominus\prime}=1.08$ V；在 3mol/L H_2SO_4 溶液中，$E^{\ominus\prime}=1.15$ V。溶液的酸度增加，$K_2Cr_2O_7$ 的条件电位亦随之增大。也即，可以通过控制溶液的酸度，将重铬酸钾的氧化性调节到适当程度，从而减少副反应、提高检测的选择性。例如，在 HCl 浓度低于 3mol/L 时，$Cr_2O_7^{2-}$ 不氧化 Cl^-，可以在 HCl 介质中用 $K_2Cr_2O_7$ 滴定 Fe^{2+}。此外，$K_2Cr_2O_7$ 容易提纯（含量 99.99%），化学性质稳定，可以直接配制使用。这些优势使 $K_2Cr_2O_7$ 在氧化还原滴定分析中具有广泛的应用。

重铬酸钾法是测定矿石、合金、硅酸盐等产品中铁含量的国家标准方法。经典的重铬酸钾法中，要在浓的热 HCl 溶液中用 $SnCl_2$ 将 Fe^{3+} 还原为 Fe^{2+}，再用 $HgCl_2$ 氧化除去过量的 $SnCl_2$。每分析测定一份试液，都将产生含汞约 480mg 的废水，需要用 9.6～10t 的水稀释才能达到国家环境部门允许的汞排放浓度（0.05mg/L）。即便稀释后达到排放标准，汞盐也会沉积在环境中，造成严重污染，形成潜在的健康危害。近年来，人们研究开发了无汞测铁的新方法，如新重铬酸钾法、硫酸铈法和 EDTA 法等。

本实验采用重铬酸钾无汞法测定铁矿石中铁含量。

先将试样用硫-磷混酸溶解，用 $SnCl_2$ 还原其中大部分的 Fe^{3+}，再用 $TiCl_3$ 还原剩余部分的 Fe^{3+}。

$$2Fe^{3+} + SnCl_4^{2-} + 2Cl^- \Longrightarrow 2Fe^{2+} + SnCl_6^{2-} \tag{3.28}$$

$$Fe^{3+} + Ti^{3+} + H_2O \Longrightarrow Fe^{2+} + TiO^{2+} + 2H^+ \tag{3.29}$$

为避免加入过多的 $TiCl_3$，需要在滴加 $TiCl_3$ 之前向溶液中加入钨酸钠。当 Fe^{3+} 定量还原为 Fe^{2+}，过量的一滴 $TiCl_3$ 使溶液中的六价钨（无色的磷钨酸）还原为蓝色的五价钨化合物（俗称"钨蓝"），故溶液呈现蓝色。为消除过量一滴 $TiCl_3$ 对滴定的影响，在正式滴定之前，滴加 $K_2Cr_2O_7$ 溶液使钨蓝刚好褪色。

之后，加入二苯胺磺酸钠指示剂，用 $K_2Cr_2O_7$ 标准溶液滴定，Fe^{2+} 被完全氧化为 Fe^{3+}，Fe^{3+} 与二苯胺磺酸钠生成稳定的紫色络合物，指示终点。

$$6Fe^{2+} + Cr_2O_7^{2-} + 14H^+ \Longrightarrow 6Fe^{3+} + 2Cr^{3+} + 7H_2O \tag{3.30}$$

实验用品

仪器：酸式滴定管、移液管、容量瓶、锥形瓶、称量瓶、量筒、烧杯等。

试剂：$K_2Cr_2O_7$ 固体试剂（分析纯）、H_2SO_4-H_3PO_4（浓酸 1∶1 混合）、3mol/L HCl

溶液、浓 HNO_3、待测铁矿石试样；

Na_2WO_4 溶液（10%）：称取 10 g Na_2WO_4 溶于适量的水中（若浑浊应过滤），加入 2～5mL 浓 H_3PO_4，加水稀释至 100mL；

$SnCl_2$ 溶液（10%）：称取 10g $SnCl_2 \cdot H_2O$ 溶于 40mL 热的浓 HCl 中，加水稀释至 100mL；

$TiCl_3$ 溶液（1.5%）：量取 10mL 原瓶装 $TiCl_3$，用 2mol/L HCl 稀释至 100mL。加少量石油醚，使之浮在 $TiCl_3$ 溶液的表层，以隔绝空气，避免 $TiCl_3$ 氧化；

二苯胺磺酸钠指示剂溶液（0.2%水溶液）。

实验步骤

（1）$K_2Cr_2O_7$ 标准溶液（0.01667mol/L）的配制　准确称取（　　）g $K_2Cr_2O_7$ 基准试剂于 100mL 小烧杯中，加入 20mL 蒸馏水溶解，然后转入 250mL 容量瓶中，加水稀释至刻度，充分摇匀后备用。

（2）铁矿石试样的预处理

① 准确称取约 0.2 g 铁矿石试样（含 Fe 量约为 60%）3 份，置于 250mL 锥形瓶中。

② 加 5mL 蒸馏水或去离子水使试样散开，轻轻摇匀后，加入 10mL 硫-磷混酸（如试样中含硫化物高时，同时加入约 1mL 浓 HNO_3）置于加热板上（在通风橱内），加热分解试样，直至冒 SO_3 白烟[1]。此时，溶液应清亮，残渣为白色或浅色时试样分解完全[2]。

③ 加入 30mL HCl 溶液（3mol/L），加热至近沸，取下。

④ 冷却至约 60℃ 趁热滴加 10% $SnCl_2$ 溶液，使大部分 Fe^{3+} 还原为 Fe^{2+}，此时溶液由黄色变为浅黄色[3]，加入 1mL10% Na_2WO_4 溶液，滴加 $TiCl_3$ 溶液至出现稳定的"钨蓝"（30s 内不褪色）为止。加入约 60mL 新鲜蒸馏水（事先煮沸除去水中的氧[4]），放置 10～20s。

（3）铁矿石试样的测定　滴加 $K_2Cr_2O_7$ 标准溶液至"钨蓝"刚好褪尽（不必记录读数）。加入 6～7 滴二苯胺磺酸钠指示剂，立即用 $K_2Cr_2O_7$ 标准溶液滴定至溶液呈现稳定的紫色为终点。平行测定 3 次。（从步骤③开始，处理一个，立即滴定一个[5]）

附注：

[1] H_2SO_4 分解温度 338℃ 比 HNO_3 分解温度 125℃ 高得多，当溶液开始冒白烟，说明浓 H_2SO_4 已开始分解，则 HNO_3 已经赶尽。若硝酸赶不尽，残余的硝酸会妨碍下一步还原剂与 Fe^{3+} 的反应。另外，只要开始冒浓厚白烟即可停止加热，如果时间过长，H_3PO_4 易形成焦磷酸盐，包夹试样，影响分析结果。

[2] 溶解时，锥形瓶底部出现白色残渣物是 $SiO_2 \cdot nH_2O$ 或 $H_3SiO_3 \cdot nH_2O$。

[3] $SnCl_2$ 不能加过量，否则测定结果偏高。如不慎过量，可滴加 2% $KMnO_4$ 溶液至呈浅黄色消除其影响。

[4] "钨蓝"是钨的低价氧化物，很不稳定。若水中的溶解氧未除尽，加入水后"钨蓝"会因被氧化而消失。

[5] 还原后的 Fe^{2+} 在磷酸介质中极易被氧化，在"钨蓝"褪色 1min 内应立即滴定。放置太久测定结果偏低。

数据记录与处理

（1）$K_2Cr_2O_7$ 标准溶液（0.01667mol/L）的配制（表 3.24）

表 3.24　$K_2Cr_2O_7$ 标准溶液的配制

$m_{K_2Cr_2O_7}$（初始）/g	
$m_{K_2Cr_2O_7}$（终点）/g	
$m_{K_2Cr_2O_7}$/g	

（2）铁矿石试样中含铁量的测定（表 3.25）

表 3.25　铁矿石试样中含铁量的测定

实验序号	1	2	3
V（初始）/mL			
V（终点）/mL			
V（消耗）/mL			
$w_{Fe_2O_3}$/%			
$\overline{w}_{Fe_2O_3}$/%			
$\lvert d_i \rvert$			
$\overline{d}_r$/%			

思考题

（1）分解试样时，为什么要加入硫-磷混酸及硝酸溶液？

（2）样品溶解后，为什么要加热至冒白烟？若滴定过程中有硝酸会对实验结果有何影响？

（3）怎样才能合理配制 $SnCl_2$ 溶液？配制过程中为什么要加入 HCl？若要久置，$SnCl_2$ 溶液应如何配制？

（4）为什么 $SnCl_2$ 溶液应趁热滴加？应滴加到什么程度为止？

（5）为什么还原处理一份试样就必须立即滴定，而不能同时预处理几份之后再一份一份地滴定？

实验 3.15　碘和硫代硫酸钠溶液的配制与标定

实验目的

（1）掌握 I_2 和 $Na_2S_2O_3$ 溶液的配制方法与保存条件；

（2）掌握标定 I_2 及 $Na_2S_2O_3$ 溶液浓度的原理和方法。

实验原理

（1）**碘量法**　碘量法是利用 I_2 的氧化性和 I^- 的还原性来进行测定的方法。由于固体 I_2 在水中的溶解度很小且易于挥发，通常将 I_2 溶解于 KI 溶液中，此时它以 I_3^- 络离子形式存在，其半反应是：

$$I_3^- + 2e \Longrightarrow 3I^- \tag{3.31}$$

为简化并强调化学计量关系，一般仍简写为 I_2。这个电对的标准电位是 0.545V，居于标准电位表中间位置，可见 I_2 是较弱的氧化剂，I^- 则是中等强度的还原剂。

用 I_2 标准溶液直接滴定 $S_2O_3^{2-}$、SO_3^{2-}、As(Ⅲ)、Sn(Ⅱ)、维生素 C 等强还原剂，称为直接碘量法（或碘滴定法）。而利用 I^- 的还原作用，可与许多氧化性物质如 MnO_4^-、

$Cr_2O_7^{2-}$、H_2O_2、Cu^{2+}、Fe^{3+} 等反应定量地析出 I_2，然后用 $Na_2S_2O_3$ 标准溶液滴定 I_2，从而间接地测定这些氧化性物质，这就是间接碘量法（或称滴定碘法）。间接碘量法应用更广。

I_3^-/I^- 电对的可逆性好，其电位在很大的 pH 范围内（pH<9）不受酸度和其他络合剂的影响，所以在选择测定条件时，只要考虑被测物质的性质就可以了。

碘量法采用淀粉为指示剂，其灵敏度高，I_2 浓度为 $1×10^{-5}$ mol/L 即显示蓝色。当溶液呈现蓝色（直接碘量法）或蓝色消失（间接碘量法）即为终点。

碘量法中的分析误差主要来自 I_2 的挥发与 I^- 被空气氧化。为防止 I_2 挥发，应加入过量 KI 使之形成 I_3^- 络离子；溶液温度切勿过高；析出 I_2 的反应最好在带塞的碘量瓶中进行，反应完全后立即滴定，滴定时切勿剧烈摇动。为防止 I^- 被空气氧化，应将析出 I_2 的反应瓶置于暗处，并提前除去 Cu^{2+}、NO_2^- 等能够催化空气氧化 I^- 的杂质。采取以上措施后，碘量法的测定误差大为减小，测定结果的准确性得到保证。

（2）碘标准溶液的制备　虽然用升华法可制得纯粹的 I_2，但 I_2 的挥发性强，准确称量困难，一般是配成大致浓度再标定。将称取的一定量的 I_2，加入过量的 KI，置于研钵中，加少量水研磨，使 I_2 全部溶解，然后再将溶液稀释，倾入棕色瓶中于暗处保存。应避免 I_2 溶液与橡胶等有机物接触，也要防止 I_2 溶液见光遇热，否则浓度会发生变化。

实验室也常用间接法制备碘溶液。如，利用 KIO_3 与 KI 在酸性溶液中发生归中反应生成 I_2，过量 I^- 与 I_2 生成 I_3^-，增加碘的溶解性。由于碘酸钾是基准物质，用这种方法配制的碘标准溶液无须后续的标定即可计算得到准确浓度。

$$KIO_3 + 5KI + 6HCl =\!=\!= 3I_2 + 3H_2O + 6KCl \tag{3.32}$$

$$I_2 + I^- =\!=\!= I_3^- \tag{3.33}$$

（3）碘标准溶液的标定　利用碘单质直接配制得到的碘标准溶液需要标定。

I_2 溶液可用 As_2O_3 作为基准物质来标定。As_2O_3 难溶于水，但易溶于碱性溶液。在 pH 为 8～9 时，I_2 快速而定量地氧化 $HAsO_2$：

$$HAsO_2 + I_2 + 2H_2O =\!=\!= HAsO_4^{2-} + 2I^- + 4H^+ \tag{3.34}$$

标定时一般先酸化试液，再加入 $NaHCO_3$，调节 pH=8 左右。由于 As_2O_3 有剧毒，这种标定方法的应用受限。实际上，也可使用 $Na_2S_2O_3$ 标准溶液标定这类 I_2 溶液。

$$2S_2O_3^{2-} + I_2 =\!=\!= S_4O_6^{2-} + 2I^- \tag{3.35}$$

（4）硫代硫酸钠溶液的配制与标定　结晶 $Na_2S_2O_3 \cdot 5H_2O$ 一般都含有少量的杂质，如 S、Na_2SO_3、Na_2SO_4、Na_2CO_3 及 NaCl 等，同时还容易风化和潮解，因此不能直接称量配制标准溶液。已配好的 $Na_2S_2O_3$ 溶液也不稳定，主要原因有三点：

① 被酸分解，即使溶解在水中的 CO_2 也能使它发生分解：

$$Na_2S_2O_3 + CO_2 + H_2O =\!=\!= NaHCO_3 + NaHSO_3 + S\downarrow \tag{3.36}$$

② 微生物的作用：这是使 $Na_2S_2O_3$ 分解的主要原因。

$$Na_2S_2O_3 =\!=\!= Na_2SO_3 + S\downarrow \tag{3.37}$$

③ 空气的氧化作用：

$$2Na_2S_2O_3 + O_2 =\!=\!= 2Na_2SO_4 + 2S\downarrow \tag{3.38}$$

因此，配制 $Na_2S_2O_3$ 溶液时，应采取下列措施：第一，应用新煮沸并冷却的蒸馏水配制溶液，以除去水中溶解的 CO_2 和 O_2 并杀死微生物；第二，加入少量 Na_2CO_3，使溶液呈弱碱性，以抑制细菌生长；第三，溶液应储存于棕色试剂瓶并置于暗处，以防光照分解。溶液放置一段时间后应重新标定，若发现溶液变浑浊表示有硫析出，应弃去重配。配制时各步操作均应非常细致，所用仪器必须洁净。

标定 $Na_2S_2O_3$ 溶液浓度的基准物有 $K_2Cr_2O_7$、KIO_3、$KBrO_3$ 和纯铜等,都采用间接法标定。以 $K_2Cr_2O_7$ 为例,在酸性溶液中与 KI 作用:

$$Cr_2O_7^{2-} + 6I^- + 14H^+ = 2Cr^{3+} + 3I_2 + 7H_2O \quad (3.39)$$

析出的 I_2 以淀粉为指示剂,用 $Na_2S_2O_3$ 标准溶液滴定:

$$I_2 + 2S_2O_3^{2-} = S_4O_6^{2-} + 2I^- \quad (3.40)$$

反应(3.39)进行较慢。为加速反应,需加入过量的 KI 并提高酸度。然而酸度过高又加速空气氧化 I^-:

$$4I^- + 4H^+ + O_2 = 2I_2 + 2H_2O \quad (3.41)$$

因此,控制反应溶液的酸度十分重要。一般控制酸度为 0.4mol/L 左右。

反应(3.41)中 I_2 与 $S_2O_3^{2-}$ 的反应是碘量法中最重要的反应。酸度控制不当会影响它们的计量关系,造成误差,因此有必要着重讨论。反应(3.41)中 I_2 与 $S_2O_3^{2-}$ 的物质的量比应为 1:2。

如果该滴定反应在酸度较高的条件下进行,会发生如下反应:

$$S_2O_3^{2-} + 2H^+ = H_2SO_3 + S\downarrow \quad (3.42)$$

$$I_2 + H_2SO_3 + H_2O = SO_4^{2-} + 4H^+ + 2I^- \quad (3.43)$$

这时,I_2 与 $S_2O_3^{2-}$ 反应的物质的量比为 1:1,由此会造成误差。

若溶液 pH 过高(即强碱性溶液中),则在滴定之前,I_2 会部分歧化生成 OI^- 和 IO_3^-,它们将部分地氧化 $S_2O_3^{2-}$ 为 SO_4^{2-}:

$$4I_2 + S_2O_3^{2-} + 10OH^- = 2SO_4^{2-} + 8I^- + 5H_2O \quad (3.44)$$

即部分 I_2 与 $S_2O_3^{2-}$ 按 4:1 物质的量比发生反应,这也会造成误差。

综上,I_2 和 $S_2O_3^{2-}$ 的滴定反应,应在中性或弱酸性条件下进行。标定时第一步反应的酸度较高,所以在用 $Na_2S_2O_3$ 滴定前先用蒸馏水稀释,一则降低酸度可减少空气对 I^- 的氧化,二则使 Cr^{3+} 的蓝绿色减弱,便于观察终点。淀粉应在近终点时加入,否则淀粉吸附部分 I_2,致使终点提前且不明显。若滴定至终点后,溶液迅速变蓝,表示 $Cr_2O_7^{2-}$ 与 I^- 的反应未定量完成,遇此情况,实验应重做。

实验用品

仪器:研钵,碘量瓶等。

试剂:$Na_2S_2O_3 \cdot 5H_2O$、I_2 固体试剂、KIO_3 或 $K_2Cr_2O_7$(干燥固体)、As_2O_3(固体,于 105℃干燥 2 h)、KI、Na_2CO_3、$NaHCO_3$、6mol/L HCl 溶液、6mol/L NaOH 溶液、淀粉溶液(0.5%)、7mol/L $NH_3 \cdot H_2O$ 溶液、20% NH_4HF_2 溶液、8mol/L HAc 溶液、10% NH_4SCN 溶液。

实验步骤

(1)0.05mol/L I_2 溶液的配制 称取(　　)g I_2 和 5 g KI 置于研钵中,在通风橱内操作。加入少量蒸馏水研磨至 I_2 全部溶解后,将溶液定量转入棕色试剂瓶中,加水稀释至 250mL,充分摇匀放置暗处过夜再标定。

(2)0.1mol/L $Na_2S_2O_3$ 溶液的配制 称取(　　)g $Na_2S_2O_3 \cdot 5H_2O$ 于烧杯中,加入 200mL 新煮沸已冷却的蒸馏水,搅拌,待 $Na_2S_2O_3$ 完全溶解后,加入 0.1 g Na_2CO_3,然后用新煮沸已冷却的蒸馏水稀释至 500mL,储存于棕色试剂瓶中,在暗处放置 3～5 天后标定。

(3) 0.05mol/L I_2 溶液的标定

① 用 As_2O_3 标定。准确称取（　　）~（　　）g 基准 As_2O_3，置于100mL小烧杯中，加入 6mol/L NaOH 溶液 10mL，温热溶解，加 2 滴酚酞指示剂，用 6mol/L HCl 中和至溶液刚好无色，然后加入 2~3g $NaHCO_3$，搅拌溶解后，将溶液定量转移到 250mL 容量瓶中，加水稀释至刻度，摇匀。

用移液管移取 25.00mL 上述试液三份，分别置于 250mL 洁净的锥形瓶中，加入 50mL 水、5g $NaHCO_3$，再加 2mL 淀粉溶液，用 I_2 标准溶液滴定至溶液呈蓝色且 30s 内不褪色即为终点，记录消耗 I_2 溶液的体积，平行标定三次，计算 I_2 标准溶液的浓度。

② 用 $Na_2S_2O_3$ 标准溶液标定。吸取 25.00mL $Na_2S_2O_3$ 标准溶液三份，分别置于 250mL 锥形瓶中，加入 50mL 水、2mL 淀粉溶液，用 I_2 标准溶液滴定至溶液呈蓝色且 30s 内不褪色即为终点。计算 I_2 标准溶液的浓度。

(4) 0.1mol/L $Na_2S_2O_3$ 溶液的标定

① 用基准 KIO_3 标定。准确称取（　　）~（　　）g 基准 KIO_3，置于 100mL 小烧杯中，加水溶解后，定量转入 250mL 容量瓶中，加水稀释至刻度，充分摇匀。

吸取 25.00mL 上述试液三份，分别置于 250mL 锥形瓶中，加入 10mL 6mol/L HCl 溶液，1g KI，溶解后，加水稀释至 120mL 左右，立即用待标定的 $Na_2S_2O_3$ 溶液滴定，当溶液由棕色变为浅黄色时，加入 2mL 淀粉溶液，继续滴定至蓝色消失即为终点，记录消耗 $Na_2S_2O_3$ 溶液的体积，计算 $Na_2S_2O_3$ 标准溶液的浓度。

② 用基准 $K_2Cr_2O_7$ 标定。准确称取（　　）~（　　）g 基准 $K_2Cr_2O_7$，置于 100mL 小烧杯中，加 30mL 蒸馏水溶解后，定量转移到 250mL 容量瓶中，加水稀释至刻度，充分摇匀。

吸取 25.00mL 上述试液三份，分别置于 250mL 碘量瓶中，加入 10mL 6mol/L HCl 溶液，1g KI，立即加盖玻璃塞，摇匀置于暗处 5~10min 后，以水冲洗瓶盖和内壁。加入 100mL 蒸馏水，用待标定的 $Na_2S_2O_3$ 溶液滴定至浅黄色，然后加入 2mL 淀粉溶液，继续用 $Na_2S_2O_3$ 溶液滴定至蓝色消失即为终点，记录消耗 $Na_2S_2O_3$ 溶液的体积，计算 $Na_2S_2O_3$ 标准溶液的浓度。

③ 用纯铜标定。准确称取（　　）~（　　）g 金属纯铜三份，分别置于 250mL 锥形瓶中，加入 2mL 6mol/L HCl 溶液，加热后，取下，慢慢滴加 30% H_2O_2 2~3mL（尽量少加，只要能使铜分解完全即可），用低温加热至试样完全溶解后，再大火加热将多余的 H_2O_2 分解赶尽，冷却后，加入 20mL 水。

滴加 1:1 氨水至溶液有沉淀生成后，加入 8mL 1:1 HAc，10mL 20% NH_4HF_2，1g KI，用 $Na_2S_2O_3$ 滴定至溶液呈淡黄色，再加 3mL 淀粉指示剂，继续滴定至浅蓝色，然后加入 10mL 10% NH_4SCN 溶液，继续用 $Na_2S_2O_3$ 溶液滴定至蓝色消失即为终点，记录消耗 $Na_2S_2O_3$ 溶液的体积，计算 $Na_2S_2O_3$ 标准溶液的浓度。

数据记录与处理

(1) 0.05mol/L I_2 溶液的配制（表 3.26）

表 3.26　0.05mol/L I_2 溶液的配制

m_{I_2}/g	

(2) 0.05mol/L I_2 溶液的标定（As_2O_3 法）（表 3.27）

表 3.27 0.05mol/L I_2 溶液的标定（As_2O_3 法）

实验序号	1	2	3
$m_{As_2O_3}$（初始）/g			
$m_{As_2O_3}$（终点）/g			
$m_{As_2O_3}$/g			
$c_{As_2O_3}$/(mol/L)			
$V_{As_2O_3}$/mL			
V_{I_2}（初始）/mL			
V_{I_2}（终点）/mL			
V_{I_2}（消耗）/mL			
c_{I_2}/(mol/L)			
$\bar{c}_{I_2}$/(mol/L)			
$|d_i|$			
$\bar{d}_r\%$			

（3）0.05mol/L I_2 溶液的标定（$Na_2S_2O_3$ 法）（表 3.28）

表 3.28 0.05mol/L I_2 溶液的标定（$Na_2S_2O_3$ 法）

实验序号	1	2	3
$V_{Na_2S_2O_3}$/mL			
V_{I_2}（初始）/mL			
V_{I_2}（终点）/mL			
V_{I_2}（消耗）/mL			
c_{I_2}/(mol/L)			
$\bar{c}_{I_2}$/(mol/L)			
$|d_i|$			
$\bar{d}_r\%$			

（4）0.1mol/L $Na_2S_2O_3$ 溶液的配制（表 3.29）

表 3.29 0.1mol/L $Na_2S_2O_3$ 溶液的配制

$m(Na_2S_2O_3)$/g	

（5）0.1mol/L $Na_2S_2O_3$ 溶液的标定（表 3.30）

表 3.30 0.1mol/L $Na_2S_2O_3$ 溶液的标定

实验序号	1	2	3
$m_{Na_2S_2O_3}$（初始）/g			
$m_{Na_2S_2O_3}$（终点）/g			
$m_{Na_2S_2O_3}$/g			
$c_{基准}$/(mol/L)			
$V_{基准}$/mL			

续表

实验序号	1	2	3
V(初始)/mL			
V(终点)/mL			
V(消耗)/mL			
$c_{Na_2S_2O_3}$/(mol/L)			
$\bar{c}_{Na_2S_2O_3}$/(mol/L)			
$\lvert d_i \rvert$			
$\bar{d}_r\%$			

思考题

(1) 为什么不能直接准确称量配制 $Na_2S_2O_3$ 标准溶液？如何配制一个相对稳定的 $Na_2S_2O_3$ 溶液？配制时为什么要用新煮沸并冷却的蒸馏水？

(2) 用 As_2O_3 标定 I_2 溶液的浓度时，为什么要加入固体 $NaHCO_3$？能否改用 Na_2CO_3，为什么？

(3) 用 $K_2Cr_2O_7$ 标准溶液标定 $Na_2S_2O_3$ 溶液的浓度时，为什么滴定前需密闭、放置暗处 5~10min？

(4) 标定 $Na_2S_2O_3$ 溶液时，为什么要加入过量的 KI？

实验3.16 葡萄糖口服液中葡萄糖含量的测定

实验目的

(1) 掌握用间接碘量法测定葡萄糖含量的原理和方法。
(2) 掌握返滴定法的原理与操作。

实验原理

葡萄糖（$C_6H_{12}O_6$）分子中含有醛基，在碱性溶液中，能被 I_2（过量）定量地氧化成相应的一元酸，反应如下：

$$I_2 + 2OH^- = IO^- + I^- + H_2O \tag{3.45}$$

$$CH_2OH(CHOH)_4CHO + IO^- + OH^- = CH_2OH(CHOH)_4COO^- + I^- + H_2O \tag{3.46}$$

碱性溶液中剩余的 IO^- 歧化为 IO_3^- 及 I^-：$3IO^- = IO_3^- + 2I^-$ (3.47)

溶液酸化后又析出 I_2：$IO_3^- + 5I^- + 6H^+ = 3I_2 \downarrow + 3H_2O$ (3.48)

然后用 $Na_2S_2O_3$ 标准溶液滴定析出的 I_2，可计算出糖氧化时所消耗的 I_2 量。

$$I_2 + 2S_2O_3^{2-} = 2I^- + S_4O_6^{2-} \tag{3.49}$$

此方法可用于测定葡萄糖口服液中葡萄糖的含量。

实验用品

仪器：酸碱滴定管、容量瓶、移液管、碘量瓶、称量瓶、锥形瓶、量筒、烧杯等。

试剂：$Na_2S_2O_3 \cdot 5H_2O$、KIO_3、KI、6mol/L 的 HCl 溶液、20% 的 NaOH 溶液、0.5% 的淀粉溶液、市售葡萄糖口服液。

实验步骤

(1) 0.1mol/L $Na_2S_2O_3$ 溶液的配制与标定　请参照实验 3.15。

(2) 葡萄糖含量的测定　用移液管准确移取市售葡萄糖口服液 25.00mL 置于 250mL 容量瓶中,加水稀释定容后待用。

准确称取 1.2~1.6g 基准 KIO_3,置于 100mL 小烧杯中,加水溶解后,定量转入 250mL 容量瓶中,加水稀释至刻度,充分摇匀。

吸取 25.00mL 上述 KIO_3 试液 3 份,分别置于 250mL 碘量瓶中,加入 10mL 6mol/L 的 HCl 溶液、1 g KI,立即盖好玻璃塞,摇动溶解后,以水冲洗瓶盖和内壁,准确加入 25.00mL 葡萄糖口服液的待测稀释液,边摇边滴加 20% 的 NaOH 溶液至溶液呈现黄色后,盖好玻璃塞,置于暗处放置 10min,以水冲洗玻璃塞并加入 4mL 6mol/L HCl 溶液,加水稀释至 120mL 左右,立即用 $Na_2S_2O_3$ 标准溶液滴定,当溶液变为浅黄色时,加入 2mL 淀粉溶液,继续滴定至蓝色消失即为终点。记录消耗的 $Na_2S_2O_3$ 溶液的体积,计算葡萄糖的含量(m/V,%)及结果的相对平均偏差。

数据记录与处理

(1) KIO_3 标准溶液的配制(表 3.31)

表 3.31　KIO_3 标准溶液的配制

m_{KIO_3}(初始)/g	
m_{KIO_3}(终点)/g	
m_{KIO_3}/g	
c_{KIO_3}/(mol/L)	

(2) 用 KIO_3 标准溶液标定 $Na_2S_2O_3$ 溶液的浓度(表 3.32)

表 3.32　用 KIO_3 标准溶液标定 $Na_2S_2O_3$ 溶液的浓度

实验序号	1	2	3
V_{KIO_3}/mL			
V(初始)/mL			
V(终点)/mL			
V(消耗)/mL			
$c_{Na_2S_2O_3}$/(mol/L)			
$\bar{c}_{Na_2S_2O_3}$/(mol/L)			
$\lvert d_i \rvert$			
$\bar{d}_r$%			

(3) 葡萄糖含量的测定(表 3.33)

表 3.33　葡萄糖含量的测定

实验序号	1	2	3
$V_{葡萄糖}$/mL			
V(初始)/mL			
V(终点)/mL			
V(消耗)/mL			

续表

实验序号	1	2	3
$m_{葡萄糖}/g$			
$w_{葡萄糖}/\%$			
$\overline{w}_{葡萄糖}/\%$			
$\lvert d_i \rvert$			
$\overline{d}_r\%$			

思考题

(1) 在测定步骤中，未加入葡萄糖待测液前加 HCl，在葡萄糖试液与碱反应后又加入 HCl，其作用分别是什么？写出它们相应的反应式。

(2) 用 $Na_2S_2O_3$ 溶液滴定前为什么要加水稀释？

(3) 为什么要在溶液呈黄色时加入淀粉溶液？淀粉过早加入有什么不好？

葡萄糖含量测定微量滴定实验

实验 3.17　铜合金中铜含量的测定

实验目的

(1) 学习铜合金试样的制备方法。

(2) 掌握用间接碘量法测定铜含量的原理和操作过程。

实验原理

铜合金是制造业的重要基础材料，主要包括铜锡合金（青铜）、铜锌合金（黄铜）、铜镍合金（白铜），分别用于生产轴承齿轮、水管阀门、耐腐蚀构件等。纯铜（又称紫铜或红铜）主要用于生产电缆电线。铜的含量对合金的力学性能、电学性能等具有重要影响。因此，测定铜合金中铜的含量对于确定铜的品位十分重要。

在溶解铜合金样品时，可用 HNO_3 溶解试样，再加入浓 H_2SO_4 在加热条件下去除过量的硝酸和低价氮的氧化物，避免其氧化 I^- 而干扰测定。也可用 HCl 和 H_2O_2 分解试样，分解完成后煮沸以除去过量的 H_2O_2 和 HCl。

$$Cu + 2HCl + H_2O_2 =\!=\!= CuCl_2 + 2H_2O \tag{3.50}$$

实验室中，可以用间接碘量法测定铜的含量。铜离子与过量的碘化钾反应定量析出 I_2，然后用 $Na_2S_2O_3$ 标准溶液滴定，反应方程式如下：

$$2Cu^{2+} + 4I^- =\!=\!= 2CuI\downarrow + I_2 \tag{3.51}$$

$$I_2 + 2S_2O_3^{2-} =\!=\!= 2I^- + S_4O_6^{2-} \tag{3.52}$$

Cu^{2+} 与 I^- 之间的反应是可逆的，为保证正向反应完全进行，需要加入过量 KI。但是生成的 CuI 沉淀会吸附部分 I_2，导致结果偏低。为减小这一误差，常在计量点前加入硫氰酸盐，使 CuI 沉淀（$K_{sp} = 1.1\times10^{-12}$）转化为溶解度更小的 CuSCN 沉淀（$K_{sp} = 4.8\times10^{-15}$）。

$$CuI + SCN^- =\!=\!= CuSCN\downarrow + I^- \tag{3.53}$$

这样不但可以释放出被吸附的 I_2，而且反应时再生出来的 I^- 与未反应的 Cu^{2+} 发生作用，可以使反应进行得更完全。但 NH_4SCN 只能在接近终点时加入，否则 SCN^- 可能直接还原 Cu^{2+} 而使结果偏低。

$$6Cu^{2+} + 7SCN^- + 4H_2O \Longrightarrow 6CuSCN\downarrow + SO_4^{2-} + CN^- + 8H^+ \tag{3.54}$$

Cu^{2+} 与 I^- 的反应在弱酸性介质中进行。酸度过低，Cu^{2+} 易水解，使反应不完全，结果偏低，而且反应速度慢，终点拖长；酸度过高，则 I^- 被空气中的氧氧化为 I_2（Cu^{2+} 催化此反应），使结果偏高。一般加入 NH_4HF_2 缓冲溶液调节 pH 为 3.0~4.0，避免 Cu^{2+} 水解以及 As(V)、Sb(V) 等干扰离子对 I^- 的氧化。加入的 F^- 还能有效地络合 Fe^{3+}，从而消除 Fe^{3+} 的干扰。

实验用品

仪器：酸碱滴定管、移液管、容量瓶、锥形瓶、称量瓶、量筒、烧杯等。

试剂：$Na_2S_2O_3 \cdot 5H_2O$ 固体试剂（分析纯）、纯铜（s）、KI（s，分析纯）、6mol/L HCl 溶液、1:1 的 $NH_3 \cdot H_2O$、1:1 的 HAc、20% 的 NH_4HF_2、10% 的 NH_4SCN、淀粉溶液、30% H_2O_2、铜合金试样。

实验步骤

(1) 0.1mol/L 的 $Na_2S_2O_3$ 标准溶液的配制与标定（用纯铜标定） 称取 12.4 g $Na_2S_2O_3 \cdot 5H_2O$ 于烧杯中，加入 200mL 新煮沸已冷却的蒸馏水，搅拌，待 $Na_2S_2O_3$ 完全溶解后，加入 0.1 g Na_2CO_3，然后用新煮沸已冷却的蒸馏水稀释至 500mL，储存于棕色试剂瓶中，在暗处放置 3~5 天后标定。

准确称取（   ）~（   ）g 金属纯铜 3 份，分别置于 250mL 锥形瓶中，加入 2mL 6mol/L HCl 溶液。加热后，取下，慢慢滴加 30% H_2O_2 2~3mL（尽量少加，只要能使铜分解完全即可），用低温加热至试样完全溶解后，再大火加热将多余的 H_2O_2 分解赶尽，冷却后，加入 20mL 水。

滴加 1:1 氨水至溶液有沉淀生成后，加入 8mL 1:1 HAc、10mL 20% NH_4HF_2、1 g KI，用 $Na_2S_2O_3$ 滴定至溶液呈淡黄色，再加入 3mL 淀粉指示剂，继续滴定至浅蓝色，然后加入 10mL 10% NH_4SCN 溶液，继续用 $Na_2S_2O_3$ 溶液滴定至蓝色消失即为终点，记录消耗 $Na_2S_2O_3$ 溶液的体积，计算 $Na_2S_2O_3$ 标准溶液的浓度。

(2) 铜合金中铜含量的测定 准确称取铜合金试样（   ）~（   ）g（含 Cu 量约 65%~70%）3 份，置于 250mL 锥形瓶中，加入 10mL 6mol/L HCl，温热后滴加约 2mL 30% H_2O_2，加热使试样溶解完全后，再大火加热赶尽 H_2O_2，冷却后加入 60mL 水。

滴加 1:1 $NH_3 \cdot H_2O$，直到溶液中刚刚有稳定的沉淀生成，再加 8mL 1:1 HAc、10mL 20% NH_4HF_2、1 g KI，立即用 $Na_2S_2O_3$ 标准溶液滴定至浅黄色，加 3mL 0.5% 淀粉溶液，继续滴定至溶液呈灰蓝色，加入 10mL 10% NH_4SCN 溶液，继续滴定至蓝色刚好消失即为终点。此时，因有白色沉淀物存在，终点颜色呈灰白色或浅肉色。记录所消耗的 $Na_2S_2O_3$ 溶液总体积。计算试样中的铜含量以及结果的相对平均偏差。

数据记录与处理

(1) $Na_2S_2O_3$ 标准溶液的配制与标定（表 3.34）

表 3.34 $Na_2S_2O_3$ 标准溶液的配制与标定

实验序号	1	2	3
$m_{Na_2S_2O_3}$（初始）/g			
$m_{Na_2S_2O_3}$（终点）/g			
$m_{Na_2S_2O_3}$/g			

续表

实验序号	1	2	3
V(初始)/mL			
V(终点)/mL			
V(消耗)/mL			
$c_{Na_2S_2O_3}$/(mol/L)			
$\bar{c}_{Na_2S_2O_3}$/(mol/L)			
$\mid d_i \mid$			
$\bar{d}_r$/%			

（2）铜合金中铜含量的测定（表3.35）

表 3.35 铜合金中铜含量的测定

实验序号	1	2	3
m(初始)/g			
m(终点)/g			
m(铜试样)/g			
V(初始)/mL			
V(终点)/mL			
V(消耗)/mL			
m_{Cu}/g			
w_{Cu}/%			
$\bar{w}_{Cu}$/%			
$\mid d_i \mid$			
$\bar{d}_r$/%			

思考题

（1）用 $HCl\text{-}H_2O_2$ 溶解铜合金试样时，若 H_2O_2 未赶尽，对结果会有什么影响？

（2）标定 $Na_2S_2O_3$ 溶液可选用的基准物质有哪些？为什么本实验选用纯铜？

（3）本实验加入过量的 KI 的作用是什么？

（4）碘量法测铜含量时，加入 NH_4HF_2 的作用是什么？为什么需在临近终点时加入 NH_4SCN？

实验 3.18 莫尔法测定自来水中的氯

实验目的

（1）掌握用莫尔法测定水中微量 Cl^- 含量的原理。

（2）学习沉淀滴定的基本操作。

实验原理

某些可溶性氯化物中氯的含量可采用银量法测定。根据加入的指示剂不同，银量法又分为莫尔法、福尔哈德法和法扬斯法。指示剂分别是铬酸钾、铁铵矾和吸附指示剂。

莫尔法中，Cl^- 以 $AgNO_3$ 标准溶液为滴定剂进行测定，K_2CrO_4 为指示剂。滴定开始时，溶液呈黄色。当 AgCl 沉淀析出完全，计量点后稍过量的 Ag^+ 与 CrO_4^{2-} 生成砖红色的

Ag_2CrO_4 沉淀而指示终点。反应如下：

$$Ag^+ + Cl^- = AgCl \downarrow （白色） \quad (3.55)$$

$$2Ag^+ + CrO_4^{2-} = Ag_2CrO_4 \downarrow （砖红色） \quad (3.56)$$

采用本法进行实验时，应注意指示剂用量和滴定酸度两个方面。

(1) 根据 AgCl 的溶度积常数，可以计算出来化学计量点时 $[Ag^+] = 10^{-5} \text{mol/L}$。根据 Ag_2CrO_4 的溶度积可以计算出此时需要的 $[Cr_2O_4^{2-}] = 0.011 \text{mol/L}$。然而，此浓度下，其本身的黄色会干扰终点颜色变化的判断。通常，将 $[CrO_4^{2-}]$ 控制在 $2.0 \times 10^{-3} \sim 5.0 \times 10^{-3} \text{mol/L}$。同时，以 K_2CrO_4 为指示剂进行空白滴定，从实验终点消耗的滴定剂中减去空白消耗的滴定剂，获得真实终点。

(2) 莫尔法应在弱碱或者中性溶液中进行。如果溶液酸度太高，如 pH<6，CrO_4^{2-} 会部分转化为 $HCrO_4^-$ 和 H_2CrO_4。此时，CrO_4^{2-} 浓度减小，指示终点的 Ag_2CrO_4 沉淀出现晚或甚至不出现，导致测定误差。若滴定溶液碱性太强，则有 AgOH 甚至 Ag_2O 沉淀析出。此时，Ag_2CrO_4 的溶解度也会增大，造成终点拖后或无法确定。一般情况下，pH 应该控制在 6~10.0。

然而，莫尔法只能测定 Cl^- 或 Br^-，不能测 I^-、SCN^-，因为 AgI、AgSCN 沉淀具有极强的吸附能力，使终点变色不明显，误差较大。

实验用品

仪器：酸碱滴定管、容量瓶、移液管、锥形瓶、称量瓶、量筒、烧杯等。

试剂：① NaCl(s)：NaCl 放在瓷坩埚中加热并不断搅拌，待爆炸声停止后再加热 15min，在干燥器中冷却后使用。

② 0.05mol/L $AgNO_3$ 溶液：称取 2.2 g $AgNO_3$ 并溶解于 250mL 不含 Cl^- 的蒸馏水中，在棕色瓶中避光保存。

③ K_2CrO_4 溶液（0.5%）。

④ 待测水样。

实验步骤

(1) $AgNO_3$ 标准溶液的标定　准确称取 0.7~0.8 g NaCl 基准试剂于小烧杯中，用蒸馏水溶解后定量转移至 250mL 容量瓶中，稀释至刻度，摇匀。

移取该溶液 25.00mL 置于锥形瓶中，加入 25mL H_2O、1.0mL K_2CrO_4 指示剂，在充分摇动下，用 $AgNO_3$ 溶液滴定至出现淡砖红色浑浊即为终点。平行测定 3 次。计算 $AgNO_3$ 溶液的平均浓度。

(2) 自来水中 Cl^- 含量的测定　量取 100mL 自来水于锥形瓶中，加 1mL K_2CrO_4 指示剂，在不断地摇动下用 $AgNO_3$ 标准溶液滴定至出现淡砖红色浑浊即为终点。记录消耗 $AgNO_3$ 溶液的体积。平行测定 3 次。计算自来水中 Cl^- 的含量和该结果的相对平均偏差。

数据记录与处理

(1) $AgNO_3$ 标准溶液的标定（表 3.36）

表 3.36　$AgNO_3$ 标准溶液的标定

实验序号	1	2	3
m_{NaCl}(初始)/g			
m_{NaCl}(终点)/g			

续表

实验序号	1	2	3
m_{NaCl}/g			
$c_{NaCl}/(mol/L)$			
V_{NaCl}/mL			
$c_{AgNO_3}/(mol/L)$			
$\bar{c}_{AgNO_3}/(mol/L)$			

(2) 自来水中 Cl^- 的测定（表3.37）

表 3.37 自来水中 Cl^- 的测定

实验序号	1	2	3
V（待测试液）/mL			
消耗硝酸银溶液体积			
V（初始）/mL			
V（终点）/mL			
V（消耗）/mL			
$c_{Cl^-}/(mol/L)$			
$\bar{c}_{Cl^-}/(mol/L)$			
$\|d_i\|$			
$\bar{d}_r/\%$			

思考题

(1) 莫尔法测定 Cl^- 时，适宜的 pH 范围是多少？为什么？

(2) 基准物质 NaCl 为什么要高温烘炒？如不烘炒对标定 $AgNO_3$ 有何影响？

实验 3.19　丁二酮肟重量法测定合金钢中的镍

实验目的

(1) 了解有机沉淀剂在重量分析中的应用。

(2) 学习烘干重量法的实验操作。

(3) 熟悉微波炉用于干燥样品方面的优点。

实验原理

通常，镍铬合金钢中含有百分之几至百分之几十的镍，可以用丁二酮肟重量法或 EDTA 络合滴定法进行测定。其中 EDTA 络合滴定法需要预先分离合金中的铁，操作比较繁琐。因此，在测定钢铁中高含量的镍时，仍常使用丁二酮肟重量法。

丁二酮肟在测定镍含量时具有较高的选择性。它在氨缓冲溶液中与镍生成红色的络合物。该络合物溶解度很小（$K_{sp}=2.3\times10^{-25}$），在溶液中形成的沉淀组成恒定，烘干后即可直接称量。丁二酮肟镍的结构如图 3.1 所示。

在酸性溶液中，丁二酮肟与钯和铂生成沉淀，在氨溶液中与镍、亚铁生成红色沉淀，故当亚铁离子存在时，必须预先氧化以消除干扰。铁（Ⅲ）、铬、钛等虽不与丁二酮肟反应，

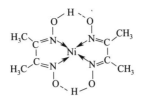

图 3.1 丁二酮肟镍的结构示意图

但在氨溶液中生成氢氧化物沉淀，亦干扰测定，故必须加入酒石酸或柠檬酸进行掩蔽。

丁二酮肟是二元酸（以 H_2D 表示），它以 HD^- 形式与 Ni^{2+} 络合，通常要控制溶液的 pH 为 7.0～8.0。若 pH 过高，丁二酮肟多以 D^{2-} 的形式存在，不与 Ni^{2+} 反应。同时，Ni^{2+} 与氨形成络合物，导致沉淀不完全，引起较大误差。

称样量以含 Ni 50～80mg 为宜。丁二酮肟的用量以过量 40%～80% 为宜，太少则沉淀不完全，过多则在沉淀冷却时析出，造成较大的正误差。

丁二酮肟的缺点之一，是试剂本身在水中的溶解度较小，必须使用乙醇溶液。但乙醇不可过量太多，否则会增大丁二酮肟镍的溶解度。在沉淀时，溶液要充分稀释，并将乙醇的浓度控制在 20% 左右，以防止过量试剂沉淀出来。

实验用品

仪器：分析天平、台秤、移液管、烧杯、容量瓶、两个玻璃坩埚（G4A 或 P16）、电动循环水真空泵及抽滤瓶、微波炉。

试剂：1% 丁二酮肟（乙醇溶液）、1:1 HCl 溶液、2mol/L HNO_3、1:1 $NH_3·H_2O$、50% 酒石酸溶液、95% 乙醇。

实验步骤

（1）坩埚恒重：以下两种方法可任选其一　微波炉加热干燥：用去离子水洗净坩埚，抽滤至水雾消失。在适宜的输出功率下，第一次加热 8min（有沉淀时 10min），第二次加热 3min，在保干器中冷却时间均为 10～12min。两次称的质量之差若不超过 0.4mg，即已恒重，否则应再次加热、冷却、称重，直至恒重。

电热恒温干燥箱加热干燥：控制温度为 (145±5)℃，第一次加热 1 h，第二次加热 30min。在保干器中冷却时间均为 30min。两次称的质量之差若不超过 0.4mg，即已恒重。

（2）溶解样品　准确称取 0.35 g 镍铬钢样，分别置于 250mL 烧杯中，盖上表面皿，从杯嘴处加入 20mL HCl 溶液和 20mL HNO_3 溶液，于通风橱中小火加热至完全溶解，再煮沸约 10min，以除去氮氧化物。稍冷，加水定容至 100mL。

（3）制备沉淀　移取 25mL 镍铬钢样于 250mL 锥形瓶中，加入 100mL 水、10mL 酒石酸溶液，在水浴中加热至 70℃，边搅拌边滴加氨水，调节 pH 为 9 左右。如有少量沉淀应用慢速滤纸过滤除去。滤液用 400mL 烧杯收集，并用热水洗涤锥形瓶三次，再用热水淋洗滤纸八次，最后使溶液总体积控制在 250～300mL。

在不断搅拌下，滴加 1:1 HCl 溶液调节 pH 为 3～4（变为深棕绿色），在水浴中加热至 70℃，再加入 20mL 乙醇和 35mL 丁二酮肟溶液。混匀之后，滴加 1:1 氨水调节 pH 为 8～9 之间，静置陈化 30min。

（4）过滤、干燥、恒重　在已知干燥恒重的玻璃坩埚中进行抽滤，将全部沉淀转移至坩埚中，先用 20% 乙醇溶液洗两次烧杯和沉淀，每次 10mL，再用温水洗涤烧杯和沉淀，少量多次，直至无 Cl^-。最后，抽干 2min 以上，直至不再产生水雾。

按照实验步骤（1）的操作条件，将沉淀干燥至恒重。称量并记录沉淀的质量。计算料液中镍的百分含量（%）。

注：丁二酮肟镍分子量为288.91；镍的原子量为58.69。

数据记录与处理

合金钢中镍含量的测定（表3.38）

表3.38 合金钢中镍含量的测定

实验序号	1	2
m（空坩埚）/g		
m（镍铬钢样）/g		
m（坩埚和沉淀）/g		
m（沉淀）/g		
w_{Ni}/%		
$\overline{w}_{Ni}$/%		

思考题

（1）为什么要先用20%的乙醇溶液洗涤烧杯和沉淀两次？
（2）如何检查Cl^-是否洗净？

实验3.20 邻二氮菲分光光度法测定微量铁

实验目的

（1）掌握分光光度计的使用方法。
（2）了解实验条件研究的一般方法，学会吸收曲线及标准曲线的绘制。
（3）掌握用分光光度法测定铁的原理。

实验原理

根据朗伯-比耳定律：$A=\varepsilon bc$，当入射光波长λ及光程b一定时，在一定浓度范围内，有色物质的吸光度A与该物质的浓度c成正比。只要绘出以吸光度A为纵坐标，浓度c为横坐标的标准曲线，测出试液的吸光度，就可以由标准曲线查得对应的浓度值，即未知样的含量。同时，还可应用相关的回归分析软件（Excel/Origin），将数据输入计算机，得到相应的分析结果。

邻二氮菲分光光度法是化工产品中测定微量铁的通用方法。在pH为2~9的溶液中，邻二氮菲和二价铁离子结合生成红色配合物，此配合物的$\lg K_{稳}=21.3$，摩尔吸光系数$\varepsilon_{510}=1.1\times10^4$ L/(mol·cm)。Fe^{3+}能与邻二氮菲生成3:1的配合物，呈淡蓝色，$\lg K_{稳}=14.1$。为了避免干扰，在加入显色剂之前，需用盐酸羟胺（$NH_2OH\cdot HCl$）将Fe^{3+}还原为Fe^{2+}，其反应式如下：

$$2Fe^{3+}+2NH_2OH\cdot HCl \longrightarrow 2Fe^{2+}+N_2+2H_2O+4H^++2Cl^- \quad (3.57)$$

邻二氮菲与铁离子的反应受到酸度的影响较大。溶液的酸度高，反应进行较慢；酸度太低，则离子易水解。本实验采用HAc-NaAc缓冲溶液控制溶液pH≈5.0。

本方法的选择性很高，相当于含铁量40倍的Sn^{2+}、Al^{3+}、Ca^{2+}、Mg^{2+}、Zn^{2+}、SiO_3^{2-}；20倍的Cr^{3+}、Mn^{2+}、VO^{3-}、PO_4^{3-}；5倍的Co^{2+}、Ni^{2+}、Cu^{2+}等离子不干扰测

定。但 Bi^{3+}、Cd^{2+}、Hg^{2+}、Zn^{2+}、Ag^+ 等离子与邻二氮菲作用生成沉淀干扰测定。

实验用品

仪器：分光光度计、酸度计、50mL 比色管、吸量管（1mL、2mL、5mL、10mL）、比色皿、洗耳球。

试剂：1.1×10^{-3} mol/L 铁标准溶液、100μg/mL 铁标准溶液、盐酸、盐酸羟胺、1mol/L 醋酸钠、0.15％邻二氮菲水溶液。

实验步骤

(1) 准备工作　打开仪器电源开关，预热，调节仪器。

(2) 测量工作

① 吸收曲线的绘制和测量波长的选择。用吸量管吸取 2.00mL 1.0×10^{-3} mol/L 铁标准溶液，注入 50mL 比色管中，加入 1.00mL 10％盐酸羟胺溶液，摇匀，加入 2.00mL 0.15％ 邻二氮菲溶液、5.00mL 1mol/L NaAc 溶液，以水稀释至刻度。在光度计上用 1cm 比色皿，采用试剂溶液为参比溶液，在 440~560 nm 间，每隔 10 nm 测量一次吸光度。以波长为横坐标，吸光度为纵坐标，绘制吸收曲线，选择测量的适宜波长。一般选用最大吸收波长 λ_{max} 为测定波长。

② 测定条件的优化

a. 邻二氮菲用量的优化。在 6 个比色管中，各加入 2.00mL 1.0×10^{-3} mol/L 铁标准溶液和 1.00mL 10％ 盐酸羟胺溶液，摇匀。分别加入 0.10mL、0.50mL、1.00mL、2.00mL、3.00mL 及 4.00mL 0.15％的邻二氮菲溶液、5.0mL 1mol/L NaAc 溶液，以水稀释至刻度，摇匀。在光度计上用 1cm 比色皿，采用试剂溶液为参比溶液，测吸光度。以邻二氮菲体积为横坐标，吸光度为纵坐标，绘制吸光度-试剂用量曲线，从而确定最佳的邻二氮菲用量。

b. 溶液 pH 的确定。取 8 个 50mL 比色管，各加入 2mL 1.00×10^{-3} mol/L 的标准铁溶液和 1mL 10％的盐酸羟胺溶液，摇匀。放置 2min，再加入特定体积（"a."中探究出的最适量）0.15％ 邻二氮菲溶液，摇匀，分别加入 0.00mL、0.20mL、0.50mL、1.00mL、1.50mL、2.00mL、3.00mL 1mol/L NaAc 溶液，用蒸馏水稀释至刻度，摇匀。用 pH 计测定各溶液的 pH。在分光光度计上，用 1cm 的比色皿选择适宜（由"①"所确定的）波长，以蒸馏水为参比，分别测其吸光度。在坐标纸上以加入的 NaAc 溶液体积数（或 pH）为横坐标，相应的吸光度为纵坐标，绘制 A-pH 曲线，确定测定过程中 pH 范围。

c. 反应时间。用吸量管吸取 2.00mL 1.00×10^{-3} mol/L 的标准铁溶液于 50mL 比色管中，加入 1mL 10％的盐酸羟胺溶液，摇匀（原则上每加入一种试剂后都要摇匀）。再加入特定体积的（"a."中探究出的）邻二氮菲溶液，加入特定体积的（"b."中探究出的）1mol/L NaAc 溶液，以水稀释至刻度，摇匀。在分光光度计上，用 1mL 的比色皿，以蒸馏水为参比溶液，每两分钟测一次，在由"①"所确定的波长下测定吸光度。以时间为横坐标，吸光度为纵坐标，绘制 A-t 吸收曲线，选择测量的最适时间。

(3) 标准曲线的制作　在 6 个 50mL 的容量瓶中，用 10mL 吸量管分别加入 0.00mL、2.00mL、4.00mL、6.00mL、8.00mL、10.00mL 100μg/mL 铁标准溶液，各加入 1mL 10％盐酸羟胺，摇匀（原则上每加入一种试剂都要摇匀）。再加入特定体积的 0.15％ 的邻二氮菲溶液和特定体积 1mol/L NaAc 溶液，以蒸馏水稀释至刻度，摇匀。放置一段时间（"c."中探究出的时间）。以试剂空白为参比，在"①"中所确定的波长下，用 1cm 的比色皿，测定各溶液的吸光度。绘制标准曲线。

(4) 试液含铁量的测定　准确吸取适量试液（如水样或工业盐酸、石灰石样品制备液等）代替标准溶液，其他步骤同上，平行3次测定其吸光度。记录光度计显示的吸光度，计算试液中铁的含量（以 mg/L 表示）。

数据记录与处理

（1）邻二氮菲-Fe^{2+} 吸收曲线的绘制（表3.39）

表3.39 吸收曲线

波长 λ/nm											
吸光度 A											
波长 λ/nm											
吸光度 A											
波长 λ/nm											
吸光度 A											

（2）实验条件的优化

① 显色剂用量与吸光度的关系。邻二氮菲用量曲线：（λ = _____ nm）

邻二氮菲的体积/mL	0.10	0.50	1.00	2.00	3.00	4.00
吸光度 A						

② 溶液的pH（醋酸钠的体积）与吸光度的关系

体积/mL	0.00	0.20	0.50	1.00	1.50	2.00	3.00
吸光度 A							

③ 反应时间

t/min	0	2	4	6	8	10	12	14	16	18	20
吸光度 A											

（3）标准曲线的制作

$V_{100\mu g/mL}$/mL	0.00	2.00	4.00	6.00	8.00	10.00
吸光度 A						

（4）样品中铁含量的测定（表3.40）

表3.40 样品中铁含量的测定

样品编号	1	2	3
吸光度 A			
w_{Fe}/%			
$\overline{w}_{Fe}$/%			

思考题

（1）邻二氮菲分光光度法测定微量铁时为何要加入盐酸羟胺溶液？
（2）参比溶液的作用是什么？在本实验中可否用蒸馏水作参比？
（3）邻二氮菲与铁的显色反应，其主要条件有哪些？
（4）在有关条件实验中，均以水为参比，为什么在测绘标准曲线和测定试液时，要以试剂空白溶液为参比？

实验 3.21　磷酸的电位滴定

实验目的

(1) 掌握电位滴定法的操作；
(2) 掌握电位滴定法中确定计量点的方法；
(3) 掌握电位滴定法测定弱酸的 pK_a 的原理及方法。

实验原理

电位滴定法是通过电极电位或 pH 值的"突跃"确定滴定终点从而分析目标物的方法。电位滴定法具有广泛的应用。在产生沉淀的反应中，或者分析混浊、有颜色的样品溶液时，通过观察指示剂的颜色判断滴定终点会产生较大误差。对此，电位滴定法显示出一定的优势。此外，电位滴定法还可用来测定某些物质的电离平衡常数。

磷酸的电位滴定实验中，以标准 NaOH 溶液为滴定剂，以 pH 计测定溶液的 pH 值或电位的变化。(pH 计的使用方法请参考 1.4.3) 以滴定剂的体积为横坐标，以溶液的 pH 值为纵坐标，绘制 NaOH-H_3PO_4 的滴定曲线（图 3.2）。在曲线上第一化学计量点的 pH 为 4.0～5.0，第二化学计量点的 pH 为 9.0～10.0。在滴定突跃处用"三切线法"作图，可以较准确地确定化学计量点。若要求更准确地确定化学计量点，则采用一级微商法 dpH/dv-V 和二级微商法 d^2pH/d^2v-V 确定（图 3.3）。

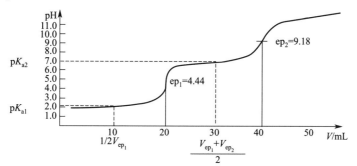

图 3.2　H_3PO_4 电位滴定曲线

磷酸为多元酸，其 pK_a 可用电位滴定法求得。

$$H_3PO_4 + H_2O \Longleftrightarrow H_3O^+ + H_2PO_4^- \tag{3.58}$$

$$K_{a1} = [H_3O^+][H_2PO_4^-]/[H_3PO_4] \tag{3.59}$$

当用 NaOH 标准液滴定至剩余 H_3PO_4 的浓度与生成 $H_2PO_4^-$ 的浓度相等，即 $[H_3PO_4] = [H_2PO_4^-]$，溶液中氢离子的浓度就是电离平衡常数 K_{a1}，即 $K_{a1} = [H_3O^+]$，$pK_{a1} = pH$。

$$H_2PO_4^- + H_2O \Longleftrightarrow H_3O^+ + HPO_4^{2-} \tag{3.60}$$

$$K_{a2} = [H_3O^+][HPO_4^{2-}]/[H_2PO_4^-] \tag{3.61}$$

当二级电离出的 H_3O^+ 被中和一半时，$[H_2PO_4^-] = [HPO_4^{2-}]$，则 $K_{a2} = [H_3O^+]$，$pK_{a2} = pH$。绘制 pH-V 滴定曲线，确定化学计量点，化学计量点一半的体积（半中和点的体积）对应的 pH 值，即为 H_3PO_4 的 pK_a。

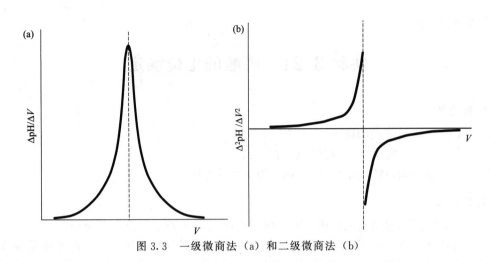

图 3.3　一级微商法（a）和二级微商法（b）

实验用品

仪器：PHSJ-4A 型精密酸度计、磁力搅拌器、50mL 滴定管、移液管、100mL 烧杯、250mL 容量瓶等。

试剂：0.1mol/L 的磷酸溶液、0.1mol/L 的 NaOH 标准溶液、pH＝4.00、6.86、9.18 的标准缓冲溶液、0.1％甲基橙指示剂（乙醇溶液）、1％酚酞指示剂（乙醇溶液）。

实验步骤

（1）0.1000mol/L NaOH 溶液的配制与标定　参见实验 3.4。

（2）用标准缓冲溶液校准 pH 计　具体操作方法见 1.4.3 pH 计的使用。

（3）按照图 3.4 所示连接滴定装置

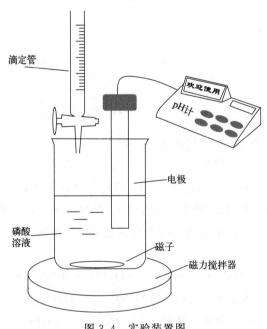

图 3.4　实验装置图

（4）测定　用滴定管精密地加入 0.1mol/L 磷酸样品溶液 20.00mL，置于 100mL 烧杯中，插入 pH 计的电极，放入搅拌磁子，滴加甲基橙和酚酞指示剂各 2~3 滴。打开磁力搅拌器的开关并调整至适当转速，用 0.1000mol/L NaOH 标准液滴定。当 NaOH 标准液体积未达到 20.00mL 之前，每加 2.00mL NaOH 记录一次 pH 值，在接近化学计量点（加入 NaOH 溶液时引起溶液的 pH 值变化逐渐增大）时，每次加入体积应逐渐减小，在化学计量点前后每加入一滴（如 0.05mL）记录一次 pH 值，尽量保证滴加的 NaOH 溶液体积相等。在"突跃"部分要多测几个点。此时，可借助甲基橙指示剂的变色来判断第一化学计量点。继续滴定，当被测试溶液中出现微红色时，测量要仔细，每次滴加的 NaOH 的体积要少，直至出现第二个化学计量点时为止。在 pH 为 7 时，用 pH 为 9.86 的缓冲溶液进行碱性校正。之后再继续滴定，直到测量的 pH 值约为 11.0 为止。

数据记录与处理

（1）NaOH 溶液的配制与标定（表 3.41）

表 3.41　NaOH 溶液的配制与标定

实验序号	1	2	3
m_{NaOH}/g			
m(初始)/g			
m(终点)/g			
$m_{基准物质}$/g			
V(初始)/mL			
V(终点)/mL			
V(消耗)/mL			
c_{NaOH}/(mol/L)			
$\bar{c}_{NaOH}$/(mol/L)			
$\lvert d_i \rvert$			
$\bar{d}_r$/%			

（2）磷酸的电位滴定（表 3.42）

表 3.42　磷酸的电位滴定

V_{NaOH}/mL								
pH 值								
V_{NaOH}/mL								
pH 值								
V_{NaOH}/mL								
pH 值								
V_{NaOH}/mL								
pH 值								
V_{NaOH}/mL								
pH 值								

（3）计算氢氧化钠与磷酸的准确浓度，需写出详细的计算过程。

（4）根据表格 3.42 中的数据，做出 pH-V 曲线、$\Delta pH/\Delta V$-V 曲线以及 $\Delta^2 pH/\Delta V^2$-V 曲线，确定化学计量点。

（5）由 pH-V 曲线找出第一个化学计量点的半中和点的 pH 以及第一个化学计量点到第二个化学计量点间的半中和点的 pH，确定 H_3PO_4 的 pK_{a1} 和 pK_{a2}。

思考题

(1) 用 NaOH 滴定 H_3PO_4，第一化学计量点和第二化学计量点所消耗的 NaOH 体积理应相等，为什么实际上并不相等？

(2) 磷酸的第三级电离常数 K_{a3} 可以从滴定曲线上求得吗？如果可以，请说明求解方法；如果不可以，说明为什么。

【拓展阅读】

(1) 三切线法 在滴定曲线上两端的平坦转折处做 AB、CD 两条切线，在曲线"突跃部分"做 EF 切线与 AB 及 CD 二线交于 Q、P 二点，通过 P、Q 二点作 PG、QH 两条平行于横轴的直线，然后在此两条平行线间作垂线，在垂线中点处作平行于横轴的直线交曲线于 O 点，此点被称为拐点，即化学计量点。通过此点分别可得到计量点时的 pH 和消耗的滴定剂的体积（图 3.5）。

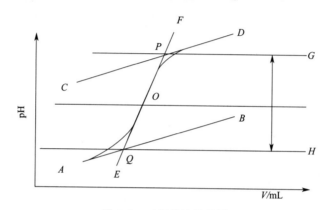

图 3.5 三切线法示意图

(2) 一级微商法和二级微商法 打开 Excel 工作表，在 A1 中输入 V，B1 中输入 pH，从第 2 行开始将采集的体积和 pH 分别输入到 A 列和 B 列。同时选中两列数据，在插入栏目下选择图表（带平滑线的散点图），即可得到电位滴定曲线。添加坐标轴标题将图完善。

另起一列，在 C1 中输入 ΔV，在 C3 中输入"＝A3－A2"可得到一个 ΔV 值。选中 C3，将鼠标拖至 C3 右下角出现黑色十字，点击鼠标左键下拉，求得所有 ΔV 值。另起一列，在 D1 中输入 ΔpH，用同样方法求得所有 ΔpH 值。在 E1 中输入 $\Delta pH/\Delta V$，在 E3 中输入"＝D3/C3"，下拉鼠标求得所有 $\Delta pH/\Delta V$ 值。选中 A 列和 E 列数据作图，得到一级微商曲线。添加坐标轴标题将图完善。

在 F1 中输入 $\Delta^2 pH$，在 F4 中输入"＝D4－D3"求得一个 $\Delta^2 pH$ 值，选中 F4，将鼠标拖至 F4 右下角出现黑色十字，点击鼠标左键下拉，求得所有 $\Delta^2 pH$ 值。在 G1 中输入 $\Delta^2 pH/\Delta^2 V$，在 G4 中输入"＝F4/C4"求得一个 $\Delta^2 pH/\Delta^2 V$ 值，再求出所有的 $\Delta^2 pH/\Delta^2 V$ 值。选中 A 列和 F 列数据作图，得到二级微商曲线。添加坐标轴标题将图完善。

第 4 章
综合设计性实验

4.1 化学类

实验 4.1 苯甲酸含量的测定

实验目的

(1) 掌握乙醇溶剂中芳香羧酸（弱酸）的滴定原理及指示剂的选择。
(2) 掌握中性乙醇试剂的配制和检测方法。

实验原理

苯甲酸又称安息香酸，一种芳香酸类有机化合物，也是最简单的芳香酸，分子式为 $C_7H_6O_2$（$M=122.12g/mol$），结构见图 4.1。常温下，它是有光泽的、白色的鳞片状或针状结晶，无气味或微有类似安息香或苯甲醛的气味。微溶于冷水、己烷，易溶于乙醇、乙醚、三氯甲烷、苯等有机溶剂。苯甲酸以游离酸、酯或其衍生物的形式广泛存在于自然界中。

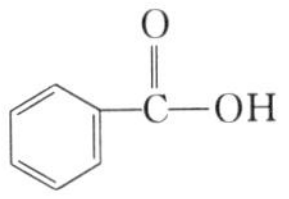

图 4.1 苯甲酸的分子结构

苯甲酸有抑制真菌、细菌、霉菌生长的作用，常用作食品或药品的防腐剂使用。适量的苯甲酸能抑制微生物的生长和繁殖，延长食品及药品的保存期。但过量摄入会对人体产生极大危害，对胃、皮肤和黏膜有一定的刺激性，对于敏感人群，较多地摄入苯甲酸及其盐会引起不良反应，如流口水、腹痛腹泻、哮喘、荨麻疹、代谢性酸中毒和抽搐等症状。因此，苯甲酸含量的测定十分重要。

目前，测定苯甲酸的方法主要有：气相色谱法、薄层色谱法、高效液相色谱法、紫外分光光度法、酸碱滴定法等。尽管色谱法具有较高的灵敏度和分离度，是检测防腐剂重要的分析手段之一，但对设备要求较高。相较而言，滴定分析法对设备要求低，是实验室中常用的苯甲酸测定方法。

苯甲酸是一元弱酸（$K_a = 6.28 \times 10^{-5}$），可用 NaOH 标准溶液直接滴定[式(4.1)]，

酚酞作指示剂，计量点时，苯甲酸钠水解溶液呈微碱性，使酚酞变红而指示终点。

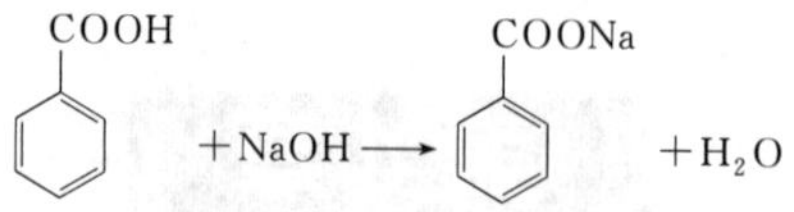

$$\text{C}_6\text{H}_5\text{COOH} + \text{NaOH} \longrightarrow \text{C}_6\text{H}_5\text{COONa} + \text{H}_2\text{O} \tag{4.1}$$

由于苯甲酸微溶于水，而易溶于乙醇，因此该滴定反应须使用乙醇作溶剂。乙醇一般显酸性，所以在该酸碱滴定中，应采用中性乙醇溶液作溶剂，所谓"中性"是指对所用指示剂显中性，以便不干扰指示剂变色。

实验用品

仪器：分析天平、台秤、烧杯、锥形瓶、量筒、滴定管；

试剂：NaOH（固体），邻苯二甲酸氢钾（KHP 优级纯），苯甲酸试样；

0.2%酚酞乙醇指示剂：0.2 g 酚酞溶于 100mL 95%的乙醇溶液；

中性乙醇溶液（对酚酞指示剂显中性）：取无水乙醇 50mL，加水 50mL，再加 2～3 滴酚酞指示剂，用 0.1mol/L NaOH 标准溶液滴定至溶液呈淡粉色备用。

实验步骤

（1）0.1mol/L NaOH 溶液配制与标定

配制：称取一定量的 NaOH 固体，用煮沸且冷却至室温的去离子水充分溶解，冷却至室温后转入聚乙烯塑料瓶中备用。

标定：根据 NaOH 溶液浓度及滴定误差要求估算基准物质 KHP 的称量质量。采用差减法准确称取计算量的 KHP 于 250mL 锥形瓶中，加水 20～30mL 使之溶解，加 2～3 滴酚酞指示剂，用待标定 NaOH 溶液滴定至溶液呈浅粉色并保持 30s 不褪色，即为终点。记录 NaOH 溶液消耗体积，平行标定 3 次（表 4.1）。

（2）试样分析　准确称取约 0.3 g 苯甲酸试样，置于 250mL 锥形瓶中，加中性乙醇溶液 25mL 使之充分溶解，加入 2～3 滴酚酞指示剂，用 NaOH 标准溶液滴定至溶液呈淡粉色（保持 30 s 不褪色）为终点。记录所消耗 NaOH 溶液的体积，平行测定 3 次（表 4.2）。

数据记录与处理

根据 NaOH 标准溶液消耗的体积，按下式[式(4.2)]计算苯甲酸的百分含量：

$$w = \frac{\bar{c}_{\text{NaOH}} V_{\text{NaOH}} M_{\text{C}_7\text{H}_6\text{O}_2}}{m_s \times 1000} \times 100\% \tag{4.2}$$

式中，$M_{\text{C}_7\text{H}_6\text{O}_2}$ 为苯甲酸的摩尔质量，122.12mol/L；m_s 为试样的质量，g。

表 4.1　NaOH 溶液浓度的标定

实验序号	1	2	3
m_{KHP}（初始）/g			
m_{KHP}（终点）/g			
m_{KHP}/g			
V_{NaOH}（初始）/mL			
V_{NaOH}（终点）/mL			
V_{NaOH}（消耗）/mL			

续表

实验序号	1	2	3
$c_{\text{NaOH}}/(\text{mol/L})$			
$\overline{c}_{\text{NaOH}}/(\text{mol/L})$			
$\lvert d_i \rvert$			
$\overline{d}_r/\%$			

表 4.2 苯甲酸含量的测定

实验序号	1	2	3
m_s/g			
$V_{\text{NaOH}}(初始)/\text{mL}$			
$V_{\text{NaOH}}(终点)/\text{mL}$			
$V_{\text{NaOH}}(消耗)/\text{mL}$			
w			
$\overline{w}$			
$\lvert d_i \rvert$			
$\overline{d}_r/\%$			

思考题

(1) 中性稀乙醇是否真正呈中性？为什么要用这种中性稀乙醇溶解而不是用水？

(2) 实验中用什么量器量取中性稀乙醇，是否需要精确量取？

(3) 本实验中，差减法称样过程中苯甲酸倾出过多（重量已超 0.7 g），是否需重称？为什么？

(4) 如果 NaOH 标准溶液吸收了空气中的 CO_2，对测定结果有何影响？

实验 4.2　苯酚含量的测定

实验目的

(1) 熟悉 $KBrO_3$ 法测定苯酚含量的原理、方法及基本操作。

(2) 了解"空白试验"的实际意义，学会"空白试验"的方法和应用。

实验原理

苯酚是一种重要的化工产品，常用于生产染料、树脂、药物、合成纤维及其他高分子材料，也广泛用于消毒、杀菌等。同时，它也具有很强的毒性，主要通过皮肤、消化道、呼吸道三种途径进入人体，使人体黏膜、心血管和中枢神经系统受到腐蚀、损害和抑制。许多化工及制药企业废水中含有苯酚成分，这些废水直接排放进入环境水体将对土壤、水资源造成严重的污染，进而对人类及其他生物造成危害。因此，苯酚含量的监测在实际应用中是十分必要的。

目前，水体中苯酚含量的测定方法主要有色谱法、电化学分析法、荧光分析法、分光光度法、化学发光法等。常量分析可采用滴定分析法，其对设备要求低，是实验室中常用的分析测试手段。

Br_2 与苯酚能够发生取代反应，生成稳定的三溴苯酚白色沉淀，反应如下：

$$\text{C}_6\text{H}_5\text{OH} + 3\text{Br}_2 \longrightarrow \text{C}_6\text{H}_2\text{Br}_3\text{OH} + 3\text{HBr} \tag{4.3}$$

由于上述反应进行较慢，且单质 Br_2 极易挥发，因此，不能采用单质 Br_2 溶液直接滴定。一般使用 $KBrO_3$ 标准溶液与过量 KBr 在酸性介质中进行反应，以产生相当量的游离 Br_2：

$$BrO_3^- + 5Br^- + 6H^+ \Longrightarrow 3Br_2(黄) + 3H_2O \tag{4.4}$$

所生成的 Br_2 与苯酚发生溴代反应后，剩余的 Br_2 可将过量的 KI 氧化置换出一定量的 I_2，再用 $Na_2S_2O_3$ 标准溶液滴定产生的 I_2，进而计算苯酚的含量，反应如下：

$$Br_2(剩余) + 2I^- \Longrightarrow 2Br^- + I_2 \tag{4.5}$$

$$I_2 + 2S_2O_3^{2-} \Longrightarrow 2I^- + S_4O_6^{2-} \tag{4.6}$$

由上可见，该反应过程存在如下计量关系，由此计算苯酚的含量。

$$KBrO_3 \approx 3Br_2 \approx C_6H_5OH \approx 3I_2 \approx 6Na_2S_2O_3 \tag{4.7}$$

本实验中，同时进行空白试验，样品测定和空白试验在相同的条件下进行，这样可以减小误差。

实验用品

仪器：分析天平、台秤、移液管、烧杯、容量瓶、滴定管、碘量瓶、量筒。

试剂：$Na_2S_2O_3$（0.05mol/L，表 4.3）、0.5％淀粉水溶液、KI 溶液（A.R.）、10％ NaOH 溶液、$KBrO_3$（基准物质）、1∶1 HCl 溶液、KBr（A.R.）和工业苯酚试样。

实验步骤

（1）$KBrO_3$-KBr 标准溶液（约 0.02mol/L）的配制　准确称取 0.25～0.30 g 干燥的 $KBrO_3$，加入 1g KBr，用适量去离子水溶解后，定容于 100mL 容量瓶备用。

（2）苯酚含量的测定

① 试样测定：准确称取约 0.2 g 苯酚试样，加入 5mL 10％ NaOH 溶液，再加少量水使溶解，并定容于 250mL 容量瓶。准确量取该苯酚试液 25.00mL 于 250mL 碘量瓶中，加入 10.00mL $KBrO_3$-KBr 标准溶液，10mL 1∶1 HCl 溶液，应立即加塞并加水封住瓶口，充分摇动 1～2min，静置 10min。加入 1g KI，加塞振摇溶解 5min 后，用少量水冲洗瓶塞及瓶颈上的附着物，加水 25mL，再用 $Na_2S_2O_3$ 标准溶液滴定至溶液呈浅黄色时，加入淀粉指示剂 1～2mL，继续滴定至蓝色恰好消失为终点，平行测定三份（表 4.4）。

② 空白试验：准确量取去离子水 25.00mL 替代苯酚试液，加入 250mL 碘量瓶中，其他操作与苯酚试样测定时相同，平行测定三份。

数据记录与处理

根据实验中的相关数据，按下式计算苯酚质量百分含量：

$$w = \frac{\left[c_{KBrO_3}V_{KBrO_3} - \frac{1}{6}\bar{c}_{Na_2S_2O_3}V_{Na_2S_2O_3}\right]M_{C_6H_5OH}}{m_s \times \frac{25}{250} \times 1000} \times 100\% \tag{4.8}$$

式中，m_s 为苯酚试样的质量，g；$M_{C_6H_5OH}$ 为苯酚的摩尔质量，94.11 g/mol。

表 4.3　$Na_2S_2O_3$ 标准溶液的标定

实验序号	1	2	3
m_{KBrO_3}(初始)/g			
m_{KBrO_3}(终点)/g			
m_{KBrO_3}/g			
c_{KBrO_3}/(mol/L)			
V_{KBrO_3}/mL			
$V_{Na_2S_2O_3}$(初始)/mL			
$V_{Na_2S_2O_3}$(终点)/mL			
$V_{Na_2S_2O_3}$(消耗)/mL			
$c_{Na_2S_2O_3}$/(mol/L)			
$\bar{c}_{Na_2S_2O_3}$/(mol/L)			
$|d_i|$			
$\bar{d}_r$/%			

表 4.4　苯酚质量含量的测定

实验序号	1	2	3
m_s(试样)/g			
V(试样)/mL			
$V_{Na_2S_2O_3}$(初始)/mL			
$V_{Na_2S_2O_3}$(终点)/mL			
$V_{Na_2S_2O_3}$(消耗)/mL			
w/%			
$\bar{w}$/%			
$|d_i|$			
$\bar{d}_r$/%			

思考题

(1) 溶解苯酚试样时，为何要加 10% NaOH 溶液？
(2) 加入 10mL 1∶1 HCl 溶液并盖塞，静置时，有何现象产生？为什么？
(3) 能否用 $Na_2S_2O_3$ 标准溶液直接滴定剩余的 Br_2？

实验 4.3　重量法测定可溶性钡盐的纯度

实验目的

(1) 了解硫酸钡重量法测定可溶性钡盐含量的原理和方法。
(2) 了解晶形沉淀的沉淀条件、原理和沉淀方法。
(3) 掌握晶形沉淀的过滤、洗涤及恒重的基本操作技术。
(4) 掌握微波炉干燥恒重样品的操作过程。

实验原理

重量分析法是通过称量物质的质量来确定被测物质组分含量的一种分析方法。分析时，一般是先采用适当方法将被测组分从试样中分离出来，转化为一定的称量形式并称重，由所称得的质量计算被测组分的含量。

Ba^{2+} 离子可与一些酸根离子作用，生成一系列微溶或难溶性化合物，如 $BaCO_3$（$K_{sp} = 2.58 \times 10^{-9}$，298.15K）、$BaC_2O_4$（$K_{sp} = 1.6 \times 10^{-7}$，298.15K）和 $BaSO_4$（$K_{sp} = 1.08 \times 10^{-10}$，298.15K）等。其中，$BaSO_4$ 溶解度最小，并且性质稳定，其化学组成与化学式相符合，摩尔质量较大，符合重量分析法对沉淀的要求。$BaSO_4$ 重量法既可以直接用于测定 Ba^{2+} 和 SO_4^{2-} 的含量，也可以用于间接测定可转化为 SO_4^{2-} 的含硫化合物中硫含量，比如水泥中的三氧化硫和铁矿石、煤、硅酸盐中的硫，等等。由于该方法的准确度较高，在分析工作中也常用重量法的测定结果作为标准，校对其他分析方法的准确度。

实验中，Ba^{2+} 与 SO_4^{2-} 形成 $BaSO_4$ 沉淀经陈化、过滤、洗涤和干燥后以 $BaSO_4$ 形式称重，进而求得 Ba^{2+} 或 SO_4^{2-} 的含量。$BaSO_4$ 沉淀初生成时一般为细小的晶体，过滤时易穿过滤纸而引起沉淀损失。为了获得颗粒较大且纯净的 $BaSO_4$ 晶形沉淀，也为了防止 $BaCO_3$ 和 $Ba(OH)_2$ 等沉淀的产生，一般在 0.05mol/L 的 HCl 介质中进行沉淀（适当提高酸度，可增加 $BaSO_4$ 在沉淀过程中的溶解度，以降低其相对过饱和度，有利于获得较好的晶形沉淀）。酸化后的溶液，加热至沸腾后，在不断搅拌下，缓慢加入热的稀 H_2SO_4 溶液，Ba^{2+} 与 SO_4^{2-} 反应形成晶形沉淀。

相较于传统的马弗炉加热，微波炉加热迅速、均匀，可快速达到很高的温度，采用玻璃砂芯坩埚不需要滤纸，避免高温下炭对 $BaSO_4$ 沉淀的还原作用。但若沉淀中包夹有 H_2SO_4 等高沸点杂质时，其在微波加热干燥 $BaSO_4$ 沉淀过程中难以分解或挥发。因此，在沉淀条件和洗涤操作方面要加以控制，比如将含 Ba^{2+} 试液进一步稀释，H_2SO_4 沉淀剂过量控制在 20%～50% 范围内。

实验用品

仪器：分析天平，电热板，烧杯（100mL，250mL），玻璃棒，微波炉，循环水真空泵（配抽滤瓶），量筒，表面皿，G4 砂芯坩埚，干燥器。

试剂：H_2SO_4 溶液（1mol/L，0.1mol/L）；HCl 溶液（2mol/L），$BaCl_2 \cdot 2H_2O$（工业品）。

实验步骤

(1) $BaSO_4$ 沉淀的制备　准确称取 0.4～0.5 g $BaCl_2 \cdot 2H_2O$ 试样于 250mL 烧杯中，加入 100mL 去离子水和 4mL HCl 溶液（2mol/L），搅拌溶解并加热接近沸腾。再取 3mL H_2SO_4 溶液（1mol/L）于 100mL 烧杯中，加 30mL 去离子后，加热接近沸腾，并在不断搅拌下将热的稀 H_2SO_4 溶液（1mol/L）滴加入 $BaCl_2$ 溶液中，直到加完为止。待 $BaSO_4$ 沉淀下沉后，在上层清液中加入 1～2 滴 0.1mol/L H_2SO_4 溶液，确定 $BaSO_4$ 是否沉淀完全。沉淀作用完全后，盖上表面皿（切勿将玻璃棒拿出杯外，以免沉淀损失）陈化 12 h（可将烧杯倾斜，使沉淀集中在烧杯一侧，以利于分离和转移），也可在热水浴或沙浴上保温 0.5 h 陈化。

(2) 坩埚恒重　用去离子水洗净坩埚，抽滤至水雾消失，然后将坩埚置于微波炉中用中高火加热（第一次加热 10min，第二次加热 4min）。之后将坩埚转移至干燥器中冷却 10～

15min，称重，两次称得质量之差若不超过 2mg 表明已恒重（否则继续加热、冷却、称重，直至恒重），坩埚质量记为 m_1。

（3）沉淀的处理　将陈化好的 $BaSO_4$ 沉淀全部转移至已经干燥恒重的坩埚（m_1）中减压过滤，并用温水洗涤沉淀多次。将盛沉淀的坩埚先置于微波炉内干燥，冷却后，称重，直至恒重[同步骤（2）的操作]，此时坩埚质量记为 m_2。

数据记录与处理

根据硫酸钡沉淀中钡的含量，计算氯化钡的质量，进而计算 $BaCl_2 \cdot 2H_2O$ 样品的纯度（表 4.5）。

$$\eta = \frac{\dfrac{M_{BaCl_2 \cdot 2H_2O}}{M_{BaSO_4}} \times (m_2 - m_1)}{m_{BaCl_2 \cdot 2H_2O}} \times 100\% \tag{4.9}$$

式中，$M_{BaCl_2 \cdot 2H_2O} = 244.26 \text{g/mol}$；$M_{BaSO_4} = 233.39 \text{g/mol}$。

表 4.5　氯化钡纯度的测定

$m_{BaCl_2 \cdot 2H_2O}/g$	m_1/g	m_2/g	$(m_2 - m_1)/g$	$\eta/\%$

思考题

（1）沉淀过程中，为什么要在稀 HCl 溶液介质中进行？
（2）沉淀过程中，滴加热 H_2SO_4 溶液时，为什么速度不能过快，并且要不断搅拌？
（3）整个过程中，为什么要求所使用的玻璃棒直至过滤、洗涤完毕后才能取出？
（4）搅拌时，为什么要求玻璃棒不要碰触杯壁及杯底？
（5）陈化过程中，将烧杯倾斜的目的是什么？

【拓展阅读】

重量分析法操作繁琐、耗时也比较长，且不适用于微量和痕量组分的测定，目前，已逐渐为其他分析方法所代替。尽管如此，因为该方法直接用分析天平称量而获得分析结果，不需要标准试样或基准物质进行比较，所以其准确度较高，在校对其他分析方法时，常用重量法的测定结果作为标准，因此仍有一定的应用价值。

实验 4.4　工业酒精中甲醇含量的测定

实验目的

（1）掌握分光光度法测定甲醇含量的原理和操作。
（2）掌握分光光度计的基本操作。

实验原理

酒精是制酒的重要原料，然而一些酒精制品中却含有一定量的甲醇，其对人体的神经系统和血液系统影响较大，经消化道、呼吸道或皮肤等途径过量摄入都会产生毒性反应。近年来，由于甲醇中毒以及甲醇积累原因造成的并发症越发频繁，因此，监测酒精制品中的甲醇含量十分必要。

目前，常用的甲醇检测方法主要有分光光度法、气相色谱法、激光拉曼光谱法、折射法等。相较而言，分光光度法具有设备要求低、适用于大批量产品检测的优点，但它对于温度、时间等条件有一定要求，需要严格控制。其中，品红亚硫酸法是检测工业酒精中甲醇含量的常用方法。

该方法的原理为：乙醇中的甲醇在 H_3PO_4 介质中被 $KMnO_4$ 氧化为甲醛（HCHO）：

$$5CH_3OH+2KMnO_4+4H_3PO_4 \Longrightarrow 5HCHO+2KH_2PO_4+2MnHPO_4+8H_2O$$

(4.10)

过量的 $KMnO_4$ 可被草酸（$H_2C_2O_4$）还原：

$$5H_2C_2O_4+2KMnO_4+3H_2SO_4 \Longrightarrow 2MnSO_4+K_2SO_4+10CO_2\uparrow+8H_2O \quad (4.11)$$

之后，采用无色的品红-亚硫酸为显色剂，与甲醛反应生成醌式结构的紫红色化合物，在590nm处测定其吸光值，并与标准系列比较定量。

仪器用品

仪器：分光光度计、具塞比色管、比色皿、移液管、容量瓶、天平、烧杯、离心机。

试剂：

高锰酸钾-磷酸溶液：70mL水中加入15mL磷酸溶液（85%）充分混合，加入3g高锰酸钾，充分溶解后用蒸馏水定容至100mL，储存于棕色试剂瓶中备用（注意存放时间不宜过长）。

草酸-硫酸溶液：称取7g 二水合草酸（$H_2C_2O_4 \cdot 2H_2O$），溶于1∶1冷硫酸中，并用该1∶1硫酸定容至100mL，混匀后储存于棕色瓶中备用。

品红-亚硫酸溶液：称取0.1g研细的碱性品红，多次加入温度为80℃的水（共60mL），边加水边研磨，充分溶解后离心取上清液于100mL试剂瓶中，冷却后加入1mL浓盐酸及10mL亚硫酸钠溶液（0.1g/mL），加水定容，储存于棕色试剂瓶中（注：若溶液呈红色时需重新配制）。

甲醇标准储备液：取1.000g甲醇于装有少量蒸馏水的100mL容量瓶中，加水定容得到浓度为10mg/mL甲醇标准储备液，低温保存备用。

甲醇标准使用液：准确移取10.00mL甲醇标准溶液加水定容于100mL容量瓶中得到1mg/mL的甲醇标准使用液。

乙醇溶液（不含甲醇与甲醛）：取500mL无水乙醇，加少许 $KMnO_4$，振摇后静止24小时后蒸馏处理，弃去最初和最后1/10的馏分，仅收集中间部分的馏分。

实验步骤

（1）**最大吸收波长的确定** 取1.00mL甲醇标准使用液于25mL具塞比色管中，加入0.5mL无甲醇甲醛的乙醇溶液、2mL高锰酸钾-磷酸溶液、3.5mL蒸馏水，充分混匀后静置15min，再加入2mL草酸-硫酸溶液混匀，褪色后加入5mL品红-亚硫酸溶液并充分混合，25℃下静置30min。在分光光度计上用1cm比色皿，以试剂空白为参比，在500～630nm之间，每隔10nm测量一次吸光度（其中在580～600nm之间，每隔2nm测量一次），以波长为横坐标，吸光度为纵坐标，绘制吸收曲线，选择测量的适宜波长（一般选用最大吸收波长 λ_{max} 为测定波长）。

（2）**实验条件的确定** 分别确定高锰酸钾-磷酸溶液、草酸-硫酸溶液、品红-亚硫酸溶液的最佳加入量。

吸取多个 1.00mL 甲醇标准使用液（1mg/mL）分别置于 25mL 具塞比色管中，加入 0.5mL 无甲醇甲醛的乙醇溶液，加蒸馏水至 5mL 刻线，混匀后加入不同量（0~3mL）的高锰酸钾-磷酸溶液，混匀后放置 10min。各管加 2mL 草酸-硫酸溶液，混匀后静置褪色，再加入 5mL 品红-亚硫酸溶液，混匀后于 25℃下静置 0.5h，以试剂空白为参比，在最大吸收波长处测吸光度，选取吸收值大且稳定时的高锰酸钾-磷酸溶液加入量。

同理，分别确定草酸-硫酸溶液及品红-亚硫酸溶液的最佳加入量。

（3）标准曲线的绘制　精确吸取 0.00mL、0.20mL、0.40mL、0.60mL、0.80mL、1.00mL 甲醇标准使用液（1mg/mL）分别置于 25mL 具塞比色管中，加入 0.5mL 无甲醇与甲醛的乙醇溶液，加水至 5mL 刻线处，混匀后加入 2mL（以优化后的量为准）高锰酸钾-磷酸溶液，混匀后放置 10min，加 2mL（以优化后的量为准）草酸-硫酸溶液，混匀后静置褪色，再加入 5mL（以优化后的量为准）品红-亚硫酸溶液，混匀后于 25℃下静置 0.5h。以 0.00mL 甲醇加入量的比色管调节零点（试剂空白），在最大吸收波长处测吸光度，以甲醇浓度为横坐标，吸光度为纵坐标，绘制标准曲线。

（4）样品测定　取 0.50mL 工业乙醇试样于 25mL 具塞比色管中，加水至 5mL，其他操作同标准曲线测定。

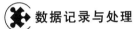
数据记录与处理

甲醇含量计算公式：

$$X = \frac{m}{V \times 1000} \times 100 \tag{4.12}$$

式中　X——样品中甲醇含量，g/(100mL)；

m——所取样品中的甲醇质量，mg；

V——样品取样体积，mL。

注：

（1）新配品红-亚硫酸溶液冰箱中放置 1~2 天再用为好，溶液呈红色时应重新配制。

（2）在上述检测过程中，工业乙醇中的醛类以及经 $KMnO_4$ 氧化其他醇生成的醛类，如乙醛、丙醛等，与品红-亚硫酸溶液作用也显色。但在一定酸性条件下，除甲醛可形成经久不变的紫红色外，其他醛所形成的色泽会慢慢消退，故无干扰。因此，操作中时间条件必须严格控制。

（3）实验中加入草酸-硫酸溶液后会放出热量，温度升高，需适当冷却后再加入品红-亚硫酸溶液。因为体系温度对吸光度有影响，因此，试样管与标准溶液管温度尽量保持一致，温度差控制在 1℃以内。

拓展阅读

甲醇中毒机理

甲醇的中毒机理是，甲醇经人体代谢产生甲醛和甲酸（俗称蚁酸），然后对人体产生伤害。常见的症状是，先是产生喝醉的感觉，数小时后头痛，恶心，呕吐，以及视线模糊。严重者会失明，乃至丧命。失明的原因：甲醇的代谢产物甲酸累积在眼睛部位，破坏视觉神经细胞。脑神经也会受到破坏，而产生永久性损害。甲酸进入血液后，会使组织酸性越来越强，损害肾脏导致肾衰竭。

实验 4.5　电位滴定法测定醋酸含量

实验目的

(1) 掌握电位滴定法的基本原理和方法。
(2) 掌握醋酸含量的测定方法，并熟练使用 pH 计。
(3) 掌握电位滴定曲线的绘制及滴定终点的计算方法。

实验原理

醋酸是有机弱酸（$K_a = 1.8 \times 10^{-5}$），可用 NaOH 标准溶液直接滴定［见式(4.13)］，由于满足 $cK_a > 10^{-8}$ 的弱酸均可被 NaOH 准确滴定，因此，该方法实际测定的是溶液的总酸量，测定结果以含量最高的醋酸 ρ_{HAc}（g/L）表示。该滴定反应产物为 NaAc，计量点时溶液 pH 值约为 8.7，可选用酚酞作指示剂，滴定终点时溶液由无色变为微粉色。

$$NaOH + CH_3COOH == CH_3COONa + H_2O \tag{4.13}$$

电位滴定法是通过测量插入溶液的两电极组成的原电池的电动势变化确定滴定终点，无需使用指示剂，因此，可用于有色或混浊溶液的滴定分析，也适用于没有合适指示剂的滴定反应。

本实验为酸碱中和反应，滴定过程中溶液的氢离子浓度（溶液 pH）不断变化，在化学计量点附近发生 pH 突跃，因此，测量溶液 pH 的变化就能确定滴定终点。采用对氢离子选择性响应的玻璃电极为指示电极，饱和甘汞电极为参比电极，与试液组成工作电池，即 pH 计。用 NaOH 标准溶液对试液进行电位滴定，测量溶液 pH 值随 NaOH 溶液加入的变化，实验装置如图 4.2，以 NaOH 溶液加入体积为横坐标，pH 值为纵坐标，绘制 pH-V 曲线［如图 4.3(a)］，确定滴定终点。此外，滴定终点的判断，也可采用一阶微商法 ［$\Delta pH/\Delta V$-V，如图 4.3(b)］或二阶微商法 ［$\Delta^2 pH/\Delta V^2$-V，如图 4.3(c)］求得。再根据 NaOH 溶液的浓度、计量点时消耗 NaOH 溶液的体积及待测试液的取用量，求出醋酸含量。

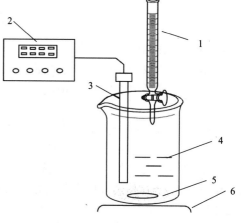

图 4.2　滴定装置示意图
1—滴定管；2—pH 计；3—复合 pH 电极；
4—待测试液；5—磁子；6—磁力搅拌器

玻璃电极（GE）、饱和甘汞电极（SCE）和待测试液组成原电池，E_M 为玻璃电极的膜电位，E_L 为液接电位。电池的电动势 E 与待测溶液的 pH 值关系如下：

$$\begin{aligned} E &= E_{SCE} - E_{GE} + E_L \\ &= E_{SCE} - (E_{AgCl/Ag} + E_M) + E_L \\ &= E_{SCE} - \left(E_{AgCl/Ag} + K + \frac{2.303RT}{F} \lg \alpha_{H^+}\right) + E_L \\ &= K' + \frac{2.303RT}{F} pH \end{aligned} \tag{4.14}$$

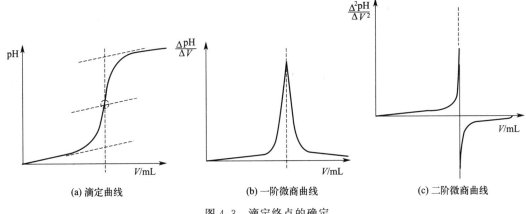

(a) 滴定曲线　　　　　　(b) 一阶微商曲线　　　　　　(c) 二阶微商曲线

图 4.3　滴定终点的确定

实验用品

仪器：pHS-3C 型精密酸度计、磁力搅拌器、分析天平、容量瓶、锥形瓶、移液管、滴定管、烧杯、量筒。

试剂：NaOH（固体）、邻苯二甲酸氢钾（KHP）基准物质、食醋（3%～5%）、0.2% 酚酞乙醇指示剂、去离子水、pH=4.00（25℃）和 pH=6.86（25℃）的标准缓冲溶液。

实验步骤

(1) 0.1mol/L NaOH 溶液配制与标定　称取一定量的 NaOH 固体，用煮沸且冷却至室温的去离子水充分溶解，冷却至室温后转入聚乙烯塑料瓶中备用。

根据 NaOH 溶液浓度及滴定误差要求估算基准物质 KHP 的称量质量。采用差减法准确称取计算量的 KHP 于 250mL 锥形瓶中，加水 20～30mL 使之溶解，加 2～3 滴酚酞指示剂，用待标定 NaOH 溶液滴定至溶液呈浅粉色并保持 30s 不褪色，即为终点。记录 NaOH 溶液消耗的体积，平行标定 3 次（表 4.6）。

(2) 醋酸含量的测定

pH 计校准：开机预热后，采用标准缓冲溶液两点法校准酸度计，电极插入去离子水中备用。

醋酸含量的粗测：准确移取 5.00mL 食醋试样，用新煮沸并冷却至室温的去离子水定容于 100mL 容量瓶中，得食醋稀释液。准确移取 25.00mL 上述食醋稀释液于 250mL 烧杯中，再加入 20mL 去离子水，将烧杯置于磁力搅拌器上，插入 pH 计（已用滤纸擦干电极头），开启搅拌，开始滴定。测量在加入 NaOH 标准溶液 0mL、1mL、2mL 直至 20mL 后各个点溶液的 pH 值，初步判断发生 pH 突跃时所需 NaOH 溶液的体积范围。

醋酸含量的精测：重复粗测时的实验步骤，在粗测时判断的化学计量点附近取较小 NaOH 滴定剂体积增量（如每加入 0.1mL 记录一次 pH 值），增加计量点附近测量点的密度。记录每次加入 NaOH 溶液后的总体积 V 和对应的 pH 值，以此绘制滴定曲线。

数据记录与处理

参照表 4.7 记录的数据，绘制滴定曲线，确定滴定终点。如图 4.3(a)，对于 pH-V 曲线，滴定终点为曲线上的转折点（斜率最大处）对应的体积 V；如图 4.3(b)，对于 $\Delta \mathrm{pH}/\Delta V$-$V$ 一级微商曲线，滴定终点为曲线尖峰处（$\Delta \mathrm{pH}/\Delta V$ 极大值处）所对应的体积 V；如图 4.3(c)，对于 $\Delta^2 \mathrm{pH}/\Delta V^2$-$V$ 二阶微商曲线，滴定终点为 $\Delta^2 \mathrm{pH}/\Delta V^2$ 由极大正值到极大负值与纵坐标零线相交处对应的体积 V。

根据滴定终点时 NaOH 标准溶液所消耗的体积、NaOH 标准溶液的浓度及待测试液的取用量，根据下式求出食醋的总酸度 ρ_{HAc}（g/L）。

$$\rho_{HAc}(g/L) = \frac{\overline{c}_{NaOH} V_{NaOH} M_{HAc} \times \frac{100}{25}}{5.00} \tag{4.15}$$

式中　M_{HAc}——HAc 的摩尔质量，60.05 g/mol；

　　　$\overline{c}_{NaOH}$——NaOH 标准滴定溶液的实际浓度，mol/L；

　　　V_{NaOH}——滴定终点消耗 NaOH 标准溶液的体积，mL。

表 4.6　NaOH 溶液浓度的标定

实验序号	1	2	3
m_{KHP}（初始）/g			
m_{KHP}（终点）/g			
m_{KHP}/g			
V_{NaOH}（初始）/mL			
V_{NaOH}（终点）/mL			
V_{NaOH}（消耗）/mL			
c_{NaOH}/(mol/L)			
$\overline{c}_{NaOH}$/(mol/L)			
$\lvert d_i \rvert$			
$\overline{d}_r$/%			

表 4.7　醋酸含量的测定

pH 值					
V_{NaOH}/mL					
pH 值					
V_{NaOH}/mL					
pH 值					
V_{NaOH}/mL					

思考题

（1）CO_2 对本实验是否有影响？如有影响应如何消除？

（2）相比于容量分析法，电位滴定分析法有哪些优缺点？

（3）若使用指示剂法进行本次测定，如何减小食醋本身颜色对结果的干扰？

实验 4.6　水泥熟料中铁、铝、钙、镁含量的测定

实验目的

（1）学习复杂物质的样品溶解和分析方法；

(2) 掌握沉淀分离 Fe、Al 的方法原理；
(3) 掌握酸效应在水泥分析中的应用原理。

实验原理

水泥主要是由硅酸盐组成，试样用盐酸分解后，铁、铝、钙、镁等组分则以 Fe^{3+}、Al^{3+}、Ca^{2+}、Mg^{2+} 等离子形式存在于溶液中，它们都能与 EDTA 形成稳定的络合物，只要控制适当的酸度，就可以用 EDTA 标准溶液将它们分别滴定。向分解后的试样溶液中加入氨水，使 Fe^{3+}、Al^{3+} 生成 $Fe(OH)_3$、$Al(OH)_3$ 沉淀与 Ca^{2+}、Mg^{2+} 分离。沉淀用盐酸溶解，调节溶液的 pH 值为 2～2.5，以磺基水杨酸钠作指示剂，用 EDTA 滴定 Fe^{3+}；然后再加入一定的过量的 EDTA，调节溶液的 pH 在 3.0～4.0，煮沸，待 Al^{3+} 与 EDTA 完全络合后再调节溶液的 pH 为 4.2，以 PAN 为指示剂，用 $CuSO_4$ 标准溶液滴定过量的 EDTA，然后分别计算 Fe_2O_3 和 Al_2O_3 的百分含量。

将分离出含有 Ca^{2+}、Mg^{2+} 的滤液，参照实验 3.9 自来水硬度的测定的分析方法进行测定。

实验用品

仪器：分析天平、台秤、移液管、烧杯、容量瓶、滴定管、量筒。

试剂：0.2%甲基红乙醇溶液；10%磺基水杨酸钠溶液；0.3% PAN 乙醇溶液；待测水泥试样；0.02mol/L EDTA 标准溶液；0.02mol/L $CuSO_4$ 标准溶液；6mol/L HCl 溶液；20% NaOH 溶液；$CaCO_3$ 基准物质；1∶1 $NH_3·H_2O$；pH＝10.0 的 NH_3-NH_4Cl 缓冲溶液（详见实验 3.8，EDTA 溶液的配制与标定）；

HAc-NaAc 缓冲溶液（pH＝4.2）：3.2g 无水 NaAc 溶于水中，加入 5mL 冰醋酸，用水稀释至 100mL；

K-B 指示剂：称取 0.2g 酸性铬蓝 K、0.4g 萘酚绿 B 于烧杯中，加水溶解后，稀释至 100mL。

实验步骤

(1) 0.02mol/L EDTA 标准溶液的配制与标定（详见实验 3.8）。

(2) 0.02mol/L $CuSO_4$ 标准溶液的配制与测定

配制：称 5.00g $CuSO_4·5H_2O$ 溶于水中，加 4～5 滴 1∶1 H_2SO_4，用水稀释至 1L。

体积比的测定：准确称取 10.00mL 0.02mol/L EDTA 溶液，加水稀释至 150mL 左右，加 10mL pH＝4.2 的 HAc-NaAc 缓冲溶液，加热至 80～90℃。加入 PAN 指示剂 5～6 滴，用 $CuSO_4$ 溶液滴定至棕红色且稳定即为终点。计算 1mL $CuSO_4$ 溶液相当于 EDTA 标准溶液的体积。

(3) 试样的溶解　准确称取 0.23～0.25 g 试样，置于 250mL 烧杯中，加少量水润湿，加 15mL 6mol/L HCl，盖上表面皿，加热煮沸，待试样分解完全后，用热水稀释至 100mL 左右。加热至沸腾停止加热，加 2 滴甲基红指示剂，在搅动下慢慢加 1∶1 $NH_3·H_2O$ 至溶液呈黄色，再加热至沸腾，停止加热，待溶液澄清后，趁热用快速定量滤纸过滤，沉淀用 0.1% NH_4NO_3 热溶液充分洗涤，至流出液中无 Cl^- 为止。滤液盛于 250mL 容量瓶中，冷至室温，用水稀释至刻度，供测定 Ca^{2+}、Mg^{2+} 时用。

(4) Fe_2O_3 的测定　滴加 6mol/L HCl 于滤纸上，使氢氧化物沉淀溶解于原烧杯中，滤纸用热水洗涤数次后弃去，将溶液煮沸以溶解可能存在的氢氧化物沉淀，加 10 滴磺基水杨酸钠指示剂，滴加 1∶1 $NH_3·H_2O$ 至溶液的 pH 值为 2～2.5（溶液呈紫红色），加热至

50~60℃，用 EDTA 标准溶液滴定至溶液由暗紫红色变为淡黄色为终点。记下 EDTA 标准溶液用量，计算试样中 Fe_2O_3 的质量分数，测 Fe^{3+} 后的溶液继续用于测定 Al^{3+}。

(5) Al_2O_3 的测定 在滴定 Fe^{3+} 后的溶液中，准确加入 20mL EDTA 标准溶液，滴加 1∶1 $NH_3·H_2O$ 至溶液 pH 值约为 4，加入 10mL HAc-NaAc 缓冲溶液，煮沸 1min，停止加热后稍冷，加 6~8 滴 PAN 指示剂。用 $CuSO_4$ 标准溶液滴定至溶液呈红色即为终点。记下 $CuSO_4$ 溶液用量，计算试样中 Al_2O_3 的质量分数。

(6) CaO 和 MgO 含量的测定 移取 25.00mL 分离 Fe^{3+}、Al^{3+} 后的滤液于锥形瓶中，加入 15mL pH=10.0 的氨缓冲溶液，再加 2~3 滴 K-B 指示剂，用 EDTA 滴定至溶液由紫红色变为纯蓝色，记录消耗体积 V_1，即为 Ca^{2+}、Mg^{2+} 消耗的总体积。

另取 25.00mL 分离 Fe^{3+}、Al^{3+} 后的滤液，依次加入 5mL 20% NaOH 溶液、20~30mL H_2O、2~3 滴 K-B 指示剂，用 EDTA 滴定至溶液变为纯蓝色，记录 Ca^{2+} 所消耗体积 V_2，求 CaO 的质量分数。

利用差减法求 MgO 的质量分数。

数据记录与处理

(1) EDTA 溶液的标定（表 4.8）

表 4.8 EDTA 溶液的标定

实验序号	1	2	3
m_{CaCO_3}（初始）/g			
m_{CaCO_3}（终点）/g			
m_{CaCO_3}/g			
$c_{Ca^{2+}}$/(mol/L)			
$V_{Ca^{2+}}$/mL			
V_{EDTA}（初始）/mL			
V_{EDTA}（终点）/mL			
V_{EDTA}（消耗）/mL			
c_{EDTA}/(mol/L)			
$\bar{c}_{EDTA}$/(mol/L)			
$|d_i|$			
$\bar{d}_r$/%			

(2) $CuSO_4$ 溶液的测定（表 4.9）

表 4.9 $CuSO_4$ 溶液的测定

实验序号	1	2	3
V_{EDTA}/mL			
V_{CuSO_4}（初始）/mL			
V_{CuSO_4}（终点）/mL			

续表

实验序号	1	2	3
V_{CuSO_4}（消耗）/mL			
V_{EDTA}/V_{CuSO_4}			
$\overline{V_{EDTA}/V_{CuSO_4}}$			

(3) Fe_2O_3 的测定（表 4.10）

表 4.10　Fe_2O_3 的测定

实验序号	1	2	3
$m_{试样}$（初始）/g			
$m_{试样}$（终点）/g			
$m_{试样}$/g			
V_{EDTA}（初始）/mL			
V_{EDTA}（终点）/mL			
V_{EDTA}（消耗）/mL			
$m_{Fe_2O_3}$/g			
$w_{Fe_2O_3}$/%			
$\overline{w}_{Fe_2O_3}$/%			
$\lvert d_i \rvert$			
$\overline{d}_r$/%			

(4) Al_2O_3 的测定（表 4.11）

表 4.11　Al_2O_3 的测定

实验序号	1	2	3
V_{CuSO_4}（初始）/mL			
V_{CuSO_4}（终点）/mL			
V_{CuSO_4}（消耗）/mL			
$w_{Al_2O_3}$/%			
$\overline{w}_{Al_2O_3}$/%			
$\lvert d_i \rvert$			
$\overline{d}_r$/%			

(5) CaO 和 MgO 的测定（表 4.12）

表 4.12　CaO 和 MgO 含量的测定

实验序号	1	2	3
$V_{试样}$/g			
$V_{1,EDTA}$（初始）/mL			
$V_{1,EDTA}$（终点）/mL			

续表

实验序号	1	2	3
$V_{1,\text{EDTA}}$(消耗)/mL			
$V_{2,\text{EDTA}}$(初始)/mL			
$V_{2,\text{EDTA}}$(终点)/mL			
$V_{2,\text{EDTA}}$(消耗)/mL			
$w_{\text{CaO}}/\%$			
$\overline{w}_{\text{CaO}}/\%$			
$|d_i|$			
$\overline{d}_r/\%$			
$w_{\text{MgO}}/\%$			
$\overline{w}_{\text{MgO}}/\%$			
$|d_i|$			
$\overline{d}_r/\%$			

思考题

（1）用 EDTA 滴定 Al^{3+} 时，为什么要采用返滴定法？

（2）在测定 Fe^{3+}、Al^{3+}、Ca^{2+}、Mg^{2+} 时，为什么要严格控制不同的 pH 值？

（3）测定 Ca^{2+}、Mg^{2+} 时，Fe^{3+}、Al^{3+} 的干扰可以采用哪些办法加以消除？

4.2 环境类

实验 4.7 水质总碱度测定

实验目的

（1）掌握酸碱滴定法测定水碱度的原理和方法。

（2）学习掌握水碱度的表示方法及计算。

实验原理

碱度是水介质与氢离子反应的定量能力。用甲基橙作指示剂，滴定的终点 pH 为 4.3～4.5，称为甲基橙碱度或 M 碱度。此时，水中的氢氧化物、碳酸盐及碳酸氢盐全部被中和，所测得的是水中各种弱酸盐类的总和，因此又称为总碱度，可换算为相应的碱度单位 mmol/100g：用 1 升水能结合的质子的量表示；mg/L（$CaCO_3$）：用 1 升水中能结合 H^+ 的物质所相当的 $CaCO_3$ 的质量来表示。M 碱度＝全部 HCO_3^{2-} ＋全部 CO_3^{2-} ＋全部 OH^-，测定结果用相当于碳酸钙的质量浓度，mg/L 为单位表示。

实验用品

仪器：酸式滴定管、移液管、250mL 锥形瓶、分析天平、量筒。

试剂：0.5g/L 甲基橙指示剂、0.10mol/L 盐酸溶液、无水碳酸钠（基准试剂）。

实验步骤

(1) 0.10mol/L 盐酸溶液的标定　用称量瓶在分析天平上准确称量（　）～（　）g 无水 Na_2CO_3（摩尔质量 $M=105.99$ g/mol），置于 250mL 锥形瓶中。用量筒加入 50mL 去离子水溶解，加入 1~2 滴甲基橙指示剂，用预标定的 HCl 溶液滴定至溶液由黄色变橙色即为终点。在接近终点时，为除去溶液中溶解的 CO_2，应剧烈振动，或加热驱赶 CO_2，冷却后再滴定至终点。平行标定 3 次（表 4.13）。根据基准物质 Na_2CO_3 的质量和消耗的盐酸体积，按如下公式计算 HCl 溶液的浓度 c_{HCl}。

$$c_{HCl} = \frac{2m \times 1000}{105.99 V} \quad (4.16)$$

式中　c_{HCl}——盐酸标准溶液的浓度，mol/L；
　　　　m——碳酸钠的质量，g；
　　　　V——消耗的盐酸体积，mL。

(2) 测定水样　吸取 50.00mL（V_1）水样于 250mL 锥形瓶中，加 1~2 滴甲基橙指示剂，用盐酸标准溶液滴定至试液由黄色突变为橙色，记录盐酸消耗的体积 V_2。平行测定 3 次（表 4.14）。

总碱度计算公式：

$$\rho_{CaCO_3} = \frac{c_{HCl} \times 100.1 \times V_2 \times 1000}{2V_1} \quad (4.17)$$

式中，ρ_{CaCO_3}：水样的总碱度，mg/L；c_{HCl}：盐酸标准溶液的浓度，mol/L；V_1：所取水样的体积，mL；V_2：滴定水样消耗标准盐酸溶液的体积，mL。

数据记录与处理

表 4.13　0.1mol/L HCl 溶液的标定

实验序号	1	2	3
$m_{Na_2CO_3}$/g			
V_{HCl}(初始)/mL			
V_{HCl}(终点)/mL			
V_{HCl}/mL			
c_{HCl}/(mol/L)			
$\bar{c}_{HCl}$/(mol/L)			
$|d_i|$			
$\bar{d}_r$/%			

表 4.14　水样中总碱度的测定

实验序号	1	2	3
$V_{水样}$/g			
V_{HCl}(初始)/mL			
V_{HCl}(终点)/mL			
V_{HCl}/mL			
ρ_{CaCO_3}/(mol/L)			

续表

实验序号	1	2	3
$\bar{\rho}_{CaCO_3}/(mol/L)$			
$|d_i|$			
$\bar{d}_r/\%$			

思考题

酸碱滴定法与配位滴定法相比有哪些不同点？

拓展阅读

常见水质 pH 指标

碱度是水介质与氢离子反应的定量能力。一些常见水质标准 pH 指标：

(1) 饮用水：通常，饮用水的 pH 值标准范围在 6.5~8.5 之间，接近中性。

(2) 美国 EPA（环境保护署）饮用水质标准：一级水，没有具体标准；二级水，pH 6.5~8.5。

(3) 欧盟（EU）饮用水水质标准：pH 6.5~9.5。对瓶装或桶装的净水，pH 最小值降至 4~5。

(4) 日本饮用水水质标准：pH 5.8~8.6。

(5) 我国生活饮用水卫生标准：pH 6.5~8.5（GB 5749—2022）。

(6) 瓶装饮用纯净水：pH 5.0~7.0。

实验 4.8 河水中化学需氧量（COD）的测定

实验目的

(1) 了解测定化学需氧量（COD）的意义。

(2) 掌握用酸性 $KMnO_4$ 法测定水样中 COD 的原理和方法。

实验原理

化学需氧量（COD）是量度水体受还原性物质（主要是有机物）污染程度的综合性指标，是 1 L 水中还原性物质（无机或有机的）在一定条件下被氧化时所消耗的氧含量（以 mg/L 计）。COD 值越高，说明水体受有机物的污染越严重。因水中除含有无机还原性物质（如 NO_2^-、S^{2-}、Fe^{2+} 等）外，还可能含有少量有机物质。有机物腐烂促使水中微生物繁殖，污染水质，影响人体健康。若水中 COD 量高则呈现黄色，并有明显的酸性，将危害水生生物；工业生产上用此水，对蒸汽锅炉有腐蚀作用，还影响印染等产品质量，所以水体中 COD 的测定是很重要的。

不同的条件得出的需氧量不同，因而必须严格控制反应条件。COD 的测定，根据氧化剂的不同，分为 $KMnO_4$ 法（COD_{Mn}）和 $K_2Cr_2O_7$ 法（COD_{Cr}）。$KMnO_4$ 法操作简便、快速，在一定程度上可以说明水体受有机物污染的状况，适合测定地面水、饮用水、河水等污染程度较轻的水样；而对于工业污水及生活污水中含有成分较多且复杂、污染较严重的水质则适宜采用 $K_2Cr_2O_7$ 法测定，此法氧化率高，重现性好。本实验采用酸性高锰酸钾法测定水中化学需氧量。

在酸性溶液中，加入过量的 $KMnO_4$ 溶液，加热使水中有机物及还原性物质充分氧化后，剩余的 $KMnO_4$ 用一定量的过量的 $Na_2C_2O_4$ 还原，再以 $KMnO_4$ 标准溶液对 $Na_2C_2O_4$ 的过量部分进行返滴定，主要反应式如下：

$$4MnO_4^-(过量)+5C+12H^+ = 4Mn^{2+}+5CO_2\uparrow+6H_2O \qquad (4.18)$$

$$2MnO_4^-(剩余)+5C_2O_4^{2-}(过量)+16H^+ = 2Mn^{2+}+10CO_2\uparrow+8H_2O \qquad (4.19)$$

$$5C_2O_4^{2-}(剩余)+2MnO_4^-+16H^+ = 2Mn^{2+}+10CO_2\uparrow+8H_2O \qquad (4.20)$$

通过计算求出水中所含有机物和无机还原性物质所消耗的 $KMnO_4$ 量。

水样中含 Cl^- 量大于 300mg/L，将使测定结果偏高，可加纯水适当稀释；或加入 Ag_2SO_4，使 Cl^- 生成沉淀；或采用碱性高锰酸钾法、重铬酸钾法，以消除干扰。水样中如有 Fe^{2+}、H_2S、NO_2^- 等还原性物质将干扰测定，但它们在室温条件下，就能被 $KMnO_4$ 氧化，因而水样在室温条件下先用 $KMnO_4$ 溶液滴定，除去干扰离子，此 $KMnO_4$ 的量不应计数。水样采集后应立即进行分析，如有特殊情况要放置时，可加入少量硫酸铜以抑制微生物对有机物的分解。若水样用蒸馏水稀释，应取与水样相同量的蒸馏水，测定空白值，加以校正。

实验用品

仪器：酸式滴定管、移液管、容量瓶、锥形瓶、称量瓶、量筒、烧杯。

试剂：$KMnO_4$（分析纯）、3mol/L H_2SO_4 溶液、$Na_2C_2O_4$（基准物质）、待测水样。

实验步骤

(1) 0.005mol/L $Na_2C_2O_4$ 标准溶液的配制　准确称取 0.15～0.17g 基准物质 $Na_2C_2O_4$（已干燥）于100mL 小烧杯中，加入 40mL 蒸馏水溶解，然后定量转入 500mL 容量瓶中，加水稀释至刻度，充分摇匀后备用。

(2) 0.002mol/L $KMnO_4$ 溶液的配制　取 25.00mL 已配好的 0.02mol/L $KMnO_4$ 溶液（提前一周配制）于 250mL 容量瓶中，加水稀释至刻度，充分摇匀后备用。用 0.005mol/L $Na_2C_2O_4$ 标准溶液标定 $KMnO_4$ 溶液的浓度（表 4.15）

(3) 水样 COD 的测定　用量筒量取 100mL 充分搅拌的水样于锥形瓶中，加入 5mL 1:3 H_2SO_4 溶液和几粒玻璃珠（防止溶液暴沸），由滴定管加入 10.00mL $KMnO_4$ 溶液，立即加热至沸腾。从冒出的第一个大气泡开始，煮沸 10.0min（红色不应褪去）。取下锥形瓶，放置 0.5～1min，趁热准确加入 $Na_2C_2O_4$ 标准溶液 25.00mL，充分摇匀，立即用 $KMnO_4$ 溶液进行滴定。滴定至试液呈微红色且 0.5min 不褪去即为终点，消耗体积为 V_1，此时试液的温度应不低于 60℃。

用移液管移取 25.00mL 待测水样于 250mL 锥形瓶中，加 75.00mL 蒸馏水（若为污染较严重的水样取 10.00mL，然后加蒸馏水至 100mL）；加入 8mL 3mol/L H_2SO_4 溶液，混合均匀；加入 Ag_2SO_4 溶液 2mL；准确加入 0.002mol/L $KMnO_4$ 标准溶液 10.00mL，摇匀，立即放入沸水浴中加热 30min，沸水浴的液面要高于反应溶液的液面（或立即加热至沸），从冒第一个大泡开始计时，煮沸 10min，从沸水浴中取出锥形瓶，稍冷；趁热（70～80℃）准确加入 10.00mL 0.005mol/L $Na_2C_2O_4$ 标准溶液，摇匀，立即用 $KMnO_4$ 标准溶液滴定至溶液呈稳定的微红色即为终点，消耗体积为 V_2。另取蒸馏水 100.00mL 代替水样进行实验，同上述操作，测定空白值。

平行测定三次，按原理中的化学反应方程式计算水样的化学需氧量（mg/L）以及相对平均偏差（表 4.16）。

数据记录与处理

表 4.15　0.002mol/L $KMnO_4$ 溶液的标定

实验序号	1	2	3
$m_{Na_2C_2O_4}/g$			
V_{KMnO_4}(初始)/mL			
V_{KMnO_4}(终点)/mL			
V_{KMnO_4}/mL			
$c_{KMnO_4}/(mol/L)$			
$\overline{c}_{KMnO_4}/(mol/L)$			
$\|d_i\|$			
$\overline{d}_r/\%$			

表 4.16　水样 COD 的测定

实验序号	1	2	3
V_s/mL			
V_1(初始)/mL			
V_1(终点)/mL			
V_1/mL			
V_2(初始)/mL			
V_2(终点)/mL			
V_2/mL			
COD/(mg/L)			
$\overline{COD}$/(mg/L)			
$\|d_i\|$			
$\overline{d}_r/\%$			

思考题

（1）水样的采集与保存应当注意哪些事项？
（2）当水样中 Cl^- 含量高时，能否用该法测定？为什么？
（3）水样中加入 $KMnO_4$ 煮沸时，若紫红色消失说明什么？应采取什么措施？
（4）测定水中化学需氧量有哪些方法？请用化学方程式表示。

实验 4.9　城市污水中硫酸盐含量的测定

实验目的

掌握重量法测定水体硫酸盐含量的原理及实验操作。

实验原理

城市污水主要为工业废水和生活污水，工业废水中多含有重金属离子 Fe^{3+}、Cu^{2+}、

Pb^{2+}、Ag^+等以及SO_4^{2-}、Cl^-、PO_4^{3-}等阴离子。测定其中SO_4^{2-}的含量可以使其定量生成$BaSO_4$沉淀，沉淀反应在接近沸腾的温度下进行，陈化一段时间后过滤出沉淀，烘干或灼烧沉淀、称重，然后根据$BaSO_4$的质量去计算SO_4^{2-}的含量，反应如下：

$$Ba^{2+} + SO_4^{2-} \rightleftharpoons BaSO_4 \downarrow \tag{4.21}$$

不溶于酸的物质均会干扰沉淀，故在用$BaCl_2$进行沉淀之前，先将样品用盐酸酸化并过滤，除去干扰。

实验用品

仪器：马弗炉、电热板、玻璃棒、烧杯、表面皿、瓷坩埚、洗瓶、坩埚钳、干燥器、慢速定量滤纸。

试剂：污水试样，1∶1盐酸溶液，100g/L $BaCl_2$溶液，0.1mol/L $AgNO_3$溶液，甲基红。

实验步骤

（1）取200mL水样（若SO_4^{2-}超过1000mg/L时相应减少取样量）置于400mL烧杯中，加两滴甲基红指示剂，用1∶1盐酸调节溶液至红色，再加2mL，盖上表面皿，将烧杯置于电热板加热至沸腾，沸腾后继续煮沸至少5min，稍冷，用蒸馏水冲洗表面皿2~3次，用定性滤纸过滤，用热水洗涤烧杯及滤纸4~5次，弃去沉淀，保留滤液。

（2）将滤液置于电热板上加热至沸腾，在不断搅拌下缓缓滴加大概10mL的$BaCl_2$溶液，直至不再产生沉淀，再多加2mL。在80~90℃保持2h，冷却后过夜陈化。

（3）用慢速定量滤纸过滤陈化后的沉淀。用热水转移并洗涤沉淀，用少量温水反复洗涤沉淀物，直至洗涤液无氯离子为止（用0.1mol/L $AgNO_3$溶液检验）。

（4）用滤纸包裹好沉淀，放入事先在800℃下灼烧至恒重的瓷坩埚中烘干后在电炉上低温灰化（不要让滤纸烧出火焰），然后将坩埚放入马弗炉中，在800℃下灼烧1h后取出，稍冷盖上盖子，放入干燥器中冷却至室温（约30min）。直至灼烧至恒重后称重（表4.17）。

数据记录与处理

计算硫酸盐含量的公式如下：

$$w = \frac{(m_2 - m_1) \times 411.6 \times 1000}{V} \tag{4.22}$$

式中　w——硫酸盐含量，mg/L；

　　　m_2——坩埚和硫酸钡的质量之和，g；

　　　m_1——坩埚质量，g；

　　　V——试样体积，mL；

411.6——$BaSO_4$质量换算为SO_4^{2-}的因素。

表4.17　SO_4^{2-}的含量测定

实验序号	1	2	3
$V_{试样}$/mL			
m_1/g			
m_2/g			
m_{BaSO_4}/g			

续表

实验序号	1	2	3
$w_{SO_4^{2-}}/(\text{mg/L})$			
$\overline{w}_{SO_4^{2-}}/(\text{mg/L})$			
$\lvert d_i \rvert$			
$\overline{d}_r/\%$			

注：
(1) 恒重是指连续两次灼烧或烘干后，称得的质量差小于 0.3mg。
(2) 硫酸钡的灰化应保证空气供应充分，否则沉淀易被滤纸烧成的炭还原。灼烧后的沉淀将会呈灰色或黑色。这时可在冷却后的沉淀中加入 2～3 滴浓硫酸，然后小心加热至 SO_2 白烟不再产生为止，再在 800℃ 灼烧至恒重。

思考题

(1) 为什么试样溶液中要加入盐酸溶液？
(2) 如何检验滤液中有无氯离子？

拓展阅读

废水的排放标准中，要求硫酸盐排放浓度＜1500mg/L，高于这一浓度，就属高硫酸盐废水。含硫酸盐废水中的硫酸盐本身虽然无害，但是它遇到厌氧环境会在硫酸盐还原菌(SRB)作用下产生 H_2S，H_2S 能严重腐蚀废水处理设施和排水管道，且气味恶臭，严重污染大气。另外硫酸盐废水排入水体会使纳水体酸化，pH 降低，危害水生生物；排入农田会破坏土壤结构，使土壤板结，减少农作物产量及降低农产品品质。目前，我国很多城市的地下水已经受到不同程度的硫酸盐污染，寻求行之有效的硫酸盐废水处理工艺早已成为环境工程界普遍关注的问题。

实验 4.10　分光光度法测定水性涂料中的甲醛

实验目的

掌握水性涂料中甲醛的提取及其乙酰丙酮显色分光光度法检测技术。

实验原理

采用蒸馏法将样品中的甲醛蒸出，在 pH=6 的乙酸-乙酸铵缓冲溶液中，馏分中的甲醛与乙酰丙酮在加热的条件下反应生成稳定的黄色络合物，冷却后在波长 413nm 处进行吸光度测试。根据标准工作曲线，计算试样中甲醛的含量。

实验用品

仪器：紫外-可见分光光度计、蒸馏装置（包括 500mL 蒸馏瓶、冷凝管、烧杯）、碘量瓶。
试剂：乙酸铵、冰醋酸、乙酰丙酮、碘溶液（0.1mol/L）、氢氧化钠溶液（0.1mol/L）、盐酸溶液（0.1mol/L）、硫代硫酸钠（$Na_2S_2O_3$，0.1mol/L）、甲醛溶液（37%）；乙酰丙酮溶液：0.25%（体积分率），称取 25g 乙酸铵，加适量水溶解，加 3mL 冰乙酸和 0.25mL 已蒸馏过的乙酰丙酮试剂，移入 100mL 容量瓶中，用水稀释至刻度，调整 pH=6。此溶液于 2~5℃储存，可稳定一个月。

实验步骤

（1）硫代硫酸钠标准溶液（0.1mol/L）的配制及标定（表 4.18）（参见实验 3.15）。

（2）甲醛储备溶液的配制与标定

甲醛标准溶液（1mg/mL）的配制：移取 2.8mL 甲醛溶液，置于 1000mL 容量瓶中，用水稀释至刻度。

甲醛标准溶液的标定：移取 20mL 待标定的甲醛标准溶液于碘量瓶中，准确加入 25mL 碘溶液，再加入 10mL 氢氧化钠溶液，摇匀，于暗处静置 15min 后，加 11mL 盐酸溶液，用硫代硫酸钠标准溶液滴定至淡黄色，加 1mL 淀粉溶液，继续滴定至蓝色刚刚消失为终点，记录所用硫代硫酸钠标准溶液体积 V_2（mL）。同时做空白样，记录所用硫代硫酸钠标准溶液体积 V_1（mL）。

（3）标准曲线的绘制　分别用移液管吸取 0.00mL、0.20mL、0.50mL、1.00mL、3.00mL、5.00mL、8.00mL 甲醛标准工作溶液（10 μg/mL），加水稀释至刻度，加入 2.5mL 乙酰丙酮溶液。在 60℃ 恒温水浴中加热 30min，取出后冷却至室温，用紫外-可见分光光度计以 1cm 吸收池（以水为参比）在 413nm 波长处测量吸光度。以甲醛质量为横坐标，相应的吸光度（A）为纵坐标，绘制标准曲线。

（4）甲醛含量的测定　称取搅拌均匀后的试样约 2g（精确至 1mg），置于 50mL 的容量瓶中，加水摇匀，稀释至刻度。再用移液管移取 10mL 容量瓶中的试样水溶液，置于已预先加入 10mL 水的蒸馏瓶中，并在蒸馏瓶中加入少量的沸石，在馏分接收器中预先加入适量的水，浸没馏分出口，馏分接收器的外部用冰水浴冷却。加热蒸馏，使试样蒸至近干，取下馏分接收器，用水稀释至刻度，待测。若待测试样在水中不易分散，则直接称取搅拌均匀后的试样约 0.4g（精确至 1mg），置于已预先加入 20mL 水的蒸馏瓶中，轻轻摇匀，再进行蒸馏过程操作。

在已定容的馏分接收器中加入 2.5mL 乙酰丙酮溶液，摇匀。在 60℃ 恒温水浴中加热 30min，取出后冷却至室温，用 1cm 比色皿（以水为参比）在紫外可见分光光度计上于 413nm 波长处测试吸光度。同时在相同条件下做空白样（水），测得空白样的吸光度。将试样的吸光度减去空白样的吸光度，在标准工作曲线上查得相应的甲醛质量。

数据记录与处理

（1）硫代硫酸钠标准溶液的标定（表 4.18）

表 4.18　$Na_2S_2O_3$ 标准溶液的标定

实验序号	1	2	3
$V_{Na_2S_2O_3}$（初始）/mL			
$V_{Na_2S_2O_3}$（终点）/mL			
$V_{Na_2S_2O_3}$/mL			
$c_{Na_2S_2O_3}$/(mg/L)			
$\bar{c}_{Na_2S_2O_3}$/(mg/L)			
$\lvert d_i \rvert$			
$\bar{d}_r$/%			

(2) 甲醛溶液的标定（表 4.19） 甲醛溶液的浓度用下式计算：

$$\rho_{HCHO} = \frac{(V_1 - V_2)c_{Na_2S_2O_3} \times 15}{20} \tag{4.23}$$

式中 ρ_{HCHO}——甲醛标准溶液的质量浓度，g/L；

V_1——空白样滴定所用的硫代硫酸钠标准溶液体积，mL；

V_2——甲醛溶液标定所用的硫代硫酸钠标准溶液体积，mL；

$c_{Na_2S_2O_3}$——硫代硫酸钠标准溶液的浓度，mol/L；

15——甲醛摩尔质量的 1/2，g/mol；

20——标定时所移取的甲醛标准溶液体积，mL。

表 4.19 甲醛溶液的标定

	1	2	3
V_2(初始)/mL			
V_2(终点)/mL			
V_2/mL			
V_1(初始)/mL			
V_1(终点)/mL			
V_1/mL			
ρ_{HCHO}/(g/L)			
$\bar{\rho}_{HCHO}$/(g/L)			
$|d_i|$			
$\bar{d}_r$/%			

(3) 甲醛标准曲线及样品测定（表 4.20） 甲醛含量（c）用下式计算：

$$c = \frac{m}{W}f \tag{4.24}$$

式中 c——甲醛含量，mg/kg；

m——从标准工作曲线上计算的甲醛质量，μg；

W——样品质量，g；

f——稀释因子。

表 4.20 甲醛标准曲线及样品测定

	HCHO 浓度/(mg/kg)	吸光度	样品中甲醛含量/%
1			—
2			—
3			—
4			—
5			—
6			—
样品			

思考题

(1) 说明分光光度法检测甲醛的方法原理。
(2) 采用哪些方法标定甲醛标准溶液？为什么？

拓展阅读

甲醛（化学式为 HCHO）为无色易溶于水的液体，挥发性很强，有刺激性气味。甲醛对人体有很大的伤害。首先，甲醛可以使人急性中毒，对皮肤有很大的损害。当甲醛达到一定程度的时候，人体会出现不适的症状，比如声音嘶哑、眼痒、咽喉不适、皮炎、呼吸困难，严重的时候会引起肺水肿、过敏性休克以及肝功能衰竭的情况。当吸入过量的时候，会导致人的死亡。对于新装修的房间可以通过通风的方式降低室内甲醛等有害物质，还可在房间内放置绿色植物通过光合作用来降低甲醛含量，活性炭、竹炭等吸附剂也可以很好地吸附室内甲醛。

实验 4.11 农达中草甘膦含量的测定

实验目的

(1) 掌握分光光度法测定草甘膦含量的原理。
(2) 掌握分光光度计的使用及标准曲线的绘制。

实验原理

草甘膦，化学名称为 N-磷酸甲基甘氨酸，又称为镇草宁、磷甘酸。化学式为 $C_3H_8NO_5P$，结构式如图 4.4 所示，是一种有机膦类除草剂。草甘膦的分子中含有羧酸基、氨基、甲基磷酸基等，可以进行酯化、羟氨基化、胺化、亚硝化和脱水等一系列典型的生物化学反应。草甘膦是一种高效、低毒、广谱的除草剂，是目前除草剂中产量最大的品种。草甘膦曾经被怀疑是一个致癌物，目前大多数认为它的致癌性已经非常小，但是食用过多的草甘膦，会引起过敏反应、损伤肝肾功能和消化道，引起恶心、呕吐、腹痛，严重的还会损伤呼吸系统和心血管系统，甚至危及生命。此外，如果孕妇长时间接触草甘膦，也可能导致畸胎、婴儿出生缺陷。草甘膦试样溶于水后，在酸性介质中能与亚硝酸钠作用生成草甘膦亚硝基衍生物，该化合物在 243nm 处有最大吸收峰，根据朗伯-比尔定律 $A=\varepsilon bc$，可以定量计算出试样中草甘膦的含量。

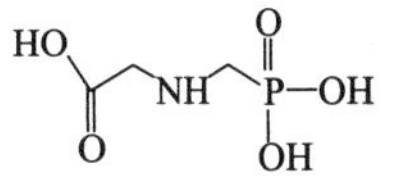

图 4.4 草甘膦的结构式

实验用品

仪器：紫外-分光光度计、石英比色皿（1cm）、容量瓶（100mL、250mL）、移液管（1.00mL、2.00mL、5.00mL）。

试剂：50%硫酸溶液（体积比）、250g/L 的溴化钾溶液、草甘膦标准样品（99.8%）、草甘膦样品（农达）、14g/L 的亚硝酸钠溶液（称取 0.28g 亚硝酸钠，溶于 20mL 水中，使用时现配）、蒸馏水。

实验步骤

(1) 草甘膦标准溶液的配制 称取 0.3000g 草甘膦标准品（精确至 0.0001g）置于小烧杯中，加 50mL 水溶解，定量转移至 250mL 容量瓶中，稀释至刻度，摇匀。

(2) 草甘膦标准曲线的绘制　用移液管分别移取适量上述草甘膦标准溶液 0.80mL，1.10mL，1.40mL，1.70mL，2.00mL 于 5 个 100mL 容量瓶中，同时另取 1 个 100mL 容量瓶作试剂空白。在上述各容量瓶中分别加入 5mL 蒸馏水，0.5mL 硫酸溶液，0.1mL 溴化钾溶液，0.5mL 亚硝酸钠溶液。加入亚硝酸钠溶液后应立即将塞子塞紧，充分摇匀，放置 20min。然后用水稀释至刻度，摇匀，最后将塞子打开，放置 15min。注意亚硝基化反应温度不能低于 15 度。

以空白作参比，在 243nm 处的波长下，用 1cm 的比色皿，测定各溶液的吸光度。以吸光度 A 为纵坐标，相应的标准样溶液的体积 V 为横坐标绘制标准曲线（表 4.21）。

(3) 农达中草甘膦含量测定　称取约 0.20g 试样（精确至 0.0001g），置于 200mL 烧杯中。加 60mL 水，缓缓加热溶解，趁热用快速滤纸过滤，仔细冲洗滤纸，将滤液接至 250mL 容量瓶中，冷至室温，稀释至刻度，摇匀。精确吸取 2.00mL 试样溶液于 100mL 容量瓶中，稀释至刻度，测定吸光度。

数据记录与处理

表 4.21　草甘膦标准曲线及样品测定

	草甘膦浓度/(mg/kg)	吸光度	样品中草甘膦含量/%
1			—
2			—
3			—
4			—
5			—
6			—
样品			

思考题

(1) 参比溶液的作用是什么？本实验中能否用蒸馏水作参比？
(2) 简述分光光度法测定草甘膦含量的原理。

4.3　生物类

实验 4.12　复合肥料中总氮含量的测定

实验目的

(1) 了解复合肥料试样的制备及处理方法。
(2) 掌握蒸馏后滴定法测定总氮的原理和方法。

实验原理

复合肥料是通过化学法或物理混合法制成的含作物营养元素氮、磷、钾中任意两种或三种的化肥。其中复合肥料中总氮含量的测定是将各种形态的氮（硝酸态、酰胺态、氰氨态）转换成氨，然后将氨吸收在过量硫酸溶液中，在甲基红-亚甲基蓝混合指示剂存在下，用氢氧化钠标准溶液返滴定。在碱性条件下，用定氮合金将硝酸态氮还原，直接蒸馏出氨。在酸性条件下还原硝酸盐成铵盐,在混合催化剂(硫酸钾和硫酸铜)存在下,用浓硫酸消化,将有机态氮或酰胺态氮和氰氨态氮转化为铵盐,从碱性溶液中蒸馏氨。

实验用品

仪器:分析天平、圆底蒸馏烧瓶(1000mL)、梨形玻璃漏斗、电炉、蒸馏设备。

试剂:

(1) 硫酸(A.R.);

(2) 盐酸(A.R.);

(3) 铬粉:细度小于$250\mu m$;

(4) 定氮合金:Cu 50%,Al 45%,Zn 5%,细度小于$250\mu m$;

(5) 硫酸钾(A.R.);

(6) 五水硫酸铜(A.R.);

(7) 混合催化剂:1000g硫酸钾和50g五水硫酸铜充分混合,并仔细研磨;

(8) 氢氧化钠溶液:400g/L;

(9) 氢氧化钠标准溶液:0.5000mol/L;

(10) 甲基红-亚甲基蓝混合指示剂;

(11) 广泛pH试纸;

(12) 硅脂。

实验步骤

(1) 复合肥料样品制备 将复合肥样品经多次缩分后取出约100g,用研磨器或研钵研磨至全部通过0.50mm孔径筛(对于潮湿肥料可通过1.00mm孔径筛),混合均匀,置于洁净、干燥瓶中,做成分分析。研磨操作要迅速,以免在研磨过程中失水或吸湿,并要防止样品过热。

从上述处理后的复合肥样品中称取总氮含量不大于235mg、硝酸态氮含量不大于60mg的试样$0.5\sim 2g$(精确至0.0002g)于圆底蒸馏烧瓶中。

(2) 试样处理与蒸馏

① 仅含铵态氮的试样。于圆底蒸馏烧瓶中加入300mL水,摇动使试样溶解,放入防爆沸物,将蒸馏烧瓶连接在蒸馏装置上。于接收器中加入40mL硫酸溶液(0.25mol/L),4~5滴甲基红-亚甲基蓝混合指示剂,并加适量水以保证封闭气体出口,将接收器连接在蒸馏装置上。蒸馏装置的磨口连接处涂硅脂密封。通过蒸馏装置的滴液漏斗加入20mL氢氧化钠溶液(400g/L),在溶液将流尽时加入20~30mL水冲洗漏斗,剩3~5mL水时关闭活塞。开通冷却水,同时开启蒸馏加热装置,沸腾时根据泡沫产生程度调节供热强度,避免泡沫溢出或液滴带出。蒸馏出至少150mL馏出液后,用pH试纸检查冷凝管出口的液滴,如果不显碱性就结束蒸馏。

② 含硝酸态氮和铵态氮的试样。于蒸馏烧瓶中加入300mL水,摇动使试样溶解,加入定氮合金3g和防爆沸物,将蒸馏烧瓶连接于蒸馏装置上。蒸馏过程除加入20mL氢氧化钠溶液(400g/L)后静置10min再加热外,其余步骤同铵态氮的试样的测定。

③ 含酰胺态氮、氰氨态氮和铵态氮的试样。将蒸馏烧瓶置于通风橱中,小心加入25mL硫酸,插上梨形玻璃漏斗,置于消化加热装置上(1500W电炉)上,加热至硫酸冒白烟15min后停止,待蒸馏烧瓶冷却至室温后小心加入250mL水。蒸馏过程除加入100mL氢氧化钠溶液(400g/L)外,其余步骤同铵态氮的试样的测定。

④ 含有机物、酰胺态氮、氰氨态氮和铵态氮的试样。将蒸馏烧瓶置于通风橱中,加入

22g 混合催化剂，小心加入 30mL 硫酸，插上梨形玻璃漏斗，置于消化加热装置上（1500W 电炉）上加热。如泡沫很多，减少供热强度至泡沫消失，继续加热至冒白烟 60min 后或直到溶液透明后停止。待烧瓶冷却至室温后小心加入 250mL 水。蒸馏过程除加入 120mL 氢氧化钠溶液（400g/L）外，其余步骤同铵态氮的试样的测定。

⑤ 含硝酸态氮、酰胺态氮、氰氨态氮和铵态氮的试样。于蒸馏烧瓶中加入 35mL 水，摇动使试样溶解，加入铬粉 1.2g，盐酸 7mL，静置 5~10min，插上梨形玻璃漏斗。置蒸馏烧瓶于通风橱内的加热装置上（1500W 电炉）上，加热至沸腾并泛起泡沫后 1min，冷却至室温，小心加入 25mL 硫酸，继续加热至冒硫酸白烟 15min，待蒸馏烧瓶冷却至室温后小心加入 400mL 水。蒸馏过程除加入 100mL 氢氧化钠溶液（400g/L）外，其余步骤同铵态氮的试样的测定。

⑥ 含有机物、硝酸态氮、酰胺态氮、氰氨态氮和铵态氮的试样或未知试样。于蒸馏烧瓶中加入 35mL 水，摇动使试样溶解，加入铬粉 1.2g，盐酸 7mL，静置 5~10min，插上梨形玻璃漏斗。置蒸馏烧瓶于通风橱内的加热装置上（1500W 电炉）上，加热至沸腾并泛起泡沫后 1min，冷却至室温，加入 22g 混合催化剂，小心加入 30mL 硫酸，继续加热。如泡沫很多，减少供热强度至泡沫消失，继续加热至冒硫酸白烟 60min 后停止，待蒸馏烧瓶冷却至室温后小心加入 400mL 水。蒸馏过程除加入氢氧化钠溶液（400g/L）为 120mL 外，其余步骤同铵态氮的试样的测定。

（3）滴定 用氢氧化钠标准溶液（0.5000mol/L）返滴定过量硫酸至混合指示剂呈现灰绿色为终点。在测定的同时，按相同的操作步骤，使用同样的试剂，但不含试样进行空白实验。

数据记录与处理

总氮含量 w，以质量分率（％）表示，按下式计算：

$$w = \frac{(V_2 - V_1) c \times 0.01401}{m} \times 100\% \tag{4.25}$$

式中 V_2——空白实验时，消耗氢氧化钠标准溶液的体积，mL；

V_1——测定复合肥料样品时，消耗氢氧化钠标准溶液的体积，mL；

c——氢氧化钠标准溶液浓度，mol/L；

0.01401——氮的毫摩尔质量值，g/mmol^{-1}；

m——试样质量，g。

计算结果保留到小数点后两位，取平行测定结果的算术平均值作为测定结果。

拓展阅读

复合肥料各营养元素的含量一般用 N-P$_2$O$_5$-K$_2$O 的相应分率表示，其总和即为肥料的浓度。根据中国标准，总含量在 25％~30％的为低浓度，30％~40％为中浓度，大于 40％为高浓度复合肥料。以复合肥料顺序的重量比来表示营养元素含量，例如 20-20-0 即为此肥料含有 20％ N，20％ P$_2$O$_5$，但不含 K$_2$O；含有两种营养元素的称作二元复合肥料（two-nutrient compound fertilizer），如含有氮和磷的磷酸二铵；含有三种营养元素的为三元复合肥料（three-nutrient compound fertilizer），如含有氮、磷、钾的硝酸磷酸钾等。根据制造方法一般将复合肥料分为化合复合肥料、混合复合肥料、掺合复合肥料 3 种类型。

实验 4.13　土壤中有机质含量的测定

实验目的

掌握土壤有机质测定方法的原理和计算方法。

实验原理

土壤有机质泛指土壤中以各种形式存在的含碳有机化合物。有机质是作物所需的 N、P、S 等各种养分的主要来源，可提高土壤的保肥性、缓冲性、生物和酶的活性。它是土壤肥力的基础，是衡量土壤肥力的重要指标之一。

用定量的重铬酸钾-硫酸溶液，在电砂浴加热条件下，使土壤中的有机碳氧化，剩余的重铬酸钾用硫酸亚铁标准溶液滴定，并以二氧化硅为添加物作试剂空白标定，根据氧化前后氧化剂质量差值，计算土壤有机质含量。标定硫酸亚铁的反应方程式如下：

$$Cr_2O_7^{2-} + 6Fe^{2+} + 14H^+ =\!=\!= 6Fe^{3+} + 2Cr^{3+} + 7H_2O \tag{4.26}$$

实验用品

仪器：分析天平、电砂浴、温度计（200～300℃）、磨口锥形瓶（150mL）、磨口简易空气冷凝管（$\phi 0.9cm \times 19cm$）、滴定管（10.00mL 和 25.00mL）、铜丝筛（1mm，0.25mm）、瓷研钵。

试剂：

(1) 重铬酸钾（G.R.）；

(2) 浓硫酸（A.R.）；

(3) 硫酸亚铁（A.R.）；

(4) 硫酸银：研成粉末；

(5) 二氧化硅：粉末状；

(6) 邻二氮菲指示剂：称取邻二氮菲 1.490g 溶于含有 0.700g 硫酸亚铁的 100mL 水溶液中，为防止变质，密闭保存于棕色瓶中备用；

(7) 重铬酸钾标准溶液（0.03330mol/L）：称取经 130℃ 烘 1.5h 的重铬酸钾 9.807g，先用少量水溶解，然后移入 1L 容量瓶内，加水定容；

(8) 重铬酸钾-硫酸溶液（0.4mol/L）：称取重铬酸钾 39.23g，溶于 600～800mL 蒸馏水中，待完全溶解后加水稀释至 1L，将溶液移入 3L 大烧杯中；另取 1L 浓硫酸，慢慢地倒入重铬酸钾水溶液内，不断搅拌，为避免溶液急剧升温，每加约 100mL 硫酸后稍停片刻，并把大烧杯放在盛有冷水的盆内冷却，待溶液的温度降到不烫手时再加另一份硫酸，直到全部加完为止；

(9) 硫酸亚铁标准溶液：称取 56g 硫酸亚铁，溶于 600～800mL 水中，加浓硫酸 20mL，搅拌均匀，加水定容至 1L（必要时过滤），贮存于棕色瓶中保存。此溶液易受空气氧化，使用时现用现标定。

实验步骤

(1) 硫酸亚铁标准溶液的标定　取 25.00mL 重铬酸钾标准溶液于 150mL 锥形瓶中，加浓硫酸 3mL 和 3～5 滴邻二氮菲指示剂，用硫酸亚铁标准溶液滴定，根据硫酸亚铁溶液的消耗量，计算硫酸亚铁标准溶液的浓度。

(2) 样品的选择和制备　选取具有代表性风干土样，用镊子挑除植物根叶等有机残体，把土块压细,使之先通过 1mm 孔径筛。充分混匀后,从中取出 10~20g 试样,磨细并全部通过 0.25mm 的孔径筛,装入 150mL 磨口锥形瓶中备用。对新采回的水稻土或长期处于渍水条件下的土壤,必须在土壤晾干压碎后,平摊成薄层,每天翻动一次,在空气中暴露一周左右才能磨样。

(3) 有机质含量的测定 根据表 4.22 有机质含量的规定称取上述制备好的风干试样 0.05~0.5g,精确到 0.0001g。置于 150mL 磨口锥形瓶中,加入 0.1g 粉末状的硫酸银,用滴定管准确加入 10.00mL 0.4mol/L 重铬酸钾-硫酸溶液,摇匀。

在上述锥形瓶瓶口上方插入简易空气冷凝管,然后移至已预热到 200~230℃ 的电砂浴上加热(见图 4.5)。当简易空气冷凝管下端落下第一滴冷凝液,开始计时,消煮 (5±0.5) min。消煮完毕后,将锥形瓶从电砂浴上取下,冷却片刻,用水冲洗冷凝管内壁及其底端外壁,使洗涤液流入原锥形瓶,瓶内溶液的总体积应控制在 60~80mL 为宜,加 3~5 滴邻二氮菲指示剂,用硫酸亚铁标准溶液滴定剩余的重铬酸钾。当溶液颜色从橙黄到蓝绿,最后变为棕红即达终点。如果试样滴定所用硫酸亚铁标准溶液的体积不到空白标定所消耗硫酸亚铁标准溶液的体积的 1/3 时,则应减少土壤称样量,重新测定。

每批试样测定必须同时做 2~3 个空白标定。取 0.500g 粉末状二氧化硅代替试样,其他步骤与试样测定相同,取其平均值。

数据记录与处理

土壤有机质含量 X(按烘干土计算)计算如下:

$$X = \frac{(V_0 - V)c \times 0.003 \times 1.724}{m} \times 100\% \tag{4.27}$$

式中 X——土壤有机质含量,%;
  V_0——空白试验消耗硫酸亚铁标准溶液体积,mL;
  V——测定试样时消耗硫酸亚铁标准溶液体积,mL;
  c——硫酸亚铁标准溶液的浓度,mol/L;
  0.003——1/4 碳原子的摩尔质量,g/mol;
  1.742——由有机碳换算为有机质的系数;
  m——烘干试样质量,g。

平行测定结果用算术平均值表示,保留三位有效数字。

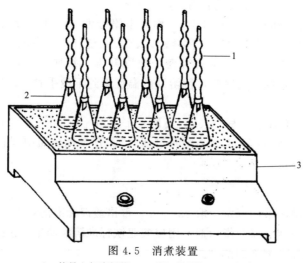

图 4.5 消煮装置
1—简易空气冷凝管;2—磨口锥形瓶;3—电砂浴

表 4.22　不同土壤有机质含量的称样量

有机质含量/%	试样质量/g
2 以下	0.4～0.5
2～7	0.2～0.3
7～10	0.1
10～15	0.05

实验 4.14　复合肥料中氯离子含量的测定

实验目的

掌握用福尔哈德法的返滴定方式测定复合肥料中氯离子含量。

实验原理

复合肥料中氯离子的含量直接影响农作物的生长，尤其是甘薯、甜菜等对氯离子比较敏感，施肥时要严格控制用量。另外长期使用氯离子含量高的复合肥料，容易使土壤酸化和盐渍化，进而会影响种子发芽、出苗，抑制农作物生长。本实验通过复合肥试样在微酸性溶液中，加入过量的硝酸银标准溶液，使氯离子转化为氯化银沉淀，用邻苯二甲酸二丁酯包裹沉淀，以硫酸铁铵为指示剂，用硫氰酸铵标准溶液滴定剩余的硝酸银。滴定反应如下：

$$Ag^+ + Cl^- =\!=\!= AgCl \downarrow \tag{4.28}$$

$$Ag^+_{(剩余)} + SCN^- =\!=\!= AgSCN \downarrow \tag{4.29}$$

实验用品

仪器：锥形瓶、容量瓶、烧杯、移液管、分析天平、滴定管。

试剂：

(1) 邻苯二甲酸二丁酯；

(2) 硝酸溶液：1∶1；

(3) 硝酸银标准溶液配制 (0.05mol/L)：称取 8.7g 硝酸银，溶解于水中，稀释至 1000mL，储存于棕色瓶中；

(4) 氯离子标准溶液 (1.000mg/mL)：准确称取 1.6487g 经 270～300℃ 烘干至质量恒定的基准氯化钠于烧杯中，用水溶解后，移入 1000mL 容量瓶中，稀释至刻度，混匀，储存于塑料瓶中；

(5) 铬酸钾指示剂 (50g/L)：称取 5.0g 分析纯铬酸钾，溶于少量蒸馏水中，加入硝酸银溶液至砖红色不褪色，搅拌均匀，放置过夜后，过滤。将滤液用蒸馏水稀释至 100mL；

(6) 硫酸铁铵指示剂 (80g/L)：溶解 8.0g 硫酸铁铵于 75mL 水中，过滤，加几滴硫酸，使棕色消失，稀释至 100mL；

(7) 硫氰酸铵标准溶液配制 (0.05mol/L)：称取 3.8g 硫氰酸铵溶解于水中，稀释至 1000mL。

实验步骤

(1) 硝酸银标准溶液的标定 (0.05mol/L)　准确移取 25.00mL 氯离子标准溶液于 250mL 锥形瓶中，加入 1mL 铬酸钾指示剂，用硝酸银标准溶液滴定，直至溶液的颜色由黄色至刚好出现砖红色沉淀即为滴定终点，同时进行空白试验。计算结果保留四位有效数字。

（2）硫氰酸铵标准溶液标定（0.05mol/L） 准确移取 25.00mL 氯离子标准溶液于 250mL 锥形瓶中，加入 5mL 硝酸溶液和 25.00mL 硝酸银标准溶液，摇动至沉淀分层，加入 5mL 邻苯二甲酸二丁酯，摇动片刻。加入水，使溶液总体积约为 100mL，加入 2mL 硫酸铁铵指示剂，用硫氰酸铵标准溶液滴定剩余的硝酸银，至出现浅砖红色为止，同时进行空白试验。计算结果保留四位有效数字。

（3）复合肥料样品制备 将复合肥样品经多次缩分后取出约 100g，用研磨器或研钵研磨至全部通过 0.50mm 孔径筛（对于潮湿肥料可通过 1.00mm 孔径筛），混合均匀，置于洁净、干燥瓶中，做成分分析。研磨操作要迅速，以免在研磨过程中失水或吸湿，并要防止样品过热。

称取上述处理过的试样约 1～10g（精确至 0.001g）（称样量范围见表 4.23）于 250mL 烧杯中，加 100mL 水，缓慢加热至沸，继续微沸 10min，冷却至室温，溶液转移到 250mL 容量瓶中，稀释至刻度，混匀。干过滤，弃去最初的部分滤液。

表 4.23 称样量范围

氯离子含量(w_1)/%	$w_1<5$	$5 \leqslant w_1 \leqslant 25$	$w_1 \geqslant 25$
称样量/g	10～5	5～1	1

（4）试样中氯离子含量测定 准确吸取一定量的滤液（含氯离子约 25mg）于 250mL 锥形瓶中，加入 5mL 硝酸溶液，加入 25.00mL 硝酸银标准溶液，摇动至沉淀分层，加入 5mL 邻苯二甲酸二丁酯，摇动片刻。加入水，使溶液总体积约为 100mL，加入 2mL 硫酸铁铵指示剂，用硫氰酸铵标准溶液滴定剩余的硝酸银，至出现浅砖红色为止。

数据记录与处理

氯离子的质量分率 w（%），按下式计算：

$$w = \frac{(c_1 V_1 - c_2 V_2) \times 0.03545}{mD} \times 100\% \tag{4.30}$$

式中 V_1——硝酸银标准溶液的体积（25.00mL）；

V_2——滴定时所消耗硫氰酸铵标准溶液的体积，mL；

c_1——硝酸银标准溶液的浓度，mol/L；

c_2——硫氰酸铵标准溶液的浓度，mol/L；

m——复合肥试样质量，g；

D——测定时吸取试液体积与试液的总体积的比值；

0.03545——氯离子的毫摩尔质量，g/mmol。

取平行测定结果的算术平均值作为测定结果。氯离子含量测定的允许差应符合表 4.24 要求。

表 4.24 氯离子含量测定的允许差

氯离子含量(w)/%	平行测定最大允许差值/%	不同实验室测定最大允许差值/%
<5	0.20	0.30
5～25	0.30	0.40
>25	0.40	0.60

实验 4.15　复合肥料中钾含量的测定

实验目的

(1) 了解复合肥料试样的制备及处理方法。
(2) 掌握以四苯硼酸钠为沉淀剂测定复合肥中钾含量的重量分析法。

实验原理

在弱碱性溶液中，四苯硼酸钠溶液与试样溶液中的钾离子生成四苯硼酸钾沉淀，将沉淀过滤、干燥及称重。如试样中含有氰氨基化合物或有机物时，可先加溴水和活性炭处理。为了防止阳离子干扰，可预先加入适量的乙二胺四乙酸二钠盐（EDTA），使阳离子与乙二胺四乙酸二钠络合。沉淀反应如下：

$$Na[B(C_6H_5)_4] + K^+ \Longrightarrow K[B(C_6H_5)_4] \downarrow + Na^+ \tag{4.31}$$

实验用品

仪器：锥形瓶、容量瓶、烧杯、移液管、分析天平、电热板、玻璃坩埚式过滤器（4号，30mL）、干燥箱。

试剂：

(1) 四苯硼酸钠溶液：15g/L；
(2) 乙二胺四乙酸二钠（EDTA）溶液：40g/L；
(3) 氢氧化钠溶液：400g/L；
(4) 溴水溶液：5%（质量分率）；
(5) 四苯硼酸钾饱和溶液；
(6) 酚酞指示剂：5g/L 乙醇溶液（95%）；
(7) 活性炭：应不吸附或不释放钾离子。

实验内容

(1) 复合肥料样品制备　将复合肥样品经多次缩分后取出约100g，用研磨器或研钵研磨至全部通过0.50mm孔径筛（对于潮湿肥料可通过1.00mm孔径筛），混合均匀，置于洁净、干燥瓶中，做成分分析。研磨操作要迅速，以免在研磨过程中失水或吸湿，并要防止样品过热。

做两份试样的平行测定。称取含氧化钾约400mg的试样2～5g（称准至0.0002g），置于250mL锥形瓶中，加约150mL水，加热煮沸30min，冷却，定量转移到250mL容量瓶中，用水稀释至刻度，混匀，干过滤，弃去最初50mL滤液。

(2) 试样溶液处理

① 试样不含氰氨基化合物或有机物。准确移取上述滤液25.00mL，置于200mL烧杯中，加EDTA溶液20mL（含阳离子较多时可加40mL），加2～3滴酚酞，滴加NaOH溶液至红色出现时，再过量1mL，在通风橱内缓慢加热煮沸15min，然后放置冷却或用流水冷却至室温，若红色消失，再用NaOH溶液调至红色。

② 试样含有氰氨基化合物或有机物。准确移取上述滤液25.00mL，置于200～250mL烧杯中，加入溴水溶液5mL，将该溶液煮沸直至所有溴水完全脱除为止（无溴颜色），若含有其他颜色，将溶液体积蒸发至小于100mL，待溶液冷却后，加0.5g活性炭，充分搅拌使

之吸附，然后过滤，并洗涤 3～5 次，每次用水约 5mL，收集全部滤液，加 EDTA 溶液 20mL（含阳离子较多时加 40mL），以下步骤同①操作。

(3) 沉淀及过滤　在不断搅拌下，于上述试样溶液（①或②）中逐滴加入 10mL 四苯硼酸钠溶液，并过量约 7mL。继续搅拌 1min，静置 15min 以上，用倾滤法将沉淀过滤于 120℃下预先恒重的 4 号玻璃坩埚式过滤器内，用四苯硼酸钾饱和溶液洗涤沉淀 5～7 次，每次用量约 5mL，最后用水洗涤 2 次，每次用量 5mL。

(4) 干燥　将盛有沉淀的坩埚置入 120℃±5℃干燥箱中，干燥 1.5h，然后放在干燥器内冷却，称重。

(5) 空白试验　除不加试样外，分析步骤及试剂用量均与上述步骤相同。

注：坩埚洗涤时，若沉淀不易洗去，可用丙酮进一步清洗。

数据记录与处理

根据四苯硼酸钾沉淀的质量计算复合肥料中 K_2O 的质量分数。以 K_2O 质量分数 w（%）表示，计算公式如下：

$$w = \frac{(m_2 - m_1) \times 0.1314}{m_0 \times 25.00/250} \times 100\% \tag{4.32}$$

式中　m_2——四苯硼酸钾沉淀的质量，g；
　　　m_1——空白试验时所得四苯硼酸钾沉淀的质量，g；
　　0.1314——四苯硼酸钾质量换算为氧化钾质量的系数；
　　　m_0——试样的质量，g；
　　25.00——吸取试样溶液体积，mL；
　　　250——试样溶液总体积，mL。

计算结果保留小数点后两位，取平行测定结果的算术平均值作为测定结果。

平行测定和不同实验室测定结果的允许差应符合表 4.25 要求。

表 4.25　平行测定和不同实验室测定结果的允许差

钾的质量分率（以 K_2O 计）/%	平行测定允许差值/%	不同实验室测定允许差值/%
<10.0	0.20	0.40
10.0～20.0	0.30	0.60
>20.0	0.40	0.80

实验 4.16　复合肥料中有效磷含量的测定

实验目的

(1) 了解复合肥料试样的制备及处理方法。
(2) 掌握磷钼酸喹啉重量法测定复合肥料中有效磷的原理和方法。

实验原理

用乙二胺四乙酸二钠振荡或柠檬酸溶液超声提取复合肥料中有效磷后，提取液中正磷酸根离子在酸性介质中与过量的喹钼柠酮试剂生成黄色磷钼酸喹啉沉淀，用磷钼酸喹啉重量法测定磷的含量。沉淀反应如下：

$$H_3PO_4+3C_9H_7N+12Na_2MoO_4+24HNO_3 =\!=\!=$$
$$(C_9H_7N)_3H_3[P(Mo_3O_{10})_4]\cdot H_2O\downarrow+11H_2O+24NaNO_3 \quad (4.33)$$

实验用品

仪器：分析天平、超声清洗仪、恒温水浴振荡器、电热恒温干燥箱（180℃±2℃）、玻璃坩埚式过滤器（4号，容积30mL）、电热板。

试剂：

(1) 硝酸溶液：1∶1；

(2) 乙二胺四乙酸二钠（EDTA）溶液：37.5g/L；

(3) 柠檬酸溶液：20g/L；

(4) 喹钼柠酮试剂：

溶液Ⅰ——溶解70g二水合钼酸钠于150mL水中。

溶液Ⅱ——溶解60g一水合柠檬酸于150mL水和85mL硝酸的混合液中。

溶液Ⅲ——在搅拌下，将溶液A加入到溶液B中。

溶液Ⅳ——溶解5mL喹啉于35mL硝酸和100mL水的混合液中。

缓慢把溶液Ⅳ加入到溶液Ⅲ中并混匀。在聚乙烯瓶中于暗处放置24h，用玻璃坩埚式过滤器过滤。量取280mL丙酮加到滤液中，加水稀释至1000mL，混匀，储存于另一洁净的聚乙烯瓶中。配制好的喹钼柠酮试剂在避光下保存不超过一周。

实验步骤

(1) 复合肥料样品制备　将复合肥样品经多次缩分后取出约100g，用研磨器或研钵研磨至全部通过0.50mm孔径筛（对于潮湿肥料可通过1.00mm孔径筛），混合均匀，置于洁净、干燥瓶中，做成分分析。研磨操作要迅速，以免在研磨过程中失水或吸湿，并要防止样品过热。

称取两份试样进行平行测定。每份试样含有100～180mg五氧化二磷，精确至0.0002g。

(2) 有效磷的提取

提取方法一（EDTA振荡提取）：按步骤(1)中平行样品测定称样量的要求，另称取试样置于滤纸上，用滤纸包裹试样，塞入250mL量瓶中，加入150mL EDTA溶液，塞紧瓶塞，摇动量瓶使滤纸破碎、试样分散于溶液中，置于60℃±2℃的恒温水浴振荡器中，保温振荡1h（振荡频率以量瓶内试样能自由翻动即可）。然后取出量瓶，冷却至室温，用水稀释至刻度，混匀。干过滤，弃去最初部分滤液，即得溶液A，供测定有效磷用。

提取方法二（柠檬酸超声提取）：以$NH_4H_2PO_4$、KH_2PO_4、硝磷复肥作为磷源的复肥，按步骤(1)中要求称取试样，置于250mL量瓶中，加入150mL柠檬酸溶液，将量瓶置于超声波清洗仪中提取6～8min（超声清洗仪液面高于容量瓶内液面），用水稀释至刻度，混匀，干过滤，弃去最初部分滤液，即得溶液B，供测定有效磷用。

注：柠檬酸超声提取法仅适用于以磷酸二氢铵、磷酸二氢钾、硝磷复肥作为磷源的肥料。

(3) 有效磷的测定　用单标线吸管吸取25.00mL溶液A或溶液B，移入500mL烧杯中，加入10mL硝酸溶液，用水稀释至100mL。在电热板上加热微沸2～3min，取下，加入35mL喹钼柠酮试剂，盖上表面皿，在电热板上微沸1min或置于近沸水浴中保温至沉淀分层，取出烧杯，冷却至室温。用预先在180℃±2℃干燥箱内干燥至恒重的玻璃坩埚式过滤器过滤，先将上层清液滤完，然后用倾泻法洗涤沉淀1～2次，每次用25mL水，将沉淀移入滤器中，再用水洗涤，所用水共125～150mL，将沉淀连同滤器置于180℃±

2℃干燥箱内，待温度达到180℃后，干燥45min，取出移入干燥器内，冷却至室温，称量。

（4）空白试验　除不加试样外，分析步骤及试剂用量均与上述步骤相同，进行平行操作。

数据记录与处理

有效磷含量（w），以五氧化二磷（P_2O_5）质量分数表示（%），按下式计算：

$$w=\frac{(m_1-m_2)\times 0.03207}{m\times \dfrac{25.00}{250}}\times 100\% \tag{4.34}$$

式中　m_1——测定有效磷所得磷钼酸喹啉沉淀的质量，g；

m_2——测定有效磷时，空白试验所得磷钼酸喹啉沉淀的质量，g；

m——测定有效磷时，试样质量，g；

0.03207——磷钼酸喹啉质量换算为五氧化二磷质量的系数；

25.00——吸取试样溶液体积，mL；

250——试样溶液总体积，mL。

计算结果保留到小数点后两位，取平行测定结果的算术平均值为测定结果。

实验4.17　土壤中有效磷的测定

实验目的

（1）了解分光光度法测定土壤中有效磷的原理和方法。

（2）掌握分光光度计的使用方法。

实验原理

磷是植物生长所必需的大量元素之一。土壤中的有效磷是指土壤中可被植物吸收利用的磷的总称。它包括全部水溶性磷、部分吸附态磷、一部分微溶性的无机磷和易矿化的有机磷等。在农业生产中一般采用土壤有效磷的指标来指导施用磷肥。

根据土壤的性质，选择不同的浸提剂提取土壤中的有效磷。利用氟化铵-盐酸溶液浸提酸性土壤中的有效磷，利用碳酸氢钠溶液浸提中性和石灰性土壤中有效磷。所提取出的磷以钼锑抗分光光度法测定，计算得出土壤样品中的有效磷含量。

钼锑抗分光光度法测定磷的原理为：在酸性条件下，正磷酸盐与钼酸铵、酒石酸锑钾反应，生成磷钼杂多酸。加入还原剂抗坏血酸还原，磷钼杂多酸中的一部分Mo（Ⅵ）被还原为Mo（Ⅴ），生成磷钼蓝（磷钼杂多蓝 $H_3PO_4\cdot 10MoO_3\cdot Mo_2O_5$ 或 $H_3PO_4\cdot 8MoO_3\cdot 2Mo_2O_5$）。在一定的浓度范围内，蓝色的深浅与磷含量成正比。

实验用品

仪器：电子天平、紫外-可见分光光度计、酸度计、恒温振荡器、容量瓶（25mL）、比色管（50mL）、吸量管（5mL）、量筒（50mL）、比色皿（1cm）、漏斗、滤纸。

试剂：

（1）硫酸（5%，体积比）：吸取5mL浓硫酸缓慢加入90mL水中，冷却后用水稀释至100mL；

(2) 酒石酸锑钾溶液（5g/L 或 3g/L）：称取酒石酸锑钾（$KSbOC_4H_4O_6 \cdot 0.5H_2O$）0.5g（或 0.3g）溶于 100mL 水中；

(3) 硫酸钼锑储备液：称取 10.0g 钼酸铵溶于 300mL 约 60℃ 的水中，冷却。另量取 126mL 浓硫酸，缓慢倒入约 400mL 水中，搅拌，冷却。然后将配制好的硫酸溶液（5%）缓缓倒入钼酸铵溶液中。再加入 100mL 酒石酸锑钾溶液，冷却后，用水定容至 1L，摇匀，储存于棕色试剂瓶中；

(4) 钼锑抗显色剂（15g/L 或 5g/L）：称取 1.5g（或 0.5g）抗坏血酸溶于 100mL 硫酸钼锑储备液，此溶液现配现用；

(5) 二硝基酚指示剂：称取 0.2g 2,4-二硝基酚或 2,6-二硝基酚溶于 100mL 水中；

(6) 氨水溶液（1:3）：按氨水与水 1:3 的体积比配制；

(7) 氟化铵-盐酸浸提剂：称取 1.11g 氟化铵溶于 400mL 水中，加入 2.1mL 浓盐酸，用水稀释至 1L，储存于塑料瓶中；

(8) 硼酸溶液（30g/L）：称取 30.0g 硼酸，在 60℃ 左右的热水中溶解，冷却后稀释至 1L；

(9) 磷标准储备液（100mg/L）：准确称取 105℃ 烘干 2h 磷酸二氢钾（优级纯）0.4394g，用水溶解后，加入 5mL 5% 硫酸，用水定容至 1 L；

(10) 磷标准溶液（5mg/L）：吸取 5.00mL 磷标准储备液于 100mL 容量瓶中，用水定容，摇匀后待用；

(11) 氢氧化钠溶液（100g/L）：称取 10g 氢氧化钠溶于 100mL 水中；

(12) 碳酸氢钠浸提剂：称取 42.0g 碳酸氢钠溶于约 950mL 水中，用氢氧化钠溶液（100g/L）调节 pH 至 8.5，用水稀释至 1L，储存于聚乙烯瓶中备用。若储存期超过 20 天，使用时必须检查并校准 pH；

(13) 钼锑储备液：称取 10.0g 钼酸铵溶于 300mL 约 60℃ 的水中，冷却。另量取 181mL 浓硫酸，缓慢倒入约 800mL 水中，搅拌，冷却。然后将配制好的硫酸溶液（5%）缓缓倒入钼酸铵溶液中。再加入 100mL 酒石酸锑钾溶液（3g/L），冷却后，用水定容至 2L，摇匀，储存于棕色试剂瓶中。

实验步骤

(1) 土壤样品预处理

① 酸性土壤：称取 5.00g 风干试样于 200mL 塑料瓶中（预先经 2mm 筛孔过筛处理），加 50.00mL NH_4F-HCl 浸提剂，在（25±1）℃ 条件下，放在振荡频率为（180±20）r/min 的振荡器里振荡 30min 后，立即用无磷滤纸干过滤。滤液待测定。

② 中性或石灰性土壤：称取 2.50g 风干试样于 200mL 塑料瓶中（预先经 2mm 筛孔过筛处理），加 50.00mL $NaHCO_3$ 浸提剂。其他步骤同上。

(2) 标准曲线的绘制

① 酸性土壤：分别准确移取 5mg/L 磷标准溶液 0.00mL、1.00mL、2.00mL、4.00mL、6.00mL、8.00mL、10.00mL 于 50mL 容量瓶中，加入 10mL NH_4F-HCl 浸提剂，再加入 10mL 硼酸溶液，摇匀，加水至 30mL，再加入二硝基酚指示剂 2 滴，用硫酸溶液或氨水溶液调节溶液刚显微黄色，加入钼锑抗显色剂 5.00mL，用水定容至刻度，充分摇匀，即得磷标准系列溶液。在室温高于 20℃ 条件下静置 30min 后，用 1cm 比色皿在 700nm 处，以标准溶液的零点调零后进行比色测定，绘制标准曲线。

② 中性或石灰性土壤：分别准确移取 5mg/L 磷标准溶液 0.00mL、0.50mL、1.00mL、

2.00mL、3.00mL、4.00mL、5.00mL 于 25mL 容量瓶中，加入 10mL NaHCO₃ 浸提剂，钼锑抗显色剂 5.00mL，慢慢摇动，排出 CO_2 后用水定容至刻度，即得磷标准系列溶液。在室温高于 20℃ 条件下静置 30min 后，用 1cm 比色皿在 880nm 处，以标准溶液的零点调零后进行比色测定，绘制标准曲线。

（3）有效磷的测定

① 酸性土壤：准确移取 10.00mL 上述酸性土壤滤液于 50mL 容量瓶中，加入 10mL 硼酸溶液，摇匀，加水至 30mL 左右，再加入二硝基酚指示剂 2 滴，用硫酸溶液或氨水溶液调节溶液刚显微黄色，加入钼锑抗显色剂 5.00mL，用水定容。在室温高于 20℃ 条件下静置 30min 后，用 1cm 比色皿在 700nm 处，以标准溶液的零点调零后进行比色测定。

② 中性或石灰性土壤：准确移取 10.00mL 上述中性或石灰性土壤滤液于 50mL 容量瓶或锥形瓶中，缓慢加入钼锑抗显色剂 5.00mL，慢慢摇动，排出 CO_2，再加入 10.00mL 水，充分摇匀，排净 CO_2。在室温高于 20℃ 条件下静置 30min 后，用 1cm 比色皿在 880nm 处，以标准溶液的零点调零后进行比色测定。

 数据记录与处理

土壤样品中有效磷 P 含量，以质量分率 w 计，数值以毫克每千克（mg/kg）表示，按下式计算：

$$w = \frac{(\rho - \rho_0)VD}{m \times 1000} \times 100\% \tag{4.35}$$

式中　ρ——从标准曲线求得的磷的浓度，mg/L；
　　　ρ_0——从标准曲线求得的空白试样中磷的浓度，mg/L；
　　　V——试液体积，mL；
　　　D——分取倍数，试样浸提剂体积与分取体积之比；
　　　m——试样质量，g。

平行测定结果以算术平均值表示，保留小数点后一位。

 拓展阅读

植物是人类赖以生存的物质财富，而植物是通过吸收土壤中养分来维持植物的生长。土壤中磷的存在对植物的营养有重要的作用，它是植物生长所必需的重要元素之一，植物用来吸收养分的根系也和磷有密切的关系。在自然界中磷大多以磷酸盐的形式存在，常见的有磷酸钙、磷灰石等。它在植物体中的含量仅次于氮和钾，几乎许多重要的有机化合物都有磷元素。磷在植物体内参与光合作用、呼吸作用、能量储备和传递、细胞分裂。磷能促进植物早期根系的形成和生长，提高植物适应外界环境条件的能力，有助于植物抗寒。磷还能提高许多水果、蔬菜和粮食作物的品质，还有助于增强一些植物的抗病性，还具有促熟作用，对丰收和作物品质是非常重要的。植物从土壤中吸收磷时，若土壤中没有足够的磷元素或没有足够供植物吸收的磷元素时，植物生长将受到极大的限制。

实验 4.18　银杏叶总黄酮含量的测定

实验目的

（1）掌握大孔树脂的使用方法。

(2) 掌握大孔树脂静态和动态吸附分离操作。
(3) 掌握紫外可见分光光度计的使用方法。

实验原理

中国是银杏的原产地，又是世界上银杏的主产区。银杏叶黄酮是银杏叶中提取的最主要的有效药用成分之一，其种类约有 20 多种。黄酮类化合物主要是指母核为 2-苯基色原酮（见图 4.6）的一类化合物，现在泛指具有 C_6—C_3—C_6 基本结构骨架（见图 4.7）的一大类天然化合物，是由两个苯环通过中央三碳相互连接而成。由于标准物的限定而很难做到直接测定 20 多种成分的含量，目前一般采用银杏叶中含量较多的芦丁为标准物分析银杏叶及其提取物中黄酮类化合物的含量。芦丁的分子式为 $C_{27}H_{30}O_{16}$，分子量为 610.51，其结构式见图 4.8。

从银杏叶浸取液中分离富集黄酮类化合物的方法主要有吸附法和萃取法。其中大孔树脂吸附法对银杏叶黄酮类的选择性比较高，因此得到的银杏叶黄酮纯度较高。大孔树脂是一种有机高分子聚合物，其吸附原理是利用分子间的范德华力，通过巨大的比表面积从溶剂中吸附黄酮类物质，再经洗脱得到纯度较高的黄酮类物质。D101 大孔弱酸型阳离子交换树脂，其功能基团羧基与吸附质黄酮类化合物上的羟基之间形成氢键，能很好地进行吸附。

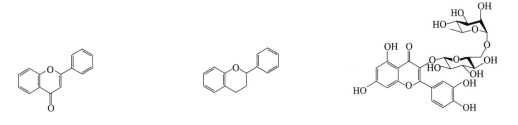

图 4.6　2-苯基色原酮　　图 4.7　C_6—C_3—C_6 结构单元　　图 4.8　芦丁结构式

紫外-可见分光光度法测定黄酮类化合物的原理是黄酮类化合物在含有亚硝酸钠的碱性溶液中，与 Al^{3+} 络合生成黄色络合物，以芦丁作为标准物质，在波长 400～600nm 内使用紫外-可见分光光度计进行测定，根据朗伯-比尔定律测定洗脱液中黄酮类化合物的含量。

实验用品

仪器：紫外-可见分光光度计、粉碎机、电子天平、电热恒温水浴锅、pH 计、容量瓶、比色管、比色皿、振荡箱、电热鼓风干燥箱、玻璃层析柱。

试剂：银杏叶、无水乙醇（A.R.）、氢氧化钠（A.R.）、硝酸铝（A.R.）、亚硝酸钠（A.R.）、盐酸（A.R.）、无水甲醇（A.R.）、芦丁标准品（G.R.）、D101 大孔树脂。

实验步骤

(1) 芦丁标准曲线的绘制　准确称取 6.2mg 芦丁标准品，用 70%乙醇溶解并定容至 25mL 容量瓶中，摇匀，即得芦丁标准品溶液，浓度为 0.248mg/mL。分别准确吸取芦丁标准品溶液 0.00mL、0.25mL、0.50mL、0.75mL、1.00mL、1.25mL、1.50mL 至 10mL 容量瓶中，分别加入 5%亚硝酸钠溶液 0.30mL，摇匀，静置 6min；再加 10%硝酸铝溶液 0.30mL，摇匀，静置 6min，再加 4%氢氧化钠溶液 4.00mL，用 70%乙醇稀释至刻度，摇匀，静置 15min。以 70%乙醇为空白，在 510nm 处测吸光度。以芦丁浓度 c 为横坐标，吸光度值 A 为纵坐标绘制芦丁标准曲线。

（2）银杏叶中黄酮类化合物的提取　取 5g 银杏叶于 50℃烘干、粉碎、过 40 目 (0.425mm) 筛后备用。称取 0.5g 银杏叶粉，加入 100mL 70%的乙醇溶液，于 70℃密闭加热 2h，抽滤取上清液，滤渣以同样的方法进行浸提，合并两次提取液，将提取液浓缩并转入 50mL 容量瓶中，用水定容至刻度，摇匀。上柱前需调节浸提液的 pH。

（3）D101 大孔树脂的预处理　先将玻璃层析柱清洗干净。在柱内加入相当于装填树脂体积 0.4~0.5 倍的无水乙醇，然后将 D101 大孔树脂通过湿法装柱加入柱中，上方液面高于树脂层约 30cm，浸泡 24h。在一定流速下，分别按照无水乙醇、水、5%盐酸、水、2%氢氧化钠、水的顺序对大孔吸附树脂进行水合和去杂处理。

注意在装柱过程中，将树脂沿玻璃棒一次性匀速加入层析柱中，同时开启柱底活塞，使树脂自然沉降而不留气泡。装柱过程中始终保持液面高出树脂层面 20cm 以上，防止液体流干。

（4）大孔树脂静态吸附-解吸实验　准确称取已处理好的大孔树脂 1g，置于带塞锥形瓶中，加入银杏叶黄酮提取液 100mL，置于振荡箱中静态吸附 24h。过滤，用蒸馏水冲洗上述吸附饱和的大孔吸附树脂至洗脱液无色，用滤纸吸干树脂表面残留的溶液后，将树脂置于另一带塞锥形瓶中，加入 80mL 70%乙醇溶液置于摇床中进行静态解吸 24h，再将树脂滤出，用分光光度计测定滤液中的黄酮浓度。

（5）大孔树脂动态吸附-解吸实验　称取 5g 大孔树脂于玻璃层析柱中（$\varphi 1.5cm \times 30cm$），径高比为 1∶8，将 50.00mL 黄酮提取液上柱，以 3mL/min 流速通过吸附柱，用分光光度计测定柱后流出液中黄酮的浓度。加入一定量蒸馏水洗脱柱中残留的浸提液，最后用 70%的乙醇作为洗脱液以 3mL/min 的流速进行解吸，收集流出液，同样用分光光度计测定其中黄酮浓度。

（6）树脂的再生　将大孔树脂于 95%乙醇中浸泡 8h，然后用蒸馏水洗至无乙醇味；再用 5% HCl 溶液浸泡 3h，用蒸馏水洗至流出液 pH 值为中性；然后用 5% NaOH 溶液浸泡 3h，用蒸馏水洗至流出液 pH 值为中性。

（7）黄酮类化合物含量测定　准确移取 2.00mL 试样溶液于 25mL 比色管中，按（1）中芦丁标准曲线的绘制步骤，以 70%乙醇为空白，在 510nm 处测吸光度，通过标准曲线和洗脱液中的黄酮样品的吸光度值，得出芦丁浓度（mg/mL），计算求出银杏叶中总黄酮得率。

数据记录与处理

总黄酮得率
$$w = \frac{y \times 25 V_1}{mV \times 1000} \times 100\% \tag{4.36}$$

式中　V_1——提取液体积，mL；
　　　V——吸取提取液的体积，mL；
　　　y——根据标准曲线求得的芦丁浓度，mg/mL；
　　　m——样品质量，g。

拓展阅读

银杏树由于其生长规律特殊，抗病能力强而受到国内外的重视。有关银杏叶提取物的有效成分及疗效的研究也日益受到了重视。以银杏叶提取物为原料，开发出的保健品、化妆品、药品、食品添加剂、功能性饮料等多达 100 多种，形成国际市场上销售额达 20 多亿美元的新兴产业。银杏叶的化学成分有黄酮类、萜类、内酯类、酚酸类以及生物碱、聚异戊二烯等化合物。黄酮类为银杏叶的主要有效成分之一，其含量随品种、产地、树龄、不同的采摘时间而不同。黄酮类化合物因其优异的抗氧化、抗病毒、防治心血管疾病、增强免疫力等作用而受世人瞩目，预计其临床价值随着研究的深入还会增加，有待加大投入，进一步加强研究，为人类卫生和健康提供自然、良好的医药资源。

实验4.19 混合氨基酸的测定

实验目的

(1) 掌握离子交换树脂分离氨基酸的基本原理；
(2) 掌握离子交换柱层析的基本操作；
(3) 掌握茚三酮显色法测定氨基酸含量的原理。

实验原理

离子交换柱层析分离混合氨基酸是基于各氨基酸电荷行为不同。氨基酸是两性电解质，分子上所带的净电荷取决于氨基酸的等电点（pI）和溶液的pH值。其中酸性氨基酸天冬氨酸的pI为2.97，碱性氨基酸赖氨酸的pI为9.74，在pH 5.3条件下，因为pH值低于赖氨酸的pI值，赖氨酸可解离成阳离子结合在732树脂上；天冬氨酸可解离成阴离子，不被树脂吸附而流出层析柱。在pH 12条件下，pH值高于赖氨酸的pI值，赖氨酸可解离成阴离子从树脂上被交换下来。这样通过改变洗脱液的pH值，可使它们被分别洗脱而达到分离的目的。

氨基酸的游离氨基与水合茚三酮生成紫色化合物，在一定范围内，该化合物颜色的深浅与氨基酸的含量成正比，因此可通过分光光度计测定氨基酸的含量。

实验用品

仪器：层析柱（20cm×ϕ1cm）、电子天平、恒流泵、收集装置、恒温水浴锅、分光光度计、pH计、电子磁力搅拌器、移液枪、具塞比色管、胶头滴管、量筒、烧杯。

试剂：

(1) 732型阳离子交换树脂[100～200目（0.147～0.074mm）]；
(2) 盐酸（A.R.）；
(3) 氢氧化钠（A.R.）；
(4) 标准氨基酸溶液：天冬氨酸、赖氨酸分别配制成2mg/mL的0.1mol/L盐酸溶液；
(5) 柠檬酸-氢氧化钠-盐酸缓冲溶液（pH 5.3）：称取柠檬酸14.25g，氢氧化钠9.30g和浓盐酸5.25mL溶于少量水后，定容至500mL，冰箱保存；
(6) 茚三酮显色剂：取0.5g茚三酮溶于75mL乙二醇甲醚中，加水定容至100mL；
(7) 混合氨基酸溶液：将2mg/mL天冬氨酸、赖氨酸溶液按1:2.5的比例混合，再用0.1mol/L盐酸溶液和水等体积稀释。

实验步骤

(1) 离子交换树脂的预处理　市售干树脂，要先在水中充分溶胀，经浮选得到颗粒大小合适的树脂。称取2g 732型阳离子交换树脂，加水倒入层析柱中。加入50mL 5% NaOH溶液，控制流速1滴/秒，结束后滴加100mL去离子水洗至中性。再加入50mL 5%HCl溶液，控制流速1滴/秒，结束后滴加100mL去离子水洗至中性，如此重复三次。

(2) 装柱　取层析柱垂直固定在铁架台上，在柱内加2～3cm高的柠檬酸-氢氧化钠-盐酸缓冲溶液。将上述处理过的离子交换树脂搅拌成悬浮状，沿柱内壁缓慢加入到层析柱内，待树脂在柱底部逐渐沉降时，慢慢打开柱底旋塞，继续加入树脂悬浮液，直至树脂层高度到层析柱高度的 3/4 处。

(3) 平衡　层析柱装好后，再缓慢沿管壁加入适量缓冲溶液至树脂床面以上 2～3cm 处，接上恒流泵，用缓冲溶液以 0.5mL/min 的流速进行平衡，用 pH 试纸检测流出液 pH 值，直至流出液 pH 值与缓冲液的 pH 值相等。

(4) 加样　关闭恒流泵，打开层析柱上端管口，缓慢打开柱底出口，小心放出层析柱内的液体至柱内液体的凹液面恰好与树脂上液面相齐，立即关闭下端出口（注意：不要使液面下降至树脂表面以下）。用移液枪吸取氨基酸混合样品 0.5mL，沿靠近树脂表面的管壁缓慢加入，注意不要冲坏树脂表面。加样后打开柱底旋塞，使液体尽可能缓慢地流下，至液体凹液面恰与树脂表面相平，立即关闭。再用胶头滴管吸取 0.5mL 缓冲液冲洗柱内壁，打开旋塞放出液体，使液体下表面与树脂表面相平，按照此法清洗 2 次。然后小心加入缓冲液离柱顶部 1cm 为止，并将层析柱与恒流泵和收集器相连。

(5) 洗脱

① 缓冲液洗脱：用柠檬酸-氢氧化钠-盐酸缓冲溶液以 6 秒/滴的流速开始洗脱，用试管收集洗脱液，每管收集 1mL，收集 1～10 号管。

② 0.01mol/L 氢氧化钠溶液（pH 12）洗脱：关闭恒流泵和旋塞，将上述缓冲液替换成 0.01mol/L 氢氧化钠溶液，打开旋塞进行洗脱，同法收集 11～40 管。收集完毕后，关闭旋塞和恒流泵。

(6) 标准曲线的绘制和样品测定

① 标准曲线的绘制。分别取 2mg/mL 的标准氨基酸溶液 0.0mL、0.2mL、0.4mL、0.6mL、0.8mL、1.0mL 于比色管中，用蒸馏水定容至 1mL。分别各加入 1.5mL 柠檬酸-氢氧化钠-盐酸缓冲溶液、1mL 茚三酮显色剂，充分混匀后，盖住试管口，100℃水浴中加热 15min，冷却至室温，用分光光度计在 570nm 测定吸光度值。以吸光度为纵坐标，氨基酸含量为横坐标绘制标准曲线。

② 氨基酸样品的测定。分别向 60 管收集液中加入 1.5mL 柠檬酸-氢氧化钠-盐酸缓冲液，混匀后再加入 1mL 茚三酮显色剂，混匀。沸水浴 15min 后取出，冷却至室温，在 570nm 下以蒸馏水为空白进行比色，测定各管的吸光度值。以吸光度值为纵坐标，收集的管数为横坐标，绘制洗脱曲线。

(7) 树脂再生　层析柱使用几次后，需将树脂取出用 1mol/L 氢氧化钠溶液洗涤，再用蒸馏水洗至中性后可继续使用。

实验 4.20　烟叶样品中还原糖的测定

实验目的

(1) 了解烟叶样品的制备及处理方法。
(2) 掌握砷钼酸比色法测定还原糖的原理和方法。

实验原理

还原糖在碱性条件下加热，可以定量地还原二价铜离子为一价铜离子，即产生砖红色氧化亚铜沉淀。氧化亚铜在酸性条件下，可将钼酸铵还原，还原后的钼酸铵再与砷酸氢二钠起作用，生成一种蓝色复合物即砷钼蓝，其颜色深浅在一定范围内与还原糖的含量成正比，用分光光度法就可测样品中还原糖的含量。

实验用品

仪器：722 型分光光度计、电子分析天平、粉碎机、水浴锅、容量瓶、锥形瓶、烧杯、量筒、比色管、移液管。

试剂：

(1) 酒精（80%）；

(2) 铜试剂：称取酒石酸钾钠 12g、无水碳酸钠 24g、碳酸氢钠 16g，将三者分别溶于 3 份 200mL 蒸馏水中，然后将 3 份溶液混匀。另称取硫酸铜 4g，溶于约 200mL 水中，完全溶解后，将此溶液慢慢地、边加边搅拌地加到上述混合液中，然后再称取 180g 无水硫酸钠溶于其中。将混合液在沸水浴中加热 20min，冷却后若出现沉淀，应过滤除去。最后将溶液稀释到 1000mL。此溶液容易出现结晶，应在 20℃ 以上保存。

(3) 砷钼酸试剂：称取 25g 钼酸铵 $[(NH_4)_6Mo_7O_{24} \cdot 4H_2O]$，溶于 450mL 蒸馏水中，再加入 22mL 浓硫酸，充分混匀。另取 3g 砷酸氢钠 $[Na_2HAsO_4 \cdot 7H_2O]$，溶于 25mL 水中，将此溶液与上述酸性钼酸铵溶液混合均匀。混合液于 37℃ 下保温 24~48h，可以用于测定，此试剂应储于棕色瓶中。

(4) 标准葡萄糖溶液（150μg/mL）：准确称取 150mg 葡萄糖，溶于 1000mL 水中。

实验步骤

(1) 标准曲线的绘制　取 6 个 10mL 比色管，分别加入 0.0mL，0.2mL，0.4mL，0.6mL，0.8mL 及 1.0mL 150μg/mL 的葡萄糖溶液，再加水补充到 1.0mL。然后准确加入 1.0mL 铜试剂，盖上玻璃塞。放入沸水浴中，煮沸 20min 后立即取出，放在冷水中冷却 5min。最后，加入 1.0mL 砷钼酸试剂，再加 7.0mL 蒸馏水稀释至刻度，摇匀后于 560nm 处比色。绘制标准曲线，求出回归方程。

(2) 烟叶样品的制备　准确称取 0.5g 粉碎的烟叶样品，放入锥形瓶中，加入 15mL 80% 酒精，在 70℃ 水浴中浸提 30min，将上清液滤入 100mL 容量瓶。残渣再分别用 10mL 酒精浸提两次，过滤，用热酒精冲洗滤渣。最后将提取液用蒸馏水定容到 100mL 容量瓶中。

(3) 测定样品还原糖　准确吸取 5.0mL 上述提取液置于 50mL 容量瓶中，加水稀释至刻度。准确量取稀释液 1.0mL，按标准曲线步骤测定 560nm 处的吸光度，计算还原糖含量。

数据记录与处理

$$还原糖含量\ w = \frac{y \times 10 \times V_1}{mV \times 10^6} \tag{4.37}$$

式中　V_1——提取液体积，mL；

　　　V——吸取提取液的体积，mL；

　　　y——根据标准曲线求得的还原糖浓度，μg/mL；

　　　m——样品质量，g。

拓展阅读

从烟草化学的角度看，糖类化合物是形成均衡烟气必不可少的成分，烟叶吸燃时，它们高温裂解成低级的醛、酮，使烟气呈酸性（pH5.3~6.5），可中和烟碱及其他含氮化合物在抽吸过程中产生的碱性化合物，对烟气的香气和吃味都有良好的作用，并能减少烟气的刺激性。因此，含糖量是烟草化学检测的必测项目。其中烟叶还原糖含量被认为是体现烟草优良品质的指标之一，在一定范围内烟叶中还原糖量越高，烟叶品质越好。

4.4 制药类

实验 4.21 中药白硇砂中 NH_4^+ 含量的测定

实验目的

(1) 掌握甲醛法测定铵盐中铵态氮含量的原理和方法。
(2) 熟练滴定操作和滴定终点的判断。

实验原理

中药白硇砂（淡硇砂），为白色结晶体，呈不规则的块状或粒状，大小不一，味咸苦而刺舌。它具有利尿、散瘀消肿、化痰、止腐、解毒、去翳、收缩子宫等功效。外用药马应龙八宝眼膏、中成药利尿八味散中就含有此药。其主要成分为 NH_4Cl，此外，也含有 Fe^{3+}、Ca^{2+}、Mg^{2+} 等。白硇砂中 NH_4^+ 的含量是用来衡量矿物药质量优劣的重要指标，准确测定其含量十分重要。

目前，测定 NH_4^+ 的方法主要有离子色谱法、比色法和电化学分析法、酸碱滴定法等。其中，酸碱滴定分析法对设备要求低，是实验室中常用的分析手段。白硇砂中 NH_4^+ 的酸性较弱（$K_a = 5.6 \times 10^{-10}$），无法用标准碱直接滴定，但一般可采用两种间接手段进行酸碱滴定。第一种是蒸馏法，在试样中加入过量的碱，加热蒸馏出来的氨吸收于已知过量的酸标准溶液中，然后用碱标准溶液返滴定过量的酸，以求出试样中 NH_4^+ 含量，该方法较准确，但操作比较麻烦。本实验采用的是第二种甲醛法，它是基于甲醛与一定量的铵盐作用，生成相当量的强酸（H^+）和质子化的六亚甲基四胺 $(CH_2)_6N_4H^+$（$K_a = 7.1 \times 10^{-6}$），反应方程式如下：

$$4NH_4^+ + 6HCHO \Longrightarrow (CH_2)_6N_4H^+ + 6H_2O + 3H^+ \quad (4.38)$$

由上式可知，$4mol\ NH_4^+$ 与甲醛作用生成了 $3mol\ H^+$ 和 $1mol$ 质子化的六亚甲基四胺，二者均可用 $NaOH$ 标准溶液准确滴定，相当于 $1mol\ NH_4^+$ 转为化 $1mol$ 较强的酸，并与 $1mol\ NaOH$ 完全反应。

因此，选用 $NaOH$ 标准溶液滴定反应中生成的酸，化学计量点时，溶液中存在的六亚甲基四胺 $(CH_2)_6N_4$ 是一种很弱的碱（$K_b = 1.49 \times 10^{-9}$），溶液 pH 约为 8.7，故选用酚酞为指示剂。

铵盐与甲醛的反应在室温下进行较慢，加入甲醛后，需放置几分钟，使反应完全。甲醛中常含有少量甲酸，使用前需先以酚酞为指示剂，用 $NaOH$ 溶液中和，否则会使测定结果偏高。若铵盐中含有游离酸，也需中和除去，即以甲基红为指示剂，用 $NaOH$ 溶液滴定至橙黄色，加以扣除。

实验用品

仪器：分析天平、台秤、称量瓶、烧杯、容量瓶（250mL）、移液管（25.00mL）、锥形瓶、滴定管。

试剂：白硇砂试样、氢氧化钠（A.R.）、40%甲醛溶液（A.R.）、0.2%酚酞乙醇溶液、0.1%甲基红乙醇溶液、邻苯二甲酸氢钾（$KHC_8H_4O_4$，KHP，$M = 204.23g/mol$）。

实验步骤

（1）**0.1mol/L NaOH 溶液的配制与标定** 用台秤称取约 1g NaOH 固体于 100mL 小烧杯中，加无 CO_2 的去离子水溶解，冷却后转移至 250mL 容量瓶中，并用去离子水定容，摇匀后用橡胶塞塞紧，备用。

以 KHP 为基准物质标定 0.1mol/L NaOH 溶液的浓度。准确称取 0.4~0.6g KHP 三份，分别置于 250mL 锥形瓶中，加入 20~30mL 去离子水溶解，加入 2~3 滴酚酞指示剂，用待标定的 NaOH 溶液滴定至溶液由无色变为浅粉色，且在 30s 内不褪色即为终点。根据称取的基准物质 KHP 的质量和消耗的 NaOH 体积，计算 NaOH 溶液的准确浓度 c_{NaOH}（表 4.26）。

（2）**甲醛溶液的处理** 甲醛常以白色聚合物状态（多聚甲醛）存在，取原瓶装甲醛（40%）上清液于烧杯中，用水稀释一倍得 20% 甲醛溶液。受空气氧化所致，甲醛中常含有微量甲酸，应除去，否则产生正误差，方法如下：在 20% 甲醛溶液中加入 1~2 滴 0.2% 酚酞指示剂，用 0.1mol/L NaOH 溶液中和至甲醛溶液呈淡红色（对酚酞显中性）。

（3）**白硇砂试样中 NH_4^+ 的测定** 由于中药白硇砂中含有不溶性杂质，因此，需充分溶解后取上清液进行测定。准确称取白硇砂试样 1.5~2.0g 于 100mL 烧杯中，加入适量去离子水溶解后定容于 250mL 容量瓶中，塞紧摇匀，静置 5min 后，准确移取上清液 25.00mL 三份分别于 250mL 锥形瓶中，加入 10mL 已中和的 20% 甲醛溶液，再加入 2 滴酚酞指示剂充分摇匀，静置 1min 后，用 0.1mol/L NaOH 标准溶液滴定至浅粉色即为终点（表 4.27）。

数据记录与处理

根据 NaOH 标准溶液所消耗的体积，按下式计算白硇砂的 NH_4Cl 含量：

$$w = \frac{\bar{c}_{NaOH} V_{NaOH} M_{NH_4Cl}}{m_s \times \frac{25}{250} \times 1000} \times 100\% \tag{4.39}$$

式中 M_{NH_4Cl}——氯化铵的摩尔质量，53.49g/mol；

m_s——白硇砂试样的质量，g。

表 4.26 NaOH 溶液浓度的标定

实验序号	1	2	3
m_{KHP}(初始)/g			
m_{KHP}(终点)/g			
m_{KHP}/g			
V_{NaOH}(初始)/mL			
V_{NaOH}(终点)/mL			
V_{NaOH}(消耗)/mL			
c_{NaOH}/(mol/L)			
$\bar{c}_{NaOH}$/(mol/L)			
$|d_i|$			
$\bar{d}_r$/%			

表 4.27　白硇砂试样中 NH_4^+ 的测定

实验序号	1	2	3
m_s/g			
V_{NaOH}(初始)/mL			
V_{NaOH}(终点)/mL			
V_{NaOH}(消耗)/mL			
w			
$\bar{w}$			
$\lvert d_i \rvert$			
$\bar{d}_r$/%			

思考题

(1) 本法测定 NH_4^+ 时为什么不能用 NaOH 标准溶液直接滴定？

(2) 本法测定 NH_4^+ 加入甲醛的作用是什么？

(3) 加入的 20% 甲醛溶液为什么预先要以酚酞为指示剂用 NaOH 标准溶液进行中和？

拓展阅读

硇砂药材有紫硇砂和白硇砂两种。紫硇砂又名碱硇砂、红硇砂、藏脑、脑砂，为块状结晶体，多数呈立方形，大小不等，有棱角或凹凸不平，有明显不规则小孔，表面暗紫色或紫红色，稍有光泽，附有少量黄白色硫磺粉末，臭气浓，味咸，可溶于水；白硇砂又名淡硇砂，为白色结晶体，呈粒状，不规则状或粉末状，有光泽，易溶于水，臭微，味咸而苦。

实验 4.22　高锰酸钾法测定补钙剂中钙含量

实验目的

(1) 掌握 $KMnO_4$ 法测定钙含量的原理及方法。

(2) 掌握沉淀分离的基本操作技术。

实验原理

钙是人体所必需的营养元素之一，也是人体中最丰富的矿物质，同时它也是人体最容易缺乏的元素之一。目前，我国居民摄入钙量严重不足，尤其是儿童和老年人缺钙比例很高。为了补充钙缺失，补钙类保健食品及补钙制品在国内外发展很快。因此，钙是保健食品、钙剂制品及乳品中常规营养分析必须检测的质量指标，而准确确定钙制品中钙的含量，为衡量钙制品质量提供了主要依据。目前，测定钙的方法主要有高锰酸钾法、分光光度法、EDTA 络合滴定法、火焰原子吸收法、离子选择电极法、电感耦合等离子质谱法等测定方法。其中，高锰酸钾法具有简单、准确、快捷、误差小等优点，是测定补钙剂中钙含量普遍采用的方法。

Ca^{2+} 在一定条件下与 $C_2O_4^{2-}$ 能形成难溶的白色 CaC_2O_4 沉淀，经过滤、洗涤后用热的稀 H_2SO_4 溶解，生成的 $H_2C_2O_4$ 用 $KMnO_4$ 标准溶液滴定，从而间接测定 Ca^{2+} 的含量，反应如下：

$$Ca^{2+} + C_2O_4^{2-} \rightleftharpoons CaC_2O_4 \downarrow \tag{4.40}$$

$$CaC_2O_4 + H_2SO_4 \rightleftharpoons CaSO_4 + H_2C_2O_4 \tag{4.41}$$

$$5H_2C_2O_4 + 2MnO_4^- + 6H^+ \rightleftharpoons 2Mn^{2+} + 10CO_2 \uparrow + 8H_2O \tag{4.42}$$

沉淀 Ca^{2+} 时，为了得到易于过滤和洗涤的粗晶形沉淀，通常在含 Ca^{2+} 的酸性溶液中加入足够量的 $(NH_4)_2C_2O_4$ 沉淀剂（$C_2O_4^{2-}$ 主要是以 $HC_2O_4^-$ 形式存在），再慢慢滴加氨水，使溶液中的 H^+ 逐渐被中和，$C_2O_4^{2-}$ 浓度缓慢增加，以得到 CaC_2O_4 粗晶形沉淀。但是，要控制沉淀完毕时，溶液 pH 值仍在 3.5~4.5 之间（防止 CaC_2O_4 溶解度太大和其他难溶性钙盐的生成），之后加热陈化 30min（陈化过程中，小颗粒晶体溶解，大颗粒晶体长大）。过滤后，沉淀表面吸附的 $C_2O_4^{2-}$ 须洗净，但为了减少 CaC_2O_4 的洗涤损失，先用稀的 $(NH_4)_2C_2O_4$ 溶液洗涤，再用微热的蒸馏水洗到不含 $C_2O_4^{2-}$ 时为止。最后，将洗净的 CaC_2O_4 沉淀溶于稀 H_2SO_4 中，加热至 70~80℃，用 $KMnO_4$ 标准溶液滴定。

市售的 $KMnO_4$ 常含有少量的杂质，如硫酸盐、卤化物、硝酸盐以及 MnO_2 等，因此不能直接配制准确浓度的标准溶液，需要进行标定。$KMnO_4$ 标准溶液常用还原剂 $Na_2C_2O_4$ 作基准物来标定（$Na_2C_2O_4$ 不含结晶水，容易提纯，性质稳定）。在 H_2SO_4 介质中，MnO_4^- 与 $C_2O_4^{2-}$ 按照反应式(4.43)进行，滴定时利用 MnO_4^- 本身的紫红色指示终点。为了使该反应能够定量且较快地进行，应严格控制反应条件：反应中溶液温度控制在 70~80℃，并趁热滴定；溶液酸度控制在 0.5~1mol/L H_2SO_4 介质中；滴定速度按照慢-快-慢的方式进行；滴定终点时溶液保持淡粉色 0.5~1min 内不褪色即为终点。

$$5C_2O_4^{2-} + 2MnO_4^- + 16H^+ \rightleftharpoons 2Mn^{2+} + 10CO_2 \uparrow + 8H_2O \tag{4.43}$$

实验用品

仪器：滴定管（50.00mL）、烧杯、漏斗、量筒（10mL、50mL）、分析天平、台秤、表面皿、电热板、水浴锅、锥形瓶、定性滤纸等。

试剂：$KMnO_4$（固体）、$Na_2C_2O_4$（基准物质）、3mol/L H_2SO_4、1mol/L H_2SO_4、6mol/L HCl、10% $NH_3 \cdot H_2O$、0.25mol/L $(NH_4)_2C_2O_4$、10%柠檬酸铵、0.1%甲基橙、0.1% $(NH_4)_2C_2O_4$、0.5mol/L $CaCl_2$。

实验步骤

(1) 0.02mol/L $KMnO_4$ 标准溶液的配制及标定　称取约 1.7g $KMnO_4$ 溶于 500mL 水中，盖上表面皿，加热至微沸，保持微沸状态 1h，并随时补充蒸发掉的水分，冷却后静置 2~3 天，用微孔玻璃漏斗过滤，滤液储于棕色瓶中置于暗处保存备用。

准确称取约 0.2g 基准物质 $Na_2C_2O_4$ 三份，置于 250mL 锥形瓶中，加入 50mL 水使之溶解，再加入 15mL 3mol/L H_2SO_4，水浴加热至 70~80℃（溶液开始冒蒸汽），趁热用待标定的 $KMnO_4$ 溶液滴定至溶液呈淡粉色，且 30s 不褪色，即为终点，记录滴定所消耗的 $KMnO_4$ 溶液体积，平行测定 3 次（表 4.28）。

(2) 补钙制剂中钙含量的测定　准确称取补钙制剂 3 份（每份含钙约 0.05g），分别置于 250mL 烧杯中，加入适量蒸馏水盖上表面皿，缓慢滴加 10mL 6mol/L HCl 溶液，加热使其溶解。冷却后，在试液中加入 5mL 10%柠檬酸铵和 50mL 蒸馏水，加入 2~3 滴甲基橙，此时溶液呈红色。溶液中再加入约 20mL 0.25mol/L $(NH_4)_2C_2O_4$ 溶液，水浴加热至 70~80℃，在不断搅拌下逐滴加入 10% $NH_3 \cdot H_2O$ 溶液，使溶液由红色转变为黄色。热水浴陈化 30min，使之形成 CaC_2O_4 粗晶形沉淀。

自然冷却后，过滤沉淀。用冷的 0.1% $(NH_4)_2C_2O_4$ 溶液洗涤沉淀几次，再用水洗涤

至滤液中不含 $C_2O_4^{2-}$ 为止（取约 1mL 滤液，加入数滴 0.5mol/L 的 $CaCl_2$ 溶液，如溶液无浑浊现象，证明已洗涤干净）。

将带有沉淀的滤纸展开贴在原烧杯内壁上，用 50mL 1mol/L H_2SO_4 溶液多次冲洗滤纸上的沉淀至烧杯中，加水至约 100mL，水浴加热至 70～80℃，用 $KMnO_4$ 标准溶液滴定至溶液呈淡粉色，再将烧杯壁上的滤纸浸入溶液中，轻轻搅动，若溶液褪色，则继续滴加 $KMnO_4$ 直至出现淡粉色，30s 内不褪色即为终点，记录消耗 $KMnO_4$ 的体积 V_1，平行测定 3 次（表 4.29）。

✳ 数据记录与处理

根据 $KMnO_4$ 标准溶液消耗的体积，计算某些补钙剂中的钙含量：

$$w = \frac{\frac{5}{2} \times \bar{c}_{KMnO_4} V_{KMnO_4} M_{Ca} \times 10^{-2}}{m_s} \times 100\% \tag{4.44}$$

式中　m_s——所测补钙剂的质量，g；
　　　M_{Ca}——钙的摩尔质量，40g/mol。

表 4.28　$KMnO_4$ 溶液浓度的标定

实验序号	1	2	3
$m_{Na_2C_2O_4}$（初始）/g			
$m_{Na_2C_2O_4}$（终点）/g			
$m_{Na_2C_2O_4}$/g			
V_{KMnO_4}（初始）/mL			
V_{KMnO_4}（终点）/mL			
V_{KMnO_4}（消耗）/mL			
c_{KMnO_4}/(mol/L)			
$\bar{c}_{KMnO_4}$/(mol/L)			
$|d_i|$			
$\bar{d}_r$/%			

表 4.29　补钙剂中钙含量测试

实验序号	1	2	3
m_s/g			
V/mL			
w/%			
$\bar{w}$/%			
$|d_i|$			
$\bar{d}_r$/%			

思考题

（1）$KMnO_4$ 法测钙含量与配位滴定法测钙含量各有何优缺点？

（2）为什么滴定至接近终点时，才将滤纸从烧杯壁上取下放入烧杯中进行滴定？

📝 拓展阅读

钙的作用是巨大的，它不仅是人体骨骼及牙齿的主要组成成分，而且在肌肉收缩、神经传导、大脑的思维活动、血液凝固、软组织的弹性和韧性、维持细胞与毛细血管的通透性、维持酸碱平衡等一系列生理反应过程当中发挥着巨大的作用。

实验 4.23　复方氢氧化铝片中 Al^{3+} 和 Mg^{2+} 含量的测定

实验目的

（1）了解成品药剂中组分含量测定的预处理方法。
（2）掌握配位滴定中返滴定法的基本原理。

实验原理

复方氢氧化铝片是一种胃药，用于缓解胃酸过多引起的胃痛、胃灼热感、反酸，也可用于治疗慢性胃炎，它是由能中和胃酸的氢氧化铝［$Al(OH)_3$］和三硅酸镁（$2MgO·3SiO_2·H_2O$），并添加解痉止痛药颠茄浸膏而成。其中，氢氧化铝中和胃酸的产物氯化铝能引起便秘，而三硅酸镁中和胃酸后的产物镁离子具有轻泻作用，二者配合使用相得益彰。然而，铝同时也是一种慢性神经性毒性物质，过多摄入会沉积在神经原纤维，使神经系统发生慢性改变，诱发老年性痴呆等疾病。因此，复方氢氧化铝片剂中铝、镁含量的测定具有重要的现实意义。

目前，测定铝、镁的方法根据含量高低主要有分光光度法、EDTA 络合滴定法、电感耦合-等离子体发射光谱法等。其中，络合滴定法是一种简便、快速且应用广泛的定量分析方法。

先将药片用酸溶解，分离除去不溶物，然后取一份试液，调节溶液 pH 3~4，定量加入过量 EDTA 溶液，煮沸数分钟使 Al^{3+} 与 EDTA 反应完全。

$$Al^{3+} + H_2Y^{2-} \Longrightarrow AlY^- + 2H^+ \tag{4.45}$$

冷却后再调节 pH 5~6，以二甲基酚橙为指示剂，用 Zn^{2+} 标准溶液返滴定过量的 EDTA，计算 Al^{3+} 的含量。

$$Zn^{2+} + H_2Y^{2-} \Longrightarrow ZnY^{2-} + 2H^+ \tag{4.46}$$

另取一份试液，调节 pH 为 8~9，将 Al^{3+} 以 $Al(OH)_3$ 沉淀的形式分离，再调节 pH=10 左右，以铬黑 T 为指示剂，用 EDTA 标准溶液滴定滤液中的 Mg^{2+}，计算 Mg^{2+} 的含量。

$$Mg^{2+} + H_2Y^{2-} \Longrightarrow MgY^{2-} + 2H^+ \tag{4.47}$$

实验用品

仪器：分析天平、电磁炉、滴定管、移液管、容量瓶、锥形瓶、量筒、烧杯、表面皿、称量瓶、研钵。

试剂：复方氢氧化铝片、0.02mol/L EDTA 标准溶液、0.02mol/L Zn^{2+} 标准溶液、0.2％二甲酚橙指示剂、20％六亚甲基四胺溶液、1∶1 HCl 溶液、1∶1 氨水、1∶2 三乙醇胺溶液、pH=10 的 $NH_3·H_2O$-NH_4Cl 缓冲溶液、0.2％甲基红乙醇溶液、铬黑 T 指示剂、NH_4Cl 固体。

实验步骤

(1) 0.02mol/L Zn^{2+} 标准溶液的配制 准确称取氧化锌约 0.4g 于 250mL 烧杯中，盖上表面皿，从烧杯嘴加入 5～10mL 1∶1 HCl，待 ZnO 全部溶解后，用去离子水定容于 250mL 容量瓶中，摇匀，计算其准确浓度 $c_{Zn^{2+}}$ （mol/L）。

(2) 0.02mol/L EDTA 标准溶液的配制与标定 称取 2g 乙二胺四乙酸二钠于 250mL 烧杯中，加入 50mL 水，加热溶解后，稀释至 250mL，存于试剂瓶中或聚乙烯塑料瓶中。

以二甲酚橙为指示剂，准确移取 Zn 标准溶液 25.00mL 于 250mL 锥形瓶中，加二甲酚橙指示剂 2～3 滴，然后滴加 20% 六亚甲基四胺至溶液呈稳定的紫红色再多加 5mL。用待标定的 EDTA 标准溶液滴定至溶液由紫红色变为亮黄色为终点，平行测定 3 次，根据 EDTA 溶液消耗的体积，计算其浓度 c_{EDTA} （mol/L）（表 4.30）。

(3) 试样的处理 取复方氢氧化铝片 10 片，研细，准确称取约 2g 于 250mL 烧杯中，加 20mL 1∶1 HCl 溶液，加去离子水至 100mL，煮沸。冷却后过滤，用去离子水洗涤沉淀，收集滤液及洗涤液于 250mL 容量瓶中，去离子水定容，摇匀，作为试样溶液。

(4) 铝的测定 准确吸取上述试样溶液 10.00mL 于 250mL 锥形瓶中，加去离子水稀释至 25mL 左右，滴加 1∶1 $NH_3 \cdot H_2O$ 至溶液刚好出现浑浊，再滴加 1∶1 HCl 溶液至沉淀恰好溶解。准确加入 EDTA 标准溶液 25.00mL，在电磁炉上加热煮沸 3min。冷却至室温后，再加入 10mL 20% 六亚甲基四胺溶液，加入 2～3 滴二甲酚橙指示剂，用锌标准溶液滴定至溶液由黄色变为紫红色为终点，记录消耗 Zn 标准溶液的体积 V_1，平行滴定 3 次（表 4.31）。

(5) 镁的测定 准确移取试样溶液 25.00mL 于小烧杯中，滴加 1∶1 $NH_3 \cdot H_2O$ 至溶液刚出现沉淀，再滴加 1∶1 HCl 溶液至沉淀恰好溶解。加入 2g 固体 NH_4Cl 后，滴加 20% 六亚甲基四胺溶液至沉淀出现，并过量 15mL。加热至 80℃并保持 10min，冷却后过滤除去 $Al(OH)_3$。用少量水洗涤沉淀数次。收集滤液和洗涤液于 250mL 锥形瓶中，加入 4mL 1∶2 三乙醇胺、10mL pH 为 10 的 $NH_3\text{-}NH_4Cl$ 缓冲溶液、1 滴甲基红指示剂和 3～5 滴铬黑 T 指示剂，用 EDTA 标准溶液滴定至溶液由暗红色变为蓝绿色，即为终点，记录消耗 EDTA 标准溶液的体积 V_2，平行滴定 3 次。

数据记录与处理

根据定量加入 EDTA 的量和消耗的 Zn 标准溶液的体积 V_1，按照式(4.48) 计算复方氢氧化铝片中 Al 的含量，以 Al_2O_3 表示。

$$w_{Al_2O_3} = \frac{(\bar{c}_{EDTA}V_{EDTA} - c_{Zn^{2+}}V_1) \times M_{Al_2O_3}}{m_s \times \dfrac{10}{250} \times 1000 \times 2} \tag{4.48}$$

式中，$w_{Al_2O_3}$——Al_2O_3 的质量分率；

$\bar{c}_{EDTA}$——EDTA 标准溶液的平均浓度，mol/L；

V_{EDTA}——加入的过量 EDTA 的体积，mL；

$c_{Zn^{2+}}$——Zn^{2+} 标准溶液的浓度，mol/L；

V_1——返滴定过量 EDTA 所消耗的 Zn^{2+} 标准溶液的体积，mL；

$M_{Al_2O_3}$——Al_2O_3 的摩尔质量，g/mol；

m_s——称取的复方氢氧化铝片试样的质量，g。

根据 EDTA 标准溶液的消耗量 V_2，按下式计算复方氢氧化铝片中 Mg 的含量，以 MgO 表示。

$$w_{MgO}=\frac{\bar{c}_{EDTA}V_2 M_{MgO}}{m_s \times \frac{25}{250} \times 1000} \times 100\% \qquad (4.49)$$

式中，w_{MgO}——MgO 的质量分率；

$\bar{c}_{EDTA}$——EDTA 标准溶液的平均浓度，mol/L；

V_2——滴定 Mg^{2+} 时消耗的 EDTA 标准溶液的体积，mL；

M_{MgO}——MgO 的摩尔质量，g/mol；

m_s——称取的复方氢氧化铝片试样的质量，g。

表 4.30　EDTA 溶液浓度的标定

实验序号	1	2	3
m_{ZnO}(初始)/g			
m_{ZnO}(终点)/g			
m_{ZnO}/g			
$c_{Zn^{2+}}$/(mol/L)			
V_{EDTA}(初始)/mL			
V_{EDTA}(终点)/mL			
V_{EDTA}(消耗)/mL			
c_{EDTA}/(mol/L)			
$\bar{c}_{EDTA}$/(mol/L)			
$|d_i|$			
$\bar{d}_r$/%			

表 4.31　药片中 Al 和 Mg 含量测定

	实验序号	1	2	3
铝的测定	m_s/g			
	V(试液)/mL			
	$\bar{c}_{EDTA}$/(mol/L)			
	V_{EDTA}/mL			
	$c_{Zn^{2+}}$/(mol/L)			
	V_1/mL			
	$w_{Al_2O_3}$/%			
	$\bar{w}_{Al_2O_3}$/%			
	$|d_i|$			
	$\bar{d}_r$/%			
	实验序号	1	2	3
镁的测定	V(试液)/mL			
	$\bar{c}_{EDTA}$/(mol/L)			
	V_2/mL			
	w_{MgO}/%			
	$\bar{w}_{MgO}$/%			
	$|d_i|$			
	$\bar{d}_r$/%			

思考题

（1）为什么不采用直接滴定法测定 Al^{3+}？

（2）测定 Al^{3+} 含量时，为什么要将试液与 EDTA 标准溶液混合后加热煮沸？

（3）测定 Mg^{2+} 时，为什么沉淀 Al^{3+} 采用 20% 六亚甲基四胺溶液调节 pH 而不是直接用氨水？

（4）测定 Mg^{2+} 时，加入三乙醇胺的作用是什么？

拓展阅读

络合滴定法是以络合反应为基础的滴定分析方法总称，自 1945 年施瓦岑巴赫提出氨羧络合剂后，络合滴定日渐发展。氨羧络合剂是一类含有氨基二乙酸—$N(CH_2COOH)_2$ 基团的有机化合物，其分子中含有氨氮和羧氧两种配位能力很强的配位原子，对许多金属有很强的络合能力。乙二胺四乙酸，简称 EDTA，常用 H_4Y 表示，是一种含有羧基和氨基的螯合剂，但其在水中的溶解度小，通常将其制成二钠盐，一般也称 EDTA 二钠盐。

实验 4.24 盐酸黄连素含量的测定

实验目的

（1）了解重量法测定盐酸黄连素的原理和方法。

（2）了解晶形沉淀的沉淀条件、原理和沉淀方法。

（3）掌握晶形沉淀的过滤、洗涤及恒重的基本操作技术。

实验原理

盐酸黄连素是季胺型檗碱的盐酸盐 $[M(C_{20}H_{18}O_4N \cdot Cl \cdot 2H_2O) = 407.85 \text{g/mol}]$，又称盐酸小檗碱，在冷水中微溶，在热水中易溶。它是一种存在于黄连、黄柏等植物中的异喹啉[类]生物碱，是一种广谱抗菌的药物，对痢疾杆菌、肺炎双球菌、金葡菌、链球菌、伤寒杆菌及阿米巴原虫有抑制作用。临床上主要用于敏感病原菌所致的胃肠炎、细菌性痢疾等肠道感染。苋莱黄连素胶囊、复方黄连素片、盐酸小檗碱片等药物的主要成分均为盐酸黄连素。药品中有效成分盐酸黄连素含量的准确测定，可以有效控制该药剂的质量。

目前，测定药品中盐酸黄连素的方法主要有紫外-可见分光光度法、高效液相色谱法、原子吸收光谱法、重量法、毛细管电泳法，等等。其中，重量分析法的准确度较高，在分析工作中也常用重量法的测定结果作为标准，校对其他分析方法的准确度。重量分析法是通过称量物质的质量来确定被测物质组分的含量。分析时，一般是先采用适当方法将被测组分从试样中分离出来并称重，由所称得的质量计算被测组分的含量。在酸性条件下，以三硝基苯酚作为沉淀剂，与盐酸黄连素形成苦味酸小檗碱沉淀 $[M(C_{20}H_{17}O_4N \cdot C_6H_3O_7N_3) = 564.56 \text{g/mol}]$ [如式(4.50)]。

$$C_{20}H_{18}O_4N \cdot Cl + C_6H_3O_7N_3 \rightleftharpoons C_{20}H_{17}O_4N \cdot C_6H_3O_7N_3 \downarrow + HCl \quad (4.50)$$

沉淀经过滤、洗涤、干燥后测定其质量，即可计算 $C_{20}H_{18}O_4N \cdot Cl$ 的含量。

实验用品

仪器：分析天平、恒温干燥箱、干燥器、称量瓶、烧杯、循环水真空泵（配抽滤瓶）、G4 砂芯坩埚。

试剂：浓盐酸（A.R.）、三硝基苯酚（A.R.）、盐酸黄连素药片、0.1mol/L HCl 溶液、三硝基苯酚饱和水溶液、三硝基苯酚黄连素饱和溶液（制备纯净的三硝基苯酚黄连素沉淀，用蒸馏水制成饱和溶液）。

实验步骤

（1）沉淀的制备 取盐酸黄连素药片 5 片，研细，精确称量约 0.2g，置于 250mL 烧杯中，加热的蒸馏水 100mL 使之溶解，加入 10mL 0.1mol/L 盐酸溶液，立即缓慢加入 30mL 三硝基苯酚饱和溶液。待沉淀下沉后，在上层清液中加入少量三硝基苯酚饱和溶液，确定是否沉淀完全。沉淀作用完全后，再水浴加热 15min，静置 2h 以上。

（2）坩埚恒重 用去离子水洗净坩埚，抽滤至水雾消失，然后将坩埚置于恒温干燥箱中加热 140℃（第一次干燥 60min，第二次干燥 30min）。之后将坩埚转移至干燥器中冷却 10~15min，称重，两次称得质量之差若不超过 2mg 表明已恒重（否则继续加热、冷却、称重，直至恒重），坩埚质量记为 m_1。

（3）沉淀的处理 将陈化好的沉淀全部转移至已经干燥恒重的坩埚（m_1）中减压过滤，并用三硝基苯酚黄连素饱和水溶液洗涤多次，再用水洗涤 3 次，每次 15mL。然后，将盛沉淀的坩埚置于恒温干燥箱中干燥、冷却、称重，直至恒重[同实验步骤(2)的操作]，此时的坩埚质量记为 m_2。

数据记录与处理

根据沉淀质量，按下式计算盐酸黄连素的含量，并填写表 4.32：

$$w_{C_{20}H_{18}O_4N\cdot Cl} = \frac{(m_2-m_1)\times \dfrac{M_{C_{20}H_{18}O_4N\cdot Cl}}{M_{C_{20}H_{17}O_4N\cdot C_6H_3O_7N_3}}}{m_s}\times 100\% \quad (4.51)$$

式中 $M_{C_{20}H_{18}O_4N\cdot Cl}$——盐酸黄连素的摩尔质量，371.85g/mol；

$M_{C_{20}H_{17}O_4N\cdot C_6H_3O_7N_3}$——沉淀物质的摩尔质量，564.56g/mol；

m_s——试样的质量，g。

表 4.32 盐酸黄连素含量的测定

m_s/g	m_1/g	m_2/g	(m_2-m_1)/g	$w_{C_{20}H_{18}O_4N\cdot Cl}$/%

思考题

（1）试样的称量量应该如何确定？是否应正好为 0.2g？

（2）加入沉淀剂前，加入 HCl 的目的是什么？

（3）为什么要在热的溶液中缓慢加入三硝基苯酚？

（4）沉淀干燥后为什么要在干燥器中冷却至室温？

（5）沉淀为什么要在水浴上加热，并静置 2h 以上？

实验 4.25 氨咖黄敏胶囊中对乙酰氨基酚含量的测定

实验目的

（1）掌握对乙酰氨基酚含量的测定原理。

（2）掌握紫外可见分光光度法测定药物含量的方法及计算。

实验原理

对乙酰氨基酚（$C_8H_9NO_2$，$M=151.16g/mol$，CAS：103-90-2）又名醋氨酚、对羟基乙酰苯胺等（结构见图 4.9），通常为白色结晶性粉末，无臭，味微苦，在热水或乙醇中易溶，在丙酮中溶解，在水中略溶。对乙酰氨基酚是最常用的解热镇痛药物，其解热作用缓慢而持久，安全有效，在临床上应用较为广泛，常用于感冒发热、关节痛、神经痛等的治疗，是扑热息痛、速效伤风胶囊、复方氨酚烷胺片、氨咖黄敏胶囊等药物的主要成分。但是，对乙酰氨基酚对人体有一定的毒副作用，过量可致肝坏死，含量控制要求严格。因此，确定药品中对乙酰氨基酚的含量，能更好地控制对乙酰氨基酚的摄入量。

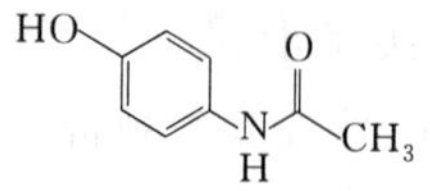

图 4.9　对乙酰氨基酚结构

目前，测定对乙酰氨基酚含量的方法有分光光度法、亚硝酸钠法、电化学法、高效液相色谱法，等等。其中，分光光度法利用对乙酰氨基酚结构中的共轭结构在紫外区的吸收，其最大吸收波长为257nm，通过测量其最大吸收波长处的吸光度即可计算其含量。

实验用品

仪器：分析天平、紫外-可见分光光度计、分刻度移液管（5.00mL、10.00mL）、烧杯、定量滤纸（直径10cm）、容量瓶（100mL、250mL）。

试剂：市售氨咖黄敏胶囊（规格：每个胶囊含对乙酰氨基酚250mg、咖啡因15mg、马来酸氯苯那敏1mg、人工牛黄10mg）、对乙酰氨基酚标准品、氢氧化钠（A.R.）、蒸馏水。

实验步骤

（1）对乙酰氨基酚标准溶液的制备　准确称取对乙酰氨基酚标准品 0.0400g 于烧杯中，分别加入 0.4% 氢氧化钠溶液 50mL、蒸馏水 50mL，振摇 15min，使对乙酰氨基酚充分溶解后，定量转移至 250mL 容量瓶中，加水定容，摇匀得标准储备液（160μg/mL）。

（2）标准曲线的绘制　取标准稀释液在 200~400nm 波长范围内测定对乙酰氨基酚紫外吸收光谱，确定其最大吸收波长（$\lambda_{max}=257nm$）。

分别准确移取一定量对乙酰氨基酚标准储备液于 100mL 容量瓶中，加 0.4% 氢氧化钠溶液 10mL，加水定容后摇匀，配制成每 1mL 含对乙酰氨基酚 2μg、4μg、6μg、8μg、10μg 的标准溶液，以水溶液做参比，在 λ_{max} 处测定其吸收值 A，绘制标准曲线，确定线性范围和标准曲线方程（表 4.33）。

（3）样品溶液的制备及测定　取市售氨咖黄敏胶囊数粒，内容物混匀。精确称取适量内容物（m_s，相当于对乙酰氨基酚约 40mg），分别加入 0.4% 氢氧化钠溶液 50mL、蒸馏水 50mL，振摇 15min，使对乙酰氨基酚充分溶解，定量转移至 250mL 容量瓶中，加水定容后摇匀，再过滤，并弃去初滤液。精确移取滤液 5.00mL 于 100mL 容量瓶中，加 0.4% 氢氧化钠溶液 10mL，加水定容后摇匀，以水溶液做参比，在 λ_{max} 处测定其吸收值 A，平行测定 3 次，根据标准曲线确定对乙酰氨基酚的浓度，进而计算氨咖黄敏胶囊中的对乙酰氨基酚的含量（表 4.34）。

数据记录与处理

根据标准曲线获得样品溶液中对乙酰氨基酚的浓度 c（μg/mL），进而计算氨咖黄敏胶囊中对乙酰氨基酚成分的含量：

$$\text{药品中对乙酰氨基酚标示量} X = \frac{c \times 20 \times 250 \times 10^{-3} \times \overline{w}}{m_s} \tag{4.52}$$

式中，$\bar{w}$——每粒氨咖黄敏胶囊的平均药重，mg；

m_s——称取氨咖黄敏胶囊内容物的量，mg。

表 4.33 标准曲线的绘制

$c/(\mu g/mL)$	2	4	6	8	10
A					
标准曲线					

表 4.34 对乙酰氨基酚含量的测定

实验序号	1	2	3
$\bar{w}$			
m_s/mg			
A			
$c/(\mu g/mL)$			
$\bar{c}/(\mu g/mL)$			
$\|d_i\|$			
$\bar{d}_r/\%$			
X/mg			

思考题

氨咖黄敏胶囊中除了主要成分对乙酰氨基酚之外，还有少量其他成分，如咖啡因、马来酸氯苯那敏、人工牛黄等等，在该紫外光谱测定中是否存在干扰？

拓展阅读

对乙酰氨基酚为苯胺类解热镇痛药，其解热、镇痛作用强度与阿司匹林类似，但抗炎作用极弱，对凝血机制无影响。布洛芬属于芳基丙酸类解热镇痛药，有明显的抗炎、解热、镇痛作用，强度也与阿司匹林相当，对血小板功能有一定的抑制作用，可延长出血时间。二者都属于非甾体类解热镇痛药，通过抑制前列腺素的合成从而发挥解热、镇痛的作用，达到使发热者退热、缓解疼痛带来的不适感等作用，但对体温正常者是无影响的，不会使其体温降低。

4.5 食品类

实验 4.26 市售柑橘中柠檬酸含量的测定

实验目的

（1）掌握配制和标定 NaOH 标准溶液的方法。
（2）掌握柠檬酸含量测定的原理和方法。

实验原理

中国是柑橘的重要原产地之一，柑橘资源丰富，品种繁多，包括橙子、柚子、柠檬、金橘、蜜橘、广柑等，秋冬时节，柑橘占据了水果市场的半壁江山。柑橘果汁丰富，酸甜适度，富含大量糖分、有机酸、维生素、矿物质等营养物质。柑橘也是很好的中药材，具有顺气、止咳、健胃、化痰、消肿、止痛、疏肝理气等多种功效。果实中糖、酸含量和糖酸比或固酸比是决定柑橘品质的最重要的指标。

大多数有机酸是固体弱酸，如果有机酸能溶于水，且解离常数 $K_a \geqslant 10^{-7}$，可称取一定量的试样，溶于水后用 NaOH 标准溶液滴定。滴定突跃在弱碱性范围内常选用酚酞为指示剂，滴定至溶液由无色变为微红色即为终点。根据 NaOH 标准溶液的浓度 c 和滴定时所消耗的体积 V 及称取有机酸的质量，计算有机酸的含量。

有机酸试样通常有柠檬酸、草酸、酒石酸、乙酰水杨酸、苯甲酸等。滴定产物是强碱弱酸盐，滴定突跃在碱性范围内，可选用酚酞作为指示剂，用 NaOH 标准溶液滴定至溶液呈粉红色（30s 不褪色）为终点。

实验用品

仪器：滴定管、锥形瓶、容量瓶、移液管（25.00mL）、烧杯。

试剂：邻苯二甲酸氢钾（基准物质，100～125℃干燥 1h，然后放入干燥器内冷却后备用）、NaOH（分析纯）、柑橘、酚酞指示剂。

实验步骤

（1）0.10mol/L NaOH 溶液的标定　准确称取 0.4～0.6g 邻苯二甲酸氢钾三份，分别置于 250mL 锥形瓶中，加入 20～30mL 水溶解，加入 2～3 滴酚酞指示剂，用待标定的 NaOH 溶液滴定至溶液呈微红色，30s 内不褪色即为终点。记录所消耗 NaOH 溶液的体积，平行测定三次（表 4.35）。

（2）柠檬酸含量测定　将柑橘样品洗净擦干，分别取可食部分于组织捣碎机中匀浆，称取 25g 浆液用水稀释，定容至 250mL。摇匀，用 0.45μm 滤膜过滤。

用 25.00mL 移液管移取上述试液于 250mL 锥形瓶中，加入酚酞指示剂 1～2 滴，用 NaOH 标准溶液滴至溶液呈微红色，保持 30s 不褪色，即为终点。记下所消耗 NaOH 溶液体积，平行测定 3 次，计算柠檬酸质量分率（表 4.36）。

$$w_{H_nA} = \frac{\frac{1}{n}cVM_{H_nA}}{m \times 1000} \times 100\% \tag{4.53}$$

其中 $n=3$ 为柠檬酸 H^+ 个数，$M_{H_3A}=210.14$g/mol 为柠檬酸的摩尔质量，m 为样品质量。

数据记录与处理

表 4.35　0.1mol/L NaOH 溶液的标定

实验序号	1	2	3
$m_{基准物质}$/g			
V_{NaOH}（初始）/mL			
V_{NaOH}（终点）/mL			
V_{NaOH}（消耗）/mL			
c_{NaOH}/(mol/L)			
$\bar{c}_{NaOH}$/(mol/L)			
$|d_i|$			
$\bar{d}_r$/%			

表 4.36 柠檬酸含量的测定

实验序号	1	2	3
m_{H_nA}/g			
V_{NaOH}(初始)/mL			
V_{NaOH}(终点)/mL			
V_{NaOH}(消耗)/mL			
$w/(mol/L)$			
$\bar{w}/(mol/L)$			
$|d_i|$			
$\bar{d}_r/\%$			

思考题

草酸、酒石酸等多元酸能否用 NaOH 溶液滴定？

拓展阅读

柠檬酸为食用酸类，可增强体内正常代谢，适当的剂量对人体无害。在某些食品中加入柠檬酸后口感好，并可促进食欲，在中国允许果酱、饮料、罐头和糖果中使用柠檬酸。

基于柠檬酸对钙的代谢可产生的影响，经常食用罐头、饮料、果酱、酸味糖果的人们，特别是孩子，要注意补钙，多喝牛奶、鱼头鱼骨汤、吃些小虾皮等，以免导致血钙不足而影响健康，胃溃疡、胃酸过多、龋齿和糖尿病患者不宜经常食用柠檬酸。柠檬酸不能加在纯奶里，否则会引起纯奶凝固。乳制品行业常把柠檬酸配成 10% 左右的溶液加入低浓度的牛奶溶液中，加入时应快速地搅拌。

实验 4.27 钙片/奶粉/菠菜中钙含量的测定

实验目的

(1) 掌握标定 EDTA 方法。
(2) 掌握 EDTA 法测定 Ca^{2+} 含量的原理和方法。

实验原理

EDTA（$Na_2H_2Y \cdot 2H_2O$）标准溶液采用间接法配制，可先配制粗略浓度，再用 Zn、ZnO、$CaCO_3$、$MgSO_4 \cdot 7H_2O$ 等基准物质来标定。样品经处理后，可以用配位滴定法测定其中钙含量，通常以 EDTA 为滴定剂，在碱性条件下以钙指示剂指示终点。钙指示剂为紫黑色粉末，在 pH=12～14 时显蓝色，与 Ca^{2+} 形成红色配合物，用于钙镁混合物中钙含量的测定，终点变色比铬黑 T 敏锐。

$$Ca^{2+} + \underset{(蓝色)}{In} = \underset{(紫红色)}{CaIn} \tag{4.54}$$

$$\underset{(紫红色)}{CaIn} + Y = CaY + \underset{(蓝色)}{In} \tag{4.55}$$

实验用品

仪器：锥形瓶（250mL）、酸式滴定管（50mL）、量筒（10mL）、烧杯（250mL）、研

钵、容量瓶（250mL）、表面皿。

试剂：$Na_2H_2Y \cdot 2H_2O$、基准$CaCO_3$、20% NaOH溶液、6mol/L HCl、钙片、奶粉、菠菜、钙指示剂。

实验步骤

（1）以$CaCO_3$为基准物标定EDTA

① EDTA标准溶液配制（0.02mol/L）：称取2.0g $Na_2H_2Y \cdot 2H_2O$于250mL的烧杯中，加入50mL水，加热溶解后，稀释至250mL，储于试剂瓶或聚乙烯瓶中备用。

② $CaCO_3$标准溶液配制（0.02mol/L）：准确称取120℃干燥过的$CaCO_3$ 0.4~0.6g于小烧杯中，加几滴水润湿，盖上表面皿，从烧杯嘴滴加6mol/L HCl 5mL直至$CaCO_3$完全溶解，用水冲洗烧杯内壁，然后将溶液移至250mL容量瓶中，再加水至刻度，摇匀。

③ EDTA标准溶液的标定：用25.00mL移液管吸取$CaCO_3$标准溶液置于250mL锥形瓶中，加水至100mL，加5~6mL 20% NaOH溶液，加2~3滴钙指示剂，用待标定的EDTA溶液滴定至溶液由紫红色变为蓝色，即为滴定终点。记录所消耗EDTA溶液的体积，平行测定三次。数据记录在表4.37中。

（2）钙含量的测定

① 样品预处理：将钙片/奶粉放入研钵中研细，准确称取0.2~0.4g研细的样品，加入10mL蒸馏水，加热，滴加6mol/L HCl，边加热边搅拌，至溶解完全。转移到250mL容量瓶中，蒸馏水定容，摇匀。

取一定量的用蒸馏水洗过的新鲜菠菜切碎，放入蒸发皿中称重，放入烘箱，烘2h，再放入坩埚中，将其置于马弗炉中灰化10h（注意升温速度不要太快，最高温500℃）。冷却后，加入2~3滴浓硝酸，1.0mL 6mol/L HCl溶解，定量转移至250mL容量瓶中，用蒸馏水稀释至刻度。

② 钙含量的测定：准确移取上述溶液25.00mL于250mL锥形瓶中，加水至100mL，加入5~6mL 20% NaOH溶液，加2~3滴钙指示剂，用待标定的EDTA溶液滴定至溶液由紫红色变为蓝色，即为滴定终点。平行滴定三次，计算样品中钙含量。数据记录在表4.38中。

数据记录与处理

表4.37 0.1mol/L EDTA溶液标定

实验序号	1	2	3
m_{CaCO_3}/g			
V_{EDTA}(初始)/mL			
V_{EDTA}(终点)/mL			
V_{EDTA}(消耗)/mL			
$c_{EDTA}/(mol/L)$			
$\bar{c}_{EDTA}/(mol/L)$			
$|d_i|$			
$\bar{d}_r/\%$			

表 4.38　钙含量的测定

实验序号	1	2	3
$m_{样品}$/g			
V_{EDTA}(初始)/mL			
V_{EDTA}(终点)/mL			
V_{EDTA}(消耗)/mL			
w_{Ca}/[mg/(100g)]			
$\bar{w}_{Ca}$/[mg/(100g)]			
$|d_i|$			
$\bar{d}_r$/%			

拓展阅读

钙是我们的生命之源，在人生长的各个阶段，都起着非常重要的作用，是人体必不可少的重要元素。钙质一旦不足，身体就无法正常运作，就会造成蛀牙、骨质疏松及骨骼软化症、幼儿发育不良、容易产生腰痛及膝痛，等等。我们很多人只知道小孩或老年人应补钙，小孩为了生长，老年人预防骨质疏松。但实际上我们每个人一生都应补钙。在我们出生后，体内的钙一直都处于一个不断累积的过程，大约到 35 岁人体的钙含量达到一生中的顶峰。以后钙流失开始加速，钙流失的量大于平时我们体内的钙积累。如果我们在 35 岁以前体内储存的钙越多，那么就可更长久地维持我们以后体内身体各种代谢的需求，因此补钙对我们的健康成长是必不可少的。

实验 4.28　酱油中 NaCl 含量的测定

实验目的

（1）掌握硝酸银标准溶液的配制和标定。
（2）掌握莫尔法测定氯化物中氯含量的原理和方法。

实验原理

莫尔法是在中性或弱碱性（pH=6.5～10）溶液中，以 K_2CrO_4 作指示剂，用 $AgNO_3$ 标准溶液直接滴定 Cl^-。由于 AgCl 的溶解度比 Ag_2CrO_4 小，根据分步沉淀原理，溶液中首先析出 AgCl 白色沉淀。当 AgCl 定量析出后，微过量的 Ag^+，即与 CrO_4^{2-} 生成砖红色的 Ag_2CrO_4 沉淀，它与 AgCl 沉淀一起，使溶液略带橙红色即为终点。反应如下：

$$Ag^+ + Cl^- = AgCl \downarrow （白色） \quad K_{sp}=1.8\times10^{-10} \quad (4.56)$$

$$2Ag^+ + CrO_4^{2-} = Ag_2CrO_4 \downarrow （砖红色） \quad K_{sp}=2.0\times10^{-12} \quad (4.57)$$

实验用品

仪器：酸式滴定管、锥形瓶（250mL）、容量瓶（250mL）、滴管、量筒、移液管、棕色试剂瓶。

试剂：$AgNO_3$、基准 NaCl、酱油、4mol/L 硝酸、4mol/L 氢氧化钠溶液、5%高锰酸钾溶液、K_2CrO_4 溶液（50g/L）。

实验步骤

（1）$AgNO_3$ 标准溶液（0.1mol/L）的配制和标定　准确称取 8.5g $AgNO_3$ 于小烧杯中，用蒸馏水溶解后，转移至棕色试剂瓶中，稀释至 500mL，盖上瓶塞，置于暗处保存。

准确称取 0.55～0.60g NaCl 基准物质于小烧杯中，用蒸馏水溶解后，转移至 100mL 容量瓶中，用水稀释至刻度，摇匀。准确移取 25.00mL NaCl 溶液于 250mL 锥形瓶中，加入 20mL 蒸馏水、1mL 的 K_2CrO_4 溶液（50g/L）。用 $AgNO_3$ 溶液滴定至呈现砖红色即为终点。记录消耗 $AgNO_3$ 溶液的体积，平行测定三份，计算 $AgNO_3$ 溶液的平均浓度（表 4.39）。

（2）酱油中 NaCl 含量测定　准确移取 10.0000mL 酱油样品于 250mL 容量瓶中，加水至刻度，定容摇匀。移取 25.00mL 稀释液于 250mL 锥形瓶中，加 20mL 水及 1mL K_2CrO_4 溶液，混匀。用 $AgNO_3$ 溶液滴定至初现砖红色沉淀即为终点。记录消耗 $AgNO_3$ 溶液的体积，平行测定三份，计算酱油中 NaCl 含量（表 4.40）。

数据记录与处理

（1）计算 $AgNO_3$ 标准溶液的浓度

$$c_{AgNO_3} = \frac{m_{NaCl} \times \frac{25.00}{100.0}}{M_{NaCl} V_{AgNO_3}} \times 1000 \tag{4.58}$$

表 4.39　0.1mol/L $AgNO_3$ 溶液标定

实验序号	1	2	3
m_{NaCl}/g			
V_{AgNO_3}（初始）/mL			
V_{AgNO_3}（终点）/mL			
V_{AgNO_3}（消耗）/mL			
c_{AgNO_3}/(mol/L)			
$\bar{c}_{AgNO_3}$/(mol/L)			
$|d_i|$			
$\bar{d}_r$/%			

（2）计算酱油中 NaCl 的含量

$$w_{NaCl} = \frac{c_{AgNO_3} V_{AgNO_3}}{m_s \times \frac{25.00}{250.0}} \times \frac{M_{NaCl}}{1000} \times 100\% \tag{4.59}$$

（$M_{NaCl} = 58.44$g/mol，$M_{AgNO_3} = 169.88$g/mol）

表 4.40　酱油中 NaCl 含量的测定

实验序号	1	2	3
m_s/g			
V_{AgNO_3}（初始）/mL			
V_{AgNO_3}（终点）/mL			

续表

实验序号	1	2	3
V_{AgNO_3}（消耗）/mL			
w_{NaCl}/%			
$\bar{w}_{NaCl}$/%			
$|d_i|$			
$\bar{d}_r$/%			

思考题

（1）莫尔法测氯时，为什么溶液的 pH 须控制在 6.5～10.5？若存在铵离子，溶液的 pH 应控制在什么范围，为什么？

（2）以 K_2CrO_4 为指示剂时，指示剂浓度过大或过小对测定有何影响？

（3）能否用莫尔法以 NaCl 标准溶液直接滴定 Ag^+？为什么？

拓展阅读

氯化钠的作用与功效：氯化钠是人所不可缺少的。成人体内所含钠离子的总量约为 60g，其中 80% 存在于细胞外液，即在血浆和细胞间液中，氯离子也主要存在于细胞外液。钠离子和氯离子的生理功能主要有：(1) 维持细胞外液的渗透压，(2) 参与体内酸碱平衡的调节，(3) 氯离子在体内参与胃酸的生成。此外，氯化钠在维持神经和肌肉的正常兴奋性上也有作用。

实验 4.29　蓝莓中花青素含量的测定

实验目的

（1）掌握花青素含量测定技术。
（2）掌握超声辅助萃取实际样品的分离提取技术。

实验原理

花青素（anthocyanidin），又称花色素，是苯并吡喃衍生物，属于酚类化合物，其结构式如图 4.10 所示。其常与一个或多个葡萄糖、鼠李糖、半乳糖、阿拉伯糖等通过糖苷键形成花色苷（anthocyanin），是自然界一类广泛存在于植物中的水溶性天然色素，也是树木叶片中的主要呈色物质，在植物细胞液泡不同的 pH 值条件下，呈现不同的颜色。以香兰素甲醇溶液与盐酸甲醇溶液混合作为显色剂，采用紫外可见分光光度法可检测花青素的含量。

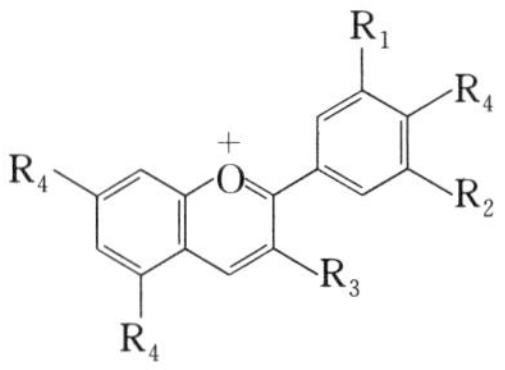

R_1 和 R_2 是 H、OH 或 OCH_3　R_3 是 H 或糖基　R_4 是 OH 或糖基

图 4.10　花青素的结构式

蓝莓是植物中花青素含量较高的一类品种，其花青素最高含量可达 4200mg/kg，蓝莓中所含花青素是目前所有植物花青素中性能最优良、副作用最低的花青素品种之一。为了进一步开发和有效利用蓝莓花青素，蓝莓中花青素提取及检测技术的研究具有十分重要的意义。植物中的花青素通常以结合态与蛋白质、纤维素结合在一起，故不易提取。超声波产生强烈振动、高加速度、强烈的空化效应、搅拌等作用，可以破坏植物的细胞壁，使溶剂渗透到细胞中，令其中的化学成分溶于溶剂中，从而提高提取效率。

实验用品

仪器：分析天平、分光光度计、烧杯、比色皿、水浴锅、超声波提取器、离心机、容量瓶、量筒。

试剂：蓝莓、花青素标准品、无水乙醇、甲醇、香兰素、盐酸。

实验步骤

（1）超声辅助萃取蓝莓中的花青素　准确称取 1g 蓝莓鲜果，将蓝莓捣碎成蓝莓糊，再置于提取瓶中，加入 15mL40％乙醇溶液，在超声波功率 200W，温度 30℃下提取 30min。重复提取 3 次，合并提取液，离心过滤，定容至 100mL。

（2）最大吸收波长的选择　取 1.0mL 浓度为 0.1mg/mL 的花青素标准溶液，加 5.0mL 混合显色剂（2％香兰素甲醇溶液：8％盐酸甲醇溶液为 1∶1），加甲醇定容到 100mL，避光反应 30min，在波长 350～600nm 之间测吸光度，绘制吸收曲线，确定花青素的最大吸收波长。

（3）标准曲线的绘制　取 0.25mL、0.50mL、0.75mL、1.00mL、1.25mL 和 1.50mL 浓度为 0.1mg/mL 的花青素标准溶液，分别加入 5.0mL 混合显色剂，加甲醇定容到 100mL，避光反应 30min，以试剂空白为参比溶液，在最大吸收波长处测定吸光度，绘制标准曲线。

（4）花青素含量的测定　取 10.00mL 超声辅助萃取蓝莓的提取液于 100mL 容量瓶中，加入 5.0mL 混合显色剂，加甲醇定容到 100mL，避光反应 30min，以试剂空白为参比溶液，在最大吸收波长处测定吸光度，计算花青素提取率（表 4.41）。

数据记录与处理

表 4.41　标准溶液的吸光度及试样中花青素含量的测定

No.	花青素的浓度/(μg/mL)	吸光度	试样中花青素的含量/％
1			—
2			—
3			—
4			—
5			—
6			—
试样溶液			

拓展阅读

近年来，作为多酚的花青素正在被人们广泛关注。科学研究表明人的寿命长短直接取决于人们抗氧化和抗自由基能力的强弱，如果解决自由基的侵害问题，那么人体细胞就可以真正自由生长。而花青素的发现为全世界的人找到了抗氧化和抗衰老的最简单有效的办法。从根本上讲，花青素是一种强有力的抗氧化剂，它能够保护人体免受自由基的损伤，还能够增强血管弹性，改善循环系统和增进皮肤的光滑度，提高视力，抑制炎症和过敏，改善关节的柔韧性，同时有助于预防多种与自由基有关的疾病，包括癌症、心脏病、过早衰老和关节炎，因此，花青素为人体带来了多种益处。

实验 4.30 奶制品中三聚氰胺的测定

实验目的

（1）掌握分光光度计的使用方法。
（2）掌握奶制品中三聚氰胺的检测原理及方法。

实验原理

三聚氰胺（Melamine）化学式为 $C_3H_6N_6$，俗称蛋白精，IUPAC 命名为 1,3,5-三嗪-2,4,6-三胺，是一种三嗪类含氮杂环有机化合物，通常被用作化工原料。根据三聚氰胺分子的结构特点，通过显色剂茜素红与三聚氰胺发生络合反应（图 4.11），利用紫外可见分光光度计测定吸光度，从而实现快速检测三聚氰胺含量。

实验用品

仪器：紫外-可见分光光度计、分析天平、称量纸、容量瓶、烧杯、玻璃棒、量筒、超声波提取器。

试剂：奶粉、乙腈、0.1%茜素红溶液（称取 0.1g 茜素红溶于水，定容至 100mL，混匀备用）；三聚氰胺标准溶液（1mg/mL）：准确称取 0.1000g 三聚氰胺于小烧杯中，加水溶解后，转移至 100mL 容量瓶中，定容至刻度，摇匀；三聚氰胺工作溶液（10μg/mL）：准确移取 1mL 三聚氰胺标准溶液（1mg/mL）于 100mL 容量瓶中，用水定容至刻度，摇匀。

实验步骤

（1）最大吸收波长的测定 在比色管中分别加入 1.0mL 浓度为 10μg/mL 的三聚氰胺工作溶液，0.1%茜素红 1mL，用水稀释至 10mL，反应 10min，以试剂空白为参比溶液，在波长 400～800nm 之间测吸光度，绘制吸收曲线，确定三聚氰胺显色反应络合物的最佳

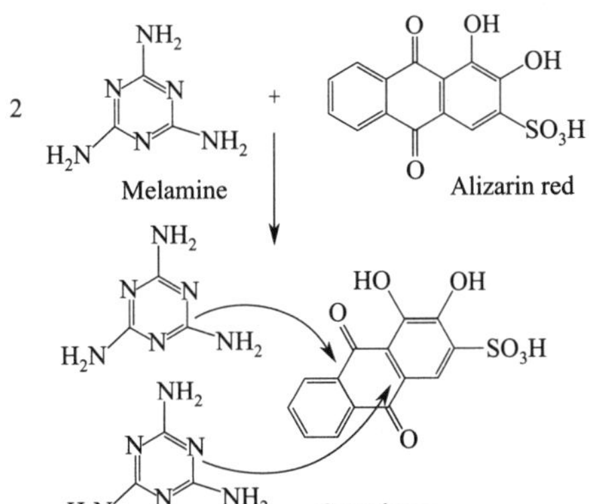

图 4.11 三聚氰胺与茜素红溶液相互作用示意图

吸收波长。

（2）显色条件优化

① 显色剂浓度的选择。在六个比色管中分别加入 1.0mL 浓度为 10μg/mL 的三聚氰胺工作溶液，0.1% 茜素红各 0.5mL、1mL、2mL、3mL、4mL、5mL，即茜素红的浓度分别为 0.005%、0.010%、0.020%、0.030%、0.040% 和 0.050%，用水稀释至 10mL，反应 10min，以试剂空白为参比溶液，在最大吸收波长处测吸光度，确定最优显色剂的浓度。

② 显色时间的选择。在六个比色管中分别加入 1.0mL 浓度为 10μg/mL 的三聚氰胺工作溶液，0.1% 茜素红 1mL，用水稀释至 10mL，分别反应 0min、5min、10min、15min、20min、30min，以试剂空白为参比溶液，在最大吸收波长处测吸光度，确定最优显色时间。

（3）标准曲线的绘制　分别取 1mL、2mL、3mL、4mL 和 5mL 浓度为 10μg/mL 的三聚氰胺溶液，分别加入 0.1% 茜素红 1mL，用水稀释到 10mL，反应 10min，以试剂空白为参比溶液，在最大吸收波长处测定吸光度，绘制标准曲线。

（4）样品分析

空白样品：取 1.0g 奶粉，加 10mL 水溶解后，加 10mL 乙腈沉淀蛋白质，超声提取 10min，离心 10min，过滤后取上清液作为提取液待用。

加标样品：取 1.0g 奶粉，加 25μL 浓度为 1mg/mL 的三聚氰胺标准溶液，10mL 水溶解后，加 10mL 乙腈沉淀蛋白质，超声提取 10min，离心 10min，过滤后取上清液作为提取液待用。

取空白和加标样品上清液 5mL，分别加入 0.1% 茜素红 1mL，用水稀释到 10mL，反应 10min 后，以试剂空白为参比溶液，在最大吸收波长处测定吸光度，利用标准曲线求得三聚氰胺浓度。平行测定三次，计算加标回收率与相对标准偏差（表 4.42）。

 数据记录与处理

表 4.42　标准溶液的吸光度及试样中三聚氰胺测定

No.	三聚氰胺浓度/(μg/mL)	吸光度	试样中三聚氰胺的含量/%
1			—
2			—
3			—
4			—
5			—
6			—
试样			
1			
2			
3			

 思考题

你还知道哪些测定三聚氰胺的其他方法？各方法有何优缺点？

实验 4.31　保健食品中维生素 B12 的测定

实验目的

（1）掌握紫外可见分光光度计的结构及基本操作。

(2) 掌握保健食品中维生素 B12 的测定。

实验原理

维生素 B12 参与体内多种代谢活动,是维持机体健康一种不可或缺的营养素。人体不能合成自身所需的维生素 B12,需要通过食物或添加维生素 B12 的营养补充剂中获取。维生素 B12 在生产提纯过程中需加入氰化钠,使天然形式的维生素 B12 转化为性质更为稳定的氰钴胺素。氰钴胺素通常作为食品添加剂,添加到保健食品复合维生素片中,达到补充维生素 B12 的目的。维生素 B12 结构如图 4.12 所示,易溶于水和乙醇,在强酸(pH<2)和碱性溶液中易分解,而在 pH 为 4.5~5.0 弱酸性条件下最稳定。

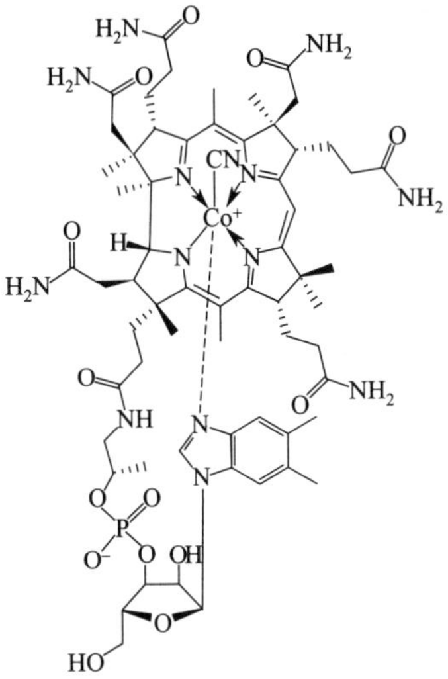

图 4.12 维生素 B12 结构式

紫外分光光度法可用于有机化合物的定性和定量分析。每种物质的吸收光谱一般都有一个最大吸收峰,这是紫外分光光度法的定性依据。在合适的测定波长下,同一化合物在不同浓度下的吸收强度不同,这是紫外吸收光谱定量分析的基础。根据维生素 B12 水溶液在 361nm 波长吸收最大的特点,采用紫外分光光度法测定复合维生素片 B12 的含量。

实验用品

仪器:紫外-可见分光光度计、比色皿、量筒、容量瓶、滤膜(水系)、超声波清洗器、离心机。

试剂:维生素 B12 标准品、保健品(复合维生素片)、乙醇。

实验步骤

(1) 维生素 B12 标准溶液(1.0mg/mL)的配制　准确称取 10.0mg 维生素 B12 标准品,用 5%乙醇溶解,定容至 10mL 棕色容量瓶中,混匀,得到维生素 B12 的标准储备液。

(2) 测定波长的选择　以水为参比,在 200~500nm 波长范围内测绘维生素 B12 标准溶液的吸收光谱,确定维生素 B12 特征吸收波长(λ_{max})。

(3) 标准曲线的制备　分别取 0.05mL、0.10mL、0.25mL、0.50mL、1.00mL、2.50mL、5.00mL 标准储备液于 10mL 棕色容量瓶内,水稀释定容,得维生素 B12 标准溶液 1~7。在所选择的 λ_{max} 下,测量 1~7 号维生素 B12 标准溶液的吸光度 A,绘制标准曲线。以吸光度 A 与质量浓度 c 绘制标准曲线,得回归方程。

（4）样品分析　将 10 粒片剂粉碎混匀，称取试样 3.0000g 于 50mL 离心管中，加 10mL 水，混匀，将其置于超声波清洗器中，超声提取 10min 后，以 4000r/min 离心 5min。重复提取 2 次，合并提取液，过 0.45μm 水系滤膜后定容于 25mL 棕色容量瓶中，得样品溶液。

在所选择的 λ_{max} 下，测定样品溶液的吸光度 A，根据标准曲线方程计算样品中维生素 B12 的含量（表 4.43）。

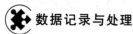数据记录与处理

表 4.43　标准溶液的吸光度及试样中维生素 B12 含量的测定

No.	维生素 B12 的浓度/(μg/mL)	吸光度	试样中维生素 B12 的含量/%
1			—
2			—
3			—
4			—
5			—
6			—
试样溶液			
1			
2			
3			

拓展阅读

　　维生素 B12 主要来源是动物性食物，存在于牛肉、猪肉、动物肝脏、蛋、奶和奶制品中，在吸收时需与钙结合才能有利于人体的机能活动。维生素 B12 能促进红细胞的形成和再生，防止贫血；促进儿童发育，增进食欲；增强体力；维持神经系统的正常功能；促使注意力集中，增强记忆力与平衡感。能使脂肪、碳水化合物、蛋白质适宜地为人体所利用。维生素 B12 为细胞分裂和维持神经组织髓鞘完整所必需，主要用于治疗恶性贫血，与叶酸合用治疗因抗叶酸药、脂肪泻等引起的巨幼红细胞性贫血。

第 5 章
探索创新实验

5.1 大学生化学竞赛（分析化学部分）

实验 5.1 稀土铕配合物的制备及配位比的测定

实验目的

(1) 稀土铕配合物 $Eu(DBM)_n(H_2O)_2$ 的制备。
(2) 用络合滴定的方法测定配合物中 Eu 的含量。
(3) 用紫外-可见分光光度法测定配合物中顺丁烯二酸二丁酯（DBM）的含量。
(4) 计算该配合物的配位数 n。

实验原理

稀土元素（REEs）由于具有特异的光、电、磁和催化等物理和化学性能，成为 21 世纪世界各国竞相研究开发的对象，是一类重要的战略资源。中国具有世界上最丰富的稀土资源。

稀土元素作为一类典型的金属元素，能够与周期表中的大多数非金属元素形成配位键。由于 β-双酮类配体对稀土离子是很好的螯合配位体，因此稀土和 β-双酮类配体形成的配合物受到人们较多的关注。下图是稀土 Eu^{m+} 与二苯甲酰甲烷（HDBM）形成配合物 $Eu(DBM)_n(H_2O)_2$ 的反应方程式：

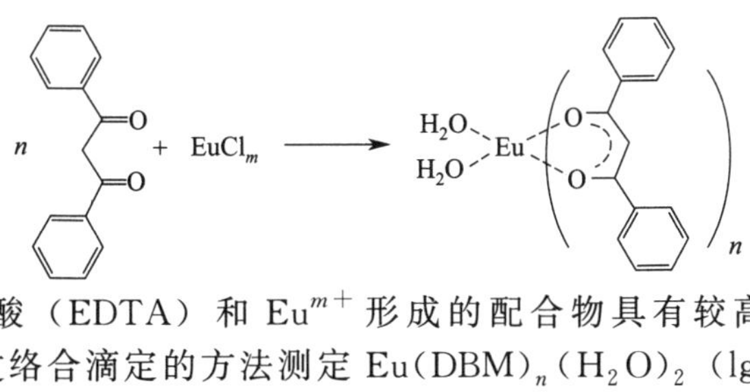

$$n \text{ [二苯甲酰甲烷]} + EuCl_m \longrightarrow \text{[}Eu(DBM)_n(H_2O)_2\text{]} \tag{5.1}$$

由于乙二胺四乙酸（EDTA）和 Eu^{m+} 形成的配合物具有较高的稳定常数（$\lg K = 17.14$），因此可以通过络合滴定的方法测定 $Eu(DBM)_n(H_2O)_2$（$\lg \beta_1 < 11$）中铕的含量。由于稀土离子特殊的电子构型，导致其络合物的光谱强弱和峰形很大程度上是由配体的性质决定，因此在一定条件下，可以通过比较配体和配合物的紫外-可见吸收光谱来确定 $Eu(DBM)_n(H_2O)_2$ 中配体的含量，从而确定配合物的化学组成。

实验用品

仪器：紫外-可见分光光度计、手提紫外灯、pH 试纸、天平、循环水泵、工具盒（含称量纸、滤纸、剪刀、标签纸、吸磁棒，记号笔，pH 值比对卡）、丁腈手套、防护眼镜、烘箱、回收产物的烧杯（400mL）、分析天平，纸条，棉布手套，磁力搅拌器（一套，附水浴缸 1 个）、铁架台、酸式滴定管（50mL）、单刻度移液管（25mL）、球形冷凝管（附橡胶管）、布氏漏斗、吸滤瓶 1 套、长颈烧瓶（100mL）、烧杯（100mL、150mL、400mL）、容量瓶（50mL、100mL、250mL）、锥形瓶（250mL）、吸量管（5mL）、量筒（10mL、50mL）、培养皿、表面皿、不锈钢刮刀、滴管、玻璃棒、洗耳球、洗瓶、试剂瓶（1L）、镊子、白瓷板、搅拌磁子。

试剂：Eu^{m+} 水溶液（0.96mol/L）、二苯甲酰甲烷（HDBM）、HDBM 储备液（无水乙醇配制，5.00×10^{-4}mol/L）、无水乙醇、95%乙醇、NaOH 溶液（2mol/L；0.5mol/L）、HCl 溶液（6mol/L；1mol/L）、氧化锌、二甲酚橙（0.2%水溶液）、六亚甲基四胺（20%水溶液）、二甲亚砜（20%乙醇溶液）、乙二胺四乙酸二钠（$Na_2H_2Y\cdot 2H_2O$）、磷酸盐缓冲溶液（pH=8.0，0.035mol/L）、去离子水。

实验步骤

（1）配合物 $Eu(DBM)_n(H_2O)_2$ 的制备　在 100mL 长颈烧瓶中加入 1.6g 二苯甲酰甲烷（HDBM，过量）和 25mL 无水乙醇，于 50℃水浴中加热，使固体溶解，从装有 0.96mol/L 氯化铕水溶液的滴定管中准确放入 1.8mL 溶液于长颈烧瓶（记录放入前后滴定管的读数，精确至 0.01mL）。先用滴管滴加 2mL 2mol/L 的 NaOH 溶液至反应瓶中（用玻棒蘸取少量溶液于滤纸上，在手提紫外灯下观察滴加 NaOH 溶液前后样品的发光情况）。然后再滴加 0.5mol/L 的 NaOH 溶液至 pH=8。升高温度至 65℃，反应 10 分钟。冷却，抽滤，用合适溶剂洗涤。产物置于 60℃烘箱中烘干，称重。

（2）$Eu(DBM)_n(H_2O)_2$ 中 Eu 的含量测定

① 锌标准溶液的配制。用减量法准确称取基准物质氧化锌 0.2100g（精确至 0.0001g）于 150mL 烧杯中，加入 5mL 6mol/L HCl 溶液。溶解后，加入适量水稀释，定量转移至 250mL 容量瓶中，定容。计算锌标准溶液的浓度（mol/L）。

② EDTA 溶液的配制及标定。称取乙二胺四乙酸二钠 3.7g，溶于 1000mL 蒸馏水中，得到浓度约为 0.01mol/L 的 EDTA 溶液。准确移取三份锌标准溶液各 25.00mL，分别置于 250mL 锥形瓶中，以二甲酚橙为指示剂，用 20%六亚甲基四胺水溶液调至紫红色后，再过量 5mL。以 0.01mol/L EDTA 溶液滴至溶液由紫红色变为亮黄色，即为终点。计算 EDTA 溶液的浓度（mol/L）及相对平均偏差。

③ $Eu(DBM)_n(H_2O)_2$ 中 Eu 的含量测定。用增量法准确称取三份自制产物（每份 0.2000g，精确至 0.0001g），分别置于 250mL 锥形瓶中（注：锥形瓶请先用无水乙醇润洗），用 25mL 二甲亚砜的乙醇溶液（20%，体积比）溶解。以二甲酚橙为指示剂，用 1mol/L HCl 溶液调至紫红色，再加入 2mL 20%六亚甲基四胺溶液，用 EDTA 标准溶液滴至终点。（注：若滴定过程中出现浑浊，可加少量乙醇使其溶解）计算产物中 Eu 的含量（mol/g）及相对平均偏差。

（3）$Eu(DBM)_n(H_2O)_2$ 中 DBM 的含量测定

① 标准曲线的绘制。用 HDBM 储备液（5.0×10^{-4}mol/L 乙醇溶液）配制浓度范围在 $0\sim5.0\times10^{-5}$mol/L 的系列标准溶液，每份标准溶液中需含 10mL 无水乙醇和 5mL pH 8.0 的磷酸盐缓冲液（0.03mol/L），用去离子水定容。于波长 300～500nm 之间，选择最优波长。测定吸光度，绘制标准曲线。

② $Eu(DBM)_n(H_2O)_2$ 中 DBM 的含量测定。准确称取自制产物 0.0200g（精确至

0.0001g），用无水乙醇溶解，定量转移至 100mL 容量瓶中，无水乙醇定容。移取上述溶液 5.00mL 于 100mL 容量瓶，用去离子水定容（需含与（3）①同等比例的无水乙醇和磷酸盐缓冲液），测量该溶液的吸光度。根据上面所得标准曲线，计算产物中 DBM 的含量（mol/g）。

数据记录与处理

（1）根据实验步骤（2）和（3）的结果，计算 $Eu(DBM)_n(H_2O)_2$ 的配位数 n。

（2）计算产物的分子量及产率（Eu 的原子量：151.96）。

思考题

（1）本实验中制备配合物时需用 NaOH 调节 pH 为 8，为什么？若用噻吩甲酰三氟丙酮（HTTA，结构式如图 5.1）为配体合成 Eu 配合物，通过计算说明 pH 至少应调节到多少？（已知 HDBM 的 $pK_a=14.17$，HTTA 的 $pK_a=6.23$，以水溶液体系为准，参考本实验浓度）

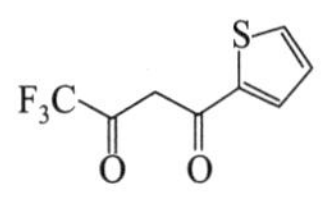

（2）一般比较不同溶液吸光能力强弱时用什么参数？用实验中所测数据，定量比较上述 DBM 和 $Eu(DBM)_n(H_2O)_2$ 溶液的吸光能力强弱。

图 5.1 HTTA 的结构式

（3）本实验中既用到了化学分析法，又用到了仪器分析法，请列举出这两类分析方法的最主要区别。

实验 5.2 氧化铁纳米颗粒的制备及用于奶制品中三聚氰胺的测定

实验目的

（1）氧化铁纳米颗粒的制备；
（2）奶制品中三聚氰胺的测定。

实验原理

氧化铁磁性纳米颗粒（Nanoparticles，Nps）是一种具有过氧化物模拟酶性质的新型纳米材料。研究结果表明，相对于辣根过氧化物酶和其他的过氧化物模拟酶纳米材料，氧化铁磁性纳米颗粒表现出良好的催化活性、稳定性、单分散性和可重复利用性。

基于氧化铁磁性纳米颗粒的过氧化物模拟酶的催化活性，利用其对过氧化氢和 2,2′-联氮-双(3-乙基苯并噻唑啉-6-磺酸)二铵盐（ABTS）之间氧化还原反应的催化作用，可建立测定乳制品中三聚氰胺的快速、简便的吸光光度分析方法。

体系中过氧化氢对三聚氰胺而言是过量的，过氧化氢首先和三聚氰胺定量反应形成 1∶1 的加合物，氧化铁磁性纳米颗粒作为催化剂催化剩余的过氧化氢和 ABTS 的氧化还原反应生成有色化合物，而有色化合物的吸光度随三聚氰胺含量的增加而降低，且具有良好的线性关系，据此可测定三聚氰胺的含量。

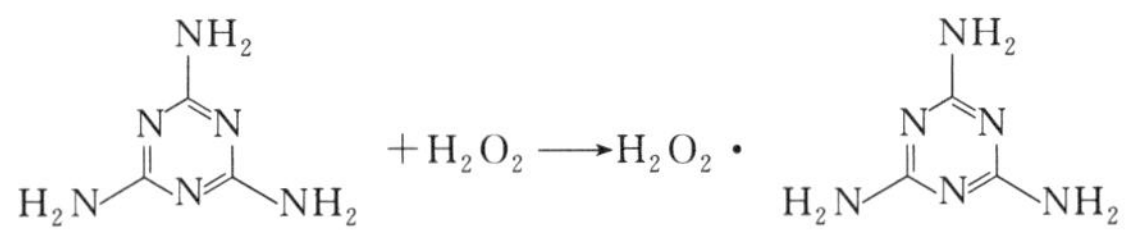

$$\text{(三聚氰胺)} + H_2O_2 \longrightarrow H_2O_2 \cdot \text{(三聚氰胺)} \tag{5.2}$$

式(5.2)为过氧化氢与三聚氰胺的反应方程式。

$$H_2O_2 \xrightarrow{\text{Nps}} HO^- + HO^+ \tag{5.3}$$

$$HO^+ + ABTS \xrightarrow{Nps} 2H_2O + ABTS^+ \tag{5.4}$$

式（5.3）和式（5.4）为氧化铁磁性纳米颗粒（Nps）催化过氧化氢与 ABTS 的氧化还原反应方程式。

实验用品

仪器：分析天平、721G 型分光光度计、氮气气袋、磁力搅拌器、磁铁、恒温水浴槽、三口烧瓶、圆底烧瓶、恒压漏斗、液封管、锥形瓶、容量瓶、手套、磁子、导气管、抽气接头、比色管、离心试管、小试管、比色皿、酸式滴定管、洗瓶、移液管、称量瓶、离心机。

试剂：$K_2Cr_2O_7$ 基准试剂、1mol/L 氯化铁溶液（含 2mol/L 盐酸）、1mol/L 氯化亚铁溶液（含 2mol/L 盐酸）、1.25mol/L 氨水、乙腈、浓盐酸、硫-磷混酸、二苯胺磺酸钠、无水乙醇、丙酮、牛奶（含 0.25mmol/L 三聚氰胺）、牛奶待测样、10% 三氯乙酸水溶液、6% $SnCl_2$ 水溶液、5% $HgCl_2$ 水溶液、ABTS（0.03mol/L）溶液、HAc-NaAc 缓冲溶液（pH=4.75）、过氧化氢（0.01mol/L）溶液。

实验步骤

（1）氧化铁纳米颗粒的制备——共沉淀法　在 100mL 三口烧瓶中将 8mL 1mol/L 氯化铁溶液和 4mL 1mol/L 氯化亚铁溶液混合，电磁搅拌，将 50mL 1.25mol/L 的氨水溶液加入到恒压漏斗中。混合溶液中通氮气 10 分钟除去溶液中的氧气后迅速加入氨水，在氮气保护下继续搅拌 20 分钟。将溶液转移至 100mL 小烧杯中，用磁铁将磁性纳米颗粒沉积于烧杯底部，倾析法倾去上清液，用蒸馏水反复洗涤至中性。加水至 60mL，分散均匀后转移出 10mL 溶液至 25mL 圆底烧瓶，电磁搅拌分散备用。将烧杯中剩余的上清液倾出，底部的纳米颗粒用乙醇洗涤两到三次，再用丙酮洗涤两到三次，每次分离时都可借助磁铁将磁性颗粒吸至烧杯底部，在空气中晾干（可借助洗耳球吹气加速干燥），称重。

（2）奶制品中三聚氰胺的测定——吸光光度法

① 样品的前处理。移取含有三聚氰胺 0.25mmol/L 的 2.00mL 牛奶于离心试管中，依次加入 2.00mL 乙腈，1.00mL 10% 的三氯乙酸溶液和 5.00mL 蒸馏水，混匀后将此溶液超声振荡 10 分钟，再放入离心机，在 4000r/min 转速下离心 10 分钟，取上清液立即进行下一步检测。

② 制作工作曲线。在小试管中分别依次加入（a）0.00mL、（b）0.10mL、（c）0.20mL、（d）0.30mL、（e）0.50mL、（f）0.80mL 样品前处理所得的清液，补加蒸馏水至 1.00mL，再分别依次加入 1.00mL HAc-NaAc 缓冲溶液、0.50mL 过氧化氢溶液，摇匀后加入 0.50mL ABTS 溶液，最后用移液枪移取 50.0μL 氧化铁磁性纳米颗粒溶液，混合均匀后在 45℃ 水浴中反应 15 分钟，然后放入冰水浴中 10 分钟以终止反应。

用磁铁将氧化铁磁性纳米颗粒沉积于试管底部，用移液枪移取 2.00mL 上清液于 10mL 比色管，并用蒸馏水稀释至刻度、摇匀，以水作参比，用分光光度计在 417nm 下测定吸光度，绘制工作曲线。

③ 待测样品中三聚氰胺的测定。移取牛奶待测样 2.00mL 于离心试管中，按前述步骤处理后，平行移取 0.50mL 上清液两份进行同上的显色反应，测定吸光度，并根据工作曲线计算样品中三聚氰胺的含量。

（3）氧化铁纳米颗粒铁含量的测定——重铬酸钾（$K_2Cr_2O_7$）法

① 重铬酸钾标准溶液的配制。配制 0.01000mol/L $K_2Cr_2O_7$ 标准溶液 250mL。

② 氧化铁纳米颗粒铁含量的测定。准确称取 0.10~0.12g 试样于 250mL 锥形瓶中，

加几滴水使试样润湿并摇动使其散开,然后加入 2mL 浓盐酸,盖上表面皿,加热至沸,待试样全溶后趁热滴加 6% $SnCl_2$ 溶液还原 Fe^{3+} 至黄色刚消失,再过量 1～2 滴 $SnCl_2$ 溶液,迅速用流水冷却至室温,立即加入 5% $HgCl_2$ 溶液 2～3mL 摇匀,放置 3～5 分钟。加水稀释至 80～100mL,加入硫-磷混酸 10mL,滴加 5～6 滴二苯胺磺酸钠指示剂,立即用 $K_2Cr_2O_7$ 标准溶液滴定至呈稳定的紫色即为终点,计算出氧化铁纳米颗粒中铁的百分含量。

数据记录与处理

根据实验步骤（2）和（3），计算样品中三聚氰胺的含量和氧化铁纳米颗粒中铁的百分含量。

实验 5.3　配合物 [Ni(Me₃en)(acac)]BPh₄ 的合成及其溶剂/热致变色行为研究

实验目的

(1) 配合物 [Ni(Me₃en)(acac)]BPh₄ 的制备；

(2) 配合物 [Ni(Me₃en)(acac)]BPh₄ 中 Ni(Ⅱ) 含量测定。

实验原理

利用乙酰丙酮和 N,N,N'-三甲基乙二胺配体可合成红色的[Ni(Me₃en)(acac)]BPh₄ 配合物,该配合物具有溶剂和热致变色效应。将它与具有层状结构的皂石（SAP）或高分子聚合物 Nafion（一种全氟磺酸聚合物）等材料混合,并经适当处理,可以制得有机溶剂识别材料（固体颗粒物或薄膜）。例如：将此材料浸入二氯甲烷、乙醚等溶剂中,材料保持红色不变；但将此材料放入甲醇、乙醇、丙酮、乙腈、N,N-二甲基甲酰胺（DMF）等溶剂中,材料则由红色变为蓝绿色；将变为蓝绿色的材料放入真空烘箱中加热干燥（80～100℃）,材料又变为红色。这类材料在一些特定溶剂的便捷检测过程中,颜色变化显著,肉眼可识别,且变化可逆,在环境、工业生产等方面具有一定的应用价值。

$$[Ni(Me_3en)(acac)]BPh_4 \text{ 红色} \xrightleftharpoons[\text{真空干燥}(80\sim100℃)]{\text{甲醇、乙醇、丙酮、乙腈、DMF}} [Ni(Me_3en)(acac)]BPh_4 \text{ 蓝绿色}$$

实验用品

仪器：移液枪（1mL）、循环水泵、旋转蒸发仪、低温浴槽、烘箱、吸磁棒、紫外-可见光谱仪、磁力搅拌器、烧瓶夹、冷凝管夹、十字夹、铁架台、瓶托、白瓷盘、圆底烧瓶（50mL）、玻璃漏斗（6cm）、固体漏斗、吸滤瓶（250mL）、布氏漏斗（6cm）、瓷滴板（白,6 孔）、锥形瓶（150mL；250mL）、磨口塞（14♯；19♯）、烧杯（50mL；100mL；400mL）、球形冷凝管（19♯）、量筒（50mL；100mL）、量杯（10mL；25mL）、培养皿（6cm）、表面皿（7cm）、容量瓶（50mL/14♯）、比色管（10mL,25mL）、单刻度移液管（25mL）、滴定管（50mL）、称量瓶（ϕ25mm×40mm）、干燥器（15cm）、滴管、玻棒、洗瓶、搅拌磁子、镍匙、漏斗板、工具盒[含洗耳球、镊子、剪刀、标签纸、记号笔、棉花、滤纸（中速、慢速）、乳胶手套、纱手套、一次性手套、护目镜]。

试剂：六水合硝酸镍[Ni(NO$_3$)$_2$·6H$_2$O]、乙酰丙酮乙醇溶液（4.0mol/L）、N,N,N'-三甲基乙二胺乙醇溶液（4.0mol/L）、碳酸钠、四苯硼钠（NaBPh$_4$）、锌、乙二胺四乙酸二钠溶液（Na$_2$H$_2$Y·2H$_2$O，0.01mol/L）、0.2%二甲酚橙溶液、HCl溶液（6mol/L，2mol/L）、30%六亚甲基四胺溶液、无水乙醇、二氯甲烷、石油醚（60~90℃）、乙腈、乙醇-水（$V_{乙醇}:V_{水}=1:1$）、二氯甲烷-石油醚（60~90℃；$V_{二氯甲烷}:V_{石油醚}=1:2$）、pH试纸（1~14）、精密pH试纸（5.4~7.0）。

实验步骤

（1）Ni(Ⅱ)配合物[Ni(Me$_3$en)(acac)]BPh$_4$的合成 在50mL圆底烧瓶中，加入0.87g六水合硝酸镍[Ni(NO$_3$)$_2$·6H$_2$O，3.0mmol，称重之前用滤纸尽可能将样品表面吸干]和15mL无水乙醇，搅拌溶解后，依次加入0.75mL乙酰丙酮、乙醇溶液（4.0mol/L，3.0mmol，用移液枪取）、0.16g碳酸钠（Na$_2$CO$_3$，1.5mmol）和0.75mL N,N,N'-三甲基乙二胺、乙醇溶液（4.0mol/L，3.0mmol，用移液枪取），继续搅拌30min。常压过滤，将溶液减压旋蒸至近干，将残留物溶解在20mL二氯甲烷中，加入1.44g四苯硼钠（NaBPh$_4$，4.2mmol），反应30min。之后，用慢速定性折叠滤纸过滤，并在滤液中加入石油醚（60~90℃）至产品析出后减压抽滤，依次用乙醇-水的混合溶剂（$V_{乙醇}:V_{水}=1:1$）、无水乙醇、二氯甲烷-石油醚（60~90℃）的混合溶剂（$V_{二氯甲烷}:V_{石油醚}=1:2$）洗涤，抽干。将此产品转移至培养皿中，置于80℃烘箱中干燥30min，称重，计算产率。

（2）Ni(Ⅱ)配合物中Ni(Ⅱ)含量的测定

① 锌标准溶液的配制。准确称取0.15~0.17g金属锌，放入100mL烧杯中，加入5mL 6mol/L HCl（必要时可微热）。待锌溶解完全后，加入适量去离子水稀释，定量转移至250mL容量瓶中，定容。计算锌标准溶液的浓度（mol/L）。

② EDTA溶液的标定。准确吸取两份25.00mL上述锌标准溶液于250mL锥形瓶中，加入50mL去离子水，3~4滴0.2%二甲酚橙指示剂，然后加入10mL 30%六亚甲基四胺溶液。用0.01mol/L EDTA标准溶液滴定至溶液由紫红色变为纯黄色，即为终点，记录读数。计算EDTA溶液的浓度（mol/L）及相对平均偏差。

③ Ni(Ⅱ)含量的测定。准确称取0.10~0.12g产品两份于250mL锥形瓶中，用20mL乙醇溶解后（若不溶，可稍加热），准确加入38~40mL 0.01mol/L EDTA标准溶液，放置5min。然后加入5mL 30%六亚甲基四胺溶液，调节溶液pH值为5.8~6.2（用pH试纸检验溶液的pH值）。以0.2%二甲酚橙溶液作指示剂（3~4滴），用0.01mol/L锌标准溶液滴定至溶液变为紫红色，记录读数。计算Ni(Ⅱ)的百分含量及相对平均偏差。

（3）Ni(Ⅱ)配合物溶剂变色现象及UV-vis吸收光谱表征 在10mL、25mL比色管中，分别以二氯甲烷和乙腈为溶剂，配制浓度约为1mg/mL（二氯甲烷为溶剂）和15mg/mL（乙腈为溶剂）的Ni(Ⅱ)配合物溶液各10mL，观察溶液颜色，并测定其UV-vis吸收光谱（波长测定范围：400~750nm）。给出每份溶液的最大吸收波长（λ_{max}）。

（4）Ni(Ⅱ)配合物的热致变色现象 取约5mg Ni(Ⅱ)配合物（约1/3药匙）于白色瓷滴板中，滴入1~2滴无水乙腈，待溶剂挥发完后，观察样品颜色。然后，将其置于80℃烘箱中10分钟，再次观察样品颜色。

数据记录与处理

根据步骤（1）和（2），计算[Ni(Me$_3$en)(acac)]BPh$_4$的产率和产物中Ni(Ⅱ)的含量。

> 思考题

(1) 画出配合物 [Ni(Me₃en)(acac)]BPh₄ 中配阳离子的结构简图,并用配合物晶体场理论解释为什么具有这种结构。

(2) 结合实验步骤 (3) 和 (4) 所观测到的现象,解释 Ni(Ⅱ) 配合物具有溶剂和热致变色现象的原因。

(3) 配位滴定是测定金属离子含量常用的方法之一,试至少列举两种其他定量测定 Ni(Ⅱ) 含量的方法。

实验5.4　多核铜（Ⅰ）配合物中铜含量的测定

> 实验目的

(1) 具有异构发光变色的 $[Cu_x(dppy)_y(CH_3CN)_{x/2}](ClO_4)_x$ 配合物的制备;

(2) 配合物中铜含量的测定。

> 实验原理

Cu(Ⅰ) 配合物具有多变的结构、优良的光/电及催化等物理和化学性能,近年来在发光材料、化学传感、生物探针和催化等方面展现了广泛的应用,在化学科学研究中备受瞩目。比如,$[Cu_x(dppy)_y(CH_3CN)_{x/2}](ClO_4)_x$ 配合物具有异构发光变色特性,它在 365nm 紫外光激发下能发射很强的蓝光,而在甲醇蒸气或含甲醇的溶剂中重结晶,结构会发生改变,导致发光颜色变为绿色。利用 $Cu(ClO_4)_2 \cdot 6H_2O$ 与铜粉在乙腈中反应得到 $[Cu(CH_3CN)_4]ClO_4$,进一步与二苯基-2-吡啶膦（dppy）反应可得到 $[Cu_x(dppy)_y(CH_3CN)_{x/2}](ClO_4)_x$ 配合物。

铜元素含量的测定有很多种手段,比如电感耦合等离子体质谱法、原子吸收法、分光光度法、碘量法、络合滴定法,等等。其中,滴定分析法,是一种简便、快速和应用广泛的定量分析方法,Cu^{2+}、Ni^{2+}、Ca^{2+}、Mg^{2+}、Al^{3+} 等均可采用乙二胺四乙酸二钠（EDTA）进行配合滴定法进行分析测定。但是,$[Cu_x(dppy)_y(CH_3CN)_{x/2}](ClO_4)_x$ 配合物中铜以 Cu(Ⅰ) 形式存在,它的含量难以用常规的化学滴定分析方法进行测定,因此,需将 Cu(Ⅰ) 氧化为 Cu(Ⅱ) 后再测定。Cu(Ⅱ) 能与 EDTA 形成较稳定的配合物（$\lg k = 18.80$）,在 pH=2～12 范围内,Cu(Ⅱ) 还能与黄色的 1-(2-吡啶偶氮)-2-奈酚（PAN）指示剂形成稳定的配合物（$\lg k = 6.70$）。

本实验将 $[Cu_x(dppy)_y(CH_3CN)_{x/2}](ClO_4)_x$ 溶于浓硝酸,在一定的 pH 条件下,以 PAN 为指示剂,用 EDTA 标准溶液进行滴定,测定产物中铜的含量。

> 实验用品

仪器：旋转蒸发仪、低温冷却液循环泵、循环水泵、365nm 紫外手电筒、超声仪、磁力搅拌器、铁架台、白陶瓷板、蝴蝶夹、安全瓶、十字夹、25.00mL 移液管、滴定管（50.00mL）、分析天平、台秤、圆底烧瓶（100mL）、玻璃板漏斗（30mL）、表面皿、烧杯（100mL）、容量瓶（250mL）、锥形瓶（250mL）、量筒（10mL、50mL）。

试剂：六水合高氯酸铜、乙腈、二苯基-2-吡啶膦（dppy,$M = 263.27$g/mol）、铜粉、无水乙醚、浓硝酸、0.2% 的 1-(2-吡啶偶氮)-2-奈酚（PAN）乙醇溶液、氧化锌（ZnO,$M = 81.38$g/mol）、约 0.01mol/L 乙二胺四乙酸二钠溶液（EDTA,$M = 372.24$g/mol）、0.2% 二甲酚橙溶液、20% 六亚甲基四胺溶液、HAc-NaAc 缓冲溶液（pH=4.2）。

实验步骤

(1) [Cu(CH$_3$CN)$_4$]ClO$_4$ 的制备　干燥的 100mL 圆底烧瓶中依次加入约 1.1g Cu(ClO$_4$)$_2$·6H$_2$O、40mL 乙腈和过量 2/3 的铜粉，塞好瓶塞，室温搅拌反应至溶液基本无色。用玻璃板漏斗将反应液快速抽滤到干燥的 100mL 圆底烧瓶中，用旋转蒸发仪减压蒸除乙腈，得到 [Cu(CH$_3$CN)$_4$]ClO$_4$ 固体。

(2) [Cu$_x$(dppy)$_y$(CH$_3$CN)$_{x/2}$](ClO$_4$)$_x$ 的制备　干燥的 100mL 圆底烧瓶中，按正确的顺序加入 20mL 乙腈、1.46g dppy、1.17g [Cu(CH$_3$CN)$_4$]ClO$_4$ 和 0.20g 铜粉，塞好瓶塞，室温搅拌反应 1.5h。反应结束后，将烧瓶中的反应液用玻璃板漏斗抽滤至干燥的 100mL 圆底烧瓶中，旋转蒸发浓缩溶液至约为 5mL，滴加无水乙醚至产生白色浑浊。静置使浑浊溶液充分结晶，后用倾析法弃去上清液，再用适当溶剂洗涤晶体，置于室温下晾干，超声使晶体从瓶壁上脱落后称重。

(3) [Cu$_x$(dppy)$_y$(CH$_3$CN)$_{x/2}$](ClO$_4$)$_x$ 中铜(Ⅰ)含量的分析

① 锌标准溶液的配制。准确称取一定量（约 0.2g）的基准试剂 ZnO 于 100mL 小烧杯中，先加入少量水润湿再滴加盐酸溶液（体积比 1∶1）使 ZnO 完全溶解，然后将溶液定量转移至 250mL 容量瓶中，稀释至刻度并摇匀，计算锌标准溶液的物质的量浓度。

② 0.01000mol/L EDTA 溶液的标定。准确吸取 25.00mL 上述锌标准溶液于 250mL 锥形瓶中，加入 25mL 去离子水，2~3 滴 0.2% 二甲酚橙指示剂，滴加 20% 六亚甲基四胺至溶液呈稳定的紫红色后过量 3mL，用 EDTA 标准溶液滴定至亮黄色为终点。平行滴定三次，计算 EDTA 溶液的物质的量浓度和相对平均偏差。

③ [Cu$_x$(dppy)$_y$(CH$_3$CN)$_{x/2}$](ClO$_4$)$_x$ 中铜的测定。准确称取 0.14~0.16g[Cu$_x$(dppy)$_y$(CH$_3$CN)$_{x/2}$](ClO$_4$)$_x$ 于 250mL 干燥的锥形瓶中，加入约 1mL 浓硝酸使样品完全溶解（用 365nm 紫外手电筒确认），加入 20mL 去离子水、20mL HAc-NaAc 缓冲溶液，滴入 6~8 滴 0.2% PAN 指示剂，用 EDTA 标准溶液滴定至溶液颜色由蓝色变为黄绿色，即为终点。平行滴定三次，计算产物中铜的质量百分含量及相对平均偏差。

数据记录与处理

根据 EDTA 标准溶液所消耗的体积，按下式计算 Cu 的含量：

$$w = \frac{\bar{c}_{EDTA} V_{EDTA} M_{Cu}}{m_s \times 1000} \times 100\% \tag{5.5}$$

式中　m_s——[Cu$_x$(dppy)$_y$(CH$_3$CN)$_{x/2}$](ClO$_4$)$_x$ 试样的质量，g；

　　　M_{Cu}——铜的摩尔质量，63.5g/mol。

思考题

合成目标物[Cu$_x$(dppy)$_y$(CH$_3$CN)$_{x/2}$](ClO$_4$)$_x$ 时铜粉的作用是什么？

5.2　创新性实验

实验 5.5　Gemini 离子液体改性氧化石墨烯薄膜在 ReO$_4^-$/TcO$_4^-$ 吸附中的应用

实验目的

(1) 学习点击化学的合成方法。

(2) 学习使用吸附法对物质进行分离回收。
(3) 掌握用电感耦合等离子体发射光谱仪（ICP-OES）分析溶液中 ReO_4^- 浓度的方法。

实验原理

锝（^{99}Tc）作为一种危险的放射性同位素，它半衰期长、毒性大，在水溶液中以 $^{99}TcO_4^-$ 形式存在。然而，在实验室中很难分析高放射性元素 ^{99}Tc，ReO_4^- 作为一种电荷密度和热力学性质与 TcO_4^- 相似的非放射性类似物，经常被用来评估 TcO_4^- 的去除性能。

点击化学是以碳-杂原子键（C—X—C）合成为基础的组合化学新方法。本实验利用巯基-烯点击反应制备了功能化氧化石墨烯柔性膜（GO-C_6），其中石墨烯网络的 C—C 键可以进一步与 Gemini 离子液体连接，随着烷基链 C_1、C_4、C_6 数量的增加，其吸附能力增强。在咪唑修饰的氧化石墨烯膜上，ReO_4^-/TcO_4^- 与 Cl^- 的离子交换反应在吸附机理中起主导作用。

实验步骤

将氧化石墨烯在 N,N-二甲基甲酰胺（DMF）中超声制备氧化石墨烯分散体，然后加入季戊四醇四-3-巯基丙酸酯（PETMP）和偶氮二异丁腈（AIBN）。将得到的混合物在 343 K 的氮气气氛下搅拌 16h，然后将功能石墨烯分散体离心。随后，将沉积物（GO-SH）分散在 10mL 甲醇中，然后加入 DMF、PETMP、DMAP 和 Gemini C_6 离子液体。超声分散 1h 后，立即倒入培养皿，在紫外光下暴露 3h。最后，将膜在室温下干燥。

将 GO-C_6 膜切割成小块，然后称重所需的质量。在不同酸浓度下，将 20mL ReO_4^- 溶液（20mg/L）与 10mg GO-C_6 振荡，进行批量吸附试验，利用 ICP-OES 测试方法检测吸附前后溶液中 ReO_4^- 的浓度。

实验 5.6　金纳米粒子比色法检测尿酸

实验目的

(1) 学习使用金纳米粒子比色法检测尿酸含量。
(2) 掌握紫外-分光光度计的使用方法。

实验原理

金纳米粒子比色法是近年来新发展起来的一种比色分析方法，其利用金纳米粒子因表面等离子体共振效应而具有高吸光系数的特点，可以实现高灵敏度的检测。使用此方法测定溶液中尿酸的含量，是利用尿酸与三聚氰胺分子间的氢键作用反应掉部分三聚氰胺，抑制三聚氰胺诱导的金纳米粒子聚集，随着尿酸浓度的增加，溶液逐渐由蓝色变为红色，其吸收曲线和吸光度也随之改变，易于观察。

实验步骤

(1) 金纳米粒子的制备　根据经典 Frens 柠檬酸钠还原氯金酸法制备金纳米粒子。将 35mL 超纯水和 0.5mL 25mmol/L $HAuCl_4$ 溶液加入 100mL 锥形瓶，摇匀，加热至沸腾后，迅速加入一定体积的 38.8mmol/L 柠檬酸钠溶液，摇匀，待溶液变色后，继续加热 5min 再关闭加热板，取下溶液冷却至室温，放入 4℃ 冰箱中保存备用。
(2) 尿酸的测定　取洁净比色管，向其中加入一定体积和浓度的三聚氰胺溶液、pH=7.0 乙酸缓冲溶液、尿酸溶液，摇匀，再加入 1.0mL 上述制备的金纳米粒子溶液，并用超纯水定容至 5mL，摇匀，取溶液加入石英比色皿中，扫描其紫外-可见吸收光谱图。

实验 5.7 电子废弃物中金元素的绿色溶解与提取

实验目的
(1) 学习配合物性质，掌握配合物降低氧化电极电势的原理。
(2) 学习电子垃圾回收的基本知识。

实验原理
金在水溶液中的电极电势很高（$E^{\ominus}=1.52\sim 1.83\text{V}$），在硝酸、硫酸、盐酸、氢氟酸以及碱中，金都不会溶解。当金与配体形成配合物后，能有效降低其电极电势。Au^+ 能与 CN^- 形成稳定的配合物，并且电极电势急剧降低至 -0.596V：

$$Au \rightleftharpoons Au^+ + e^- \quad E^{\ominus}=+1.83\text{V} \tag{5.6}$$

$$Au \rightleftharpoons Au^{3+} + 3e^- \quad E^{\ominus}=+1.52\text{V} \tag{5.7}$$

$$4Au^+ + 8CN^- + O_2 + 2H_2O \rightleftharpoons 4[Au(CN)_2]^- + 4OH^- \tag{5.8}$$

由于 $E^{\ominus}([Au(CN)_2]^-/Au)=-0.596\text{V}<E^{\ominus}(O_2/OH^-)=0.401\text{V}$，因此可以溶金，这就是利用配合物来溶金的原理。在王水法中，盐酸实际上是起到了配体的作用，经过硝酸氧化，可以实现溶金：

$$Au + HNO_3 + 4HCl \rightleftharpoons HAuCl_4 + 2H_2O + NO\uparrow \tag{5.9}$$

实验步骤
根据实验原理，合理设计选择其他绿色环保溶剂作为配体，实现溶金的效果。例如使用氯化钠实现配体的作用，Cl^- 作为配体与金形成配合物后，有效降低其电极电势，选择醋酸提供质子，在硝酸的氧化下，可以使金溶解。溶解后使用维生素 C（抗坏血酸）将金还原，达到以更安全、更环保的方式实现固体废物资源化。

实验 5.8 天然色素甜菜红的提取、含量分析及其美妆应用

实验目的
(1) 了解甜菜红素的提取方法。
(2) 了解食品添加剂甜菜红素的性质与应用。
(3) 掌握紫外-可见分光光度计的使用方法。
(4) 学会吸收曲线及标准曲线的绘制。

实验原理
甜菜红素是由甜菜醛氨酸与环多巴或环多巴的葡萄糖基化衍生物结合形成的一种紫红色物质，因最早从红甜菜中被提取鉴定而得名。其分子的骨架结构如图 5.2 所示。R1 或者 R2 的糖基化或酰基化可以产生多种结构的甜菜红素。甜菜红素分子中含有的酚羟基和胺基，使其具有较强的抗氧化能力，也促进了利用甜菜红素治疗氧化应激相关疾病，如癌症、炎症、糖尿病、高血压、高血脂、肥胖症、阿尔茨海默病等。利用甜菜红素作为添加剂，开发甜菜红素在食品、保健品、医药、美妆领域的应用具有较强的商业价值。建立甜菜红素的分析检测方法也具有重要的意义。

甜菜红素分子中含有大量的羧基、羟基等亲水性官能团，使其可以溶解在水中。这一性

图 5.2 甜菜根及甜菜红素分子结构

质为开发清洁无毒的提取方法提供了便利。甜菜红素分子结构中含有的多元环结构，使其具有明显的紫外/可见吸收，其最大吸收峰一般位于 470~550nm。根据光互补定律，甜菜红素本身显示的颜色为紫红色。利用其紫外-可见吸收性质，可以建立基于朗伯-比尔定律的测定方法。

口红是现代生活中使用最普遍的化妆品之一。然而，人类将使用口红的历史可以追溯到距今五千年前的新石器时代。考古发现的最早的口红，是公元前三千多年前由铅粉和红色矿物制作的。《齐民要术》记载了我国古代制作口红的方法，即：先用温酒浸泡香料得到香酒，再加入动物油脂、朱砂、精油，冷却凝固，即可得到颜色鲜艳细腻芳香的口红。后人还利用各种植物花卉代替朱砂制作口红。现代化学工业的发展为口红制作提供了丰富的原料和方法，促进了口红的普及。随着人们对美妆用品个性化要求的提升，古法口红、天然口红成为潮妆新品。广东深圳、湖北恩施还将古法口红的制作技艺列入了非物质文化遗产名录。本实验改良古法，以甜菜红素为色源制作口红，既是对非遗技艺的传承，也是对甜菜红素在美妆应用中的拓展。

实验步骤

（1）甜菜红素的提取　称取 10g 已打碎的新鲜红甜菜根于 50mL 玻璃烧杯中，加入 20mL 蒸馏水，搅拌均匀后放入超声清洗仪，360W 功率下超声 10min，将提取液及残渣一并倒入 50mL 离心管中，4000r/min 离心 5min，收集上清液于烧杯中，用移液管准确移取 1mL 上清液至 25mL 容量瓶中，蒸馏水定容，待用。

（2）测量波长的选择

① 甜菜红素储备溶液的配制。用分析天平准确称量 0.5000g 甜菜红素，蒸馏水溶解后转移至 50mL 容量瓶中，蒸馏水定容，得到浓度为 10mg/mL 的甜菜红素储备溶液。

② 最佳吸收波长的确定。用移液管移取 1.0mL 甜菜红素储备液，加入 25mL 比色管中，用水稀释至 10mL，振荡，得到浓度为 1mg/mL 的甜菜红素溶液。取该溶液装入比色皿中，用紫外可见分光光度计在 460~550nm 测量其最佳吸收波长。在 460~520nm 间，每隔 10nm 测定一次，在 520~550nm 间，每隔 2nm 测定一次。

（3）甜菜红素的含量分析

① 标准曲线的制作。向 5 个 25mL 比色管中分别加入 0.5mL、1.0mL、1.5mL、2.0mL、2.5mL 甜菜红储备溶液（10mg/mL），并用蒸馏水稀释至 10mL，得到一系列不同浓度的标准溶液。以试剂空白为参比，在最佳的吸收波长下，用 1cm 的比色皿测定各溶液的吸光度，绘制标准曲线。

② 实际样品中甜菜红素的测定。吸取定容后的甜菜红素提取液于比色皿中，按照标准曲线制作的测定步骤，在最佳吸收波长下测定吸光度值，从标准曲线计算提取液中甜菜红素

的含量（以 mg/mL 表示）。

(4) 制作手工口红　在台秤上称取 0.50g 甜菜红素置于 25mL 塑料烧杯中，用胶头滴管向其中滴加 1mL 甘油，搅拌使其均匀混合后待用。称取 4.00g 蜂蜡于 100mL 塑料烧杯中，向其中加入 10mL 橄榄油，水浴加热溶解呈黄色透明溶液，趁热倒入盛有甜菜红素的 25mL 塑料烧杯中，搅拌均匀，此时溶液随着温度降低成糊状，再次水浴加热呈流体状态，趁热倒入唇膏模具中，待冷却后检查并试用。

数据记录与处理

1. 最佳吸收波长的确定

波长/nm											
吸光度 A											

2. 甜菜红素的分析

浓度/(mg/mL)	0.5	1.0	1.5	2.0	2.5	未知样品
吸光度 A						

以吸光度值 A 为纵坐标，以浓度值为横坐标绘制标准曲线。计算未知样品中甜菜红素的浓度。

思考题

(1) 不同提取剂对甜菜红素的提取有什么影响？产生这一影响的原因是什么？

(2) 本实验中仅用粗提取的方法获得含有甜菜红素的产物。查阅资料回答，怎样提纯本实验获得的粗产物？

(3) 不同的溶液环境对甜菜红素的吸光度有什么影响？这可能是由什么原因引起的？

(4) 紫外-可见分光光度法进行定量分析的依据是什么？

(5) 手工制作口红时，加入了橄榄油、甜菜红素、蜂蜡、维生素 E。各个成分有什么作用？

拓展阅读

T6-新世纪紫外可见分光光度计操作指南

(1) 开机自检　依次打开打印机、仪器主机电源，仪器开始初始化，约 3min 时间初始化完成。

```
初始化 ███░░░ 43%
1.样品池电机    OK
2.滤光片        OK
3.光源电机      OK
```

初始化完成后，仪器进入主菜单。

(2) 进入光度测量状态　按 Enter 键，进入光度测量界面。

```
光度测量：
  0.000Abs
  250 nm
```

（3）进入测量界面　按 Start/Stop 键进入样品测定界面。

```
250 nm          0.002 Abs
No.    Abs      Conc
```

（4）设置测量波长　按 GOTOλ 键，在下图界面输入测量的波长，如需要在 460nm 测量，输入 460，按 Enter 键确认，仪器将自动调整波长。

```
请输入波长：
```

调整波长后界面如下：

```
460 nm          0.002 Abs
No.    Abs      Conc
```

（5）进入设置参数　在这个步骤中主要设置样品池。按 GOTOλ 键进入参数设定界面，按 ▼ 键使光标移动到"样式设定"。按 Enter 键确认，进入设定界面。

```
○ 测光方式
○ 数学计算
● 样式设定
```

（6）设定使用样品池个数　按 ▼ 键使光标移动到"使用样池数"，如图显示。按 Enter 键循环选择需要使用的样品池个数（主要根据使用比色皿数量确定，如使用 2 个比色皿，则修改为 2）。

```
○ 试样室：八联池
● 样池数：2
○ 空白溶液校正：否
○ 样池空白校正：否
```

（7）样品测量　按 RETURN 键返回到参数设定界面，再按 RETURN 键返回到光度测量界面。在 1 号样品池内放入空白溶液，2 号池内放入待测样品。关闭好样品池盖后按 ZERO 键进行空白校正，再按 Start/Stop 键进行样品测量。测量结果如下图显示：

460 nm		0.002 Abs
No.	Abs	Conc
1-1	0.012	1.000
2-1	0.032	2.000

如果需要更换波长，可以直接按 GOTO λ 键，调整波长。

注意：更换波长后必须重新按 ZERO 进行空白校正。

如果每次使用的比色皿数量是固定个数，下一次使用仪器时可以跳过第（5）、（6）步骤直接进入样品测量。

（8）结束测量　测量完成后按 PRINT 键打印数据，如果没有打印机请记录数据。退出程序或关闭仪器后测量数据将消失。确保已从样品池中取出所有比色皿，清洗干净以便下一次使用。按 RETURN 键直到返回到仪器主菜单界面后再关闭仪器电源。

附录

附录1 常用酸碱的密度、含量和浓度

试剂名称	密度/(g/mL)	含量/%	浓度/(mol/L)
盐酸	1.18~1.19	36~38	11.6~12.4
硝酸	1.39~1.40	65.0~68.0	14.4~15.2
硫酸	1.83~1.84	95~98	17.8~18.4
磷酸	1.69	85	14.6
高氯酸	1.68	70.0~72.0	11.7~12.0
乙酸	1.05	99.8(优级纯) 99.0(分析纯、化学纯)	17.4
氢氟酸	1.13	40	22.5
氢溴酸	1.49	47.0	8.6
氨水	0.88~0.90	25.0~28.0	13.3~14.8

附录2 常用基准物质的干燥条件和应用

基准物 名称	分子式	干燥后的组成	干燥条件/℃	标定对象
碳酸氢钠	$NaHCO_3$	Na_2CO_3	270~300	酸
无水碳酸钠	Na_2CO_3	Na_2CO_3	270~300	酸
硼砂	$Na_2B_4O_7 \cdot 10H_2O$	$Na_2B_4O_7 \cdot 10H_2O$	放在装有 NaCl 和蔗糖饱和溶液的干燥器中	酸
碳酸氢钾	$KHCO_3$	K_2CO_3	270~300	酸
二水合草酸	$H_2C_2O_4 \cdot 2H_2O$	$H_2C_2O_4 \cdot 2H_2O$	室温空气中干燥	碱或 $KMnO_4$
邻苯二甲酸氢钾	$KHC_8H_4O_4$	$KHC_8H_4O_4$	105~110(1h)	碱
重铬酸钾	$K_2Cr_2O_7$	$K_2Cr_2O_7$	120(1h)	还原剂
溴酸钾	$KBrO_3$	$KBrO_3$	120(1~2h)	还原剂
碘酸钾	KIO_3	KIO_3	110	还原剂
铜	Cu	Cu	室温干燥器中保存	还原剂
三氧化二砷	As_2O_3	As_2O_3	室温干燥器中保存	氧化剂
碳酸钙	$CaCO_3$	$CaCO_3$	110	EDTA
草酸钠	$Na_2C_2O_4$	$Na_2C_2O_4$	110(2h)	$KMnO_4$
锌	Zn	Zn	室温干燥器中保存	EDTA

续表

基准物		干燥后的组成	干燥条件/℃	标定对象
名称	分子式			
氧化锌	ZnO	ZnO	900～1000	EDTA
氯化钠	NaCl	NaCl	500～600	$AgNO_3$
氯化钾	KCl	KCl	500～600	$AgNO_3$
硝酸银	$AgNO_3$	$AgNO_3$	280～290	氯化物
氨基磺酸	$HOSO_2NH_2$	$HOSO_2NH_2$	在真空 H_2SO_4 干燥器中保存48h	碱

附录3 常见缓冲溶液的配制

缓冲溶液组成	pH	缓冲溶液配制方法
磷酸盐缓冲液	2.0	甲液:取磷酸16.6mL,加水至1000mL,摇匀。乙液:取磷酸氢二钠71.63g,加水至1000mL。取上述甲液72.5mL与乙液27.5mL混合,摇匀
磷酸盐缓冲液	2.5	取磷酸二氢钾100g,加水800mL,用盐酸调节pH至2.5,再用水稀释至1000mL
醋酸-锂盐	3.0	取冰醋酸50mL,加水800mL混合后,用氢氧化锂调节pH值至3.0,再加水稀释至1000mL
磷酸-三乙胺	3.2	取磷酸约4mL与三乙胺约7mL,加50%甲醇稀释至1000mL,用磷酸调节pH值至3.2
甲酸钠	3.3	取2mol/L甲酸溶液25mL,加酚酞指示液1滴,用2mol/L氢氧化钠溶液中和,再加入2mol/L甲酸溶液75mL,用水稀释至200mL,调节pH值至3.25～3.30
醋酸盐	3.5	取醋酸铵25g,加水25mL溶解后,加7mol/L盐酸溶液38mL,用2mol/L盐酸溶液或5mol/L氨溶液准确调节pH值至3.5(电位法指示),用水稀释至100mL
醋酸-醋酸钠	3.6	取醋酸钠5.1g,加冰醋酸20mL,再加水稀释至250mL
醋酸-醋酸钠	3.7	取无水醋酸钠20g,加水300mL溶解后,加溴酚蓝指示液1mL及冰醋酸60～80mL,至溶液从蓝色转变为纯绿色,再加水稀释至1000mL
乙醇-醋酸铵	3.7	取5mol/L醋酸溶液15.0mL,加乙醇60mL和水20mL,用10mol/L氢氧化铵溶液调节pH值至3.7,用水稀释至1000mL
醋酸-醋酸钠	3.8	取2mol/L醋酸钠溶液13mL与2mol/L醋酸溶液87mL,加每1mL含铜1mg的硫酸铜溶液0.5mL,再加水稀释至1000mL
枸橼酸-磷酸氢二钠	4.0	甲液:取枸橼酸21g或无水枸橼酸19.2g,加水使溶解成1000mL,置冰箱内保存。乙液:取磷酸氢二钠71.63g,加水使溶解成1000mL。取上述甲液61.45mL与乙液38.55mL混合,摇匀
醋酸-醋酸钾	4.3	取醋酸钾14g,加冰醋酸20.5mL,再加水稀释至1000mL
醋酸-醋酸铵	4.5	取醋酸铵7.7g,加水50mL溶解后,加冰醋酸6mL,加水稀释至1000mL
醋酸-醋酸钠	4.5	取醋酸钠18g,加冰醋酸9.8mL,再加水稀释至1000mL
醋酸-醋酸钠	4.6	取醋酸钠5.4g,加水50mL溶解,用冰醋酸调节pH值至4.6,再加水稀释至100mL
磷酸盐	5.0	取一定量0.2mol/L磷酸二氢钠溶液,用氢氧化钠试液调节pH值至5.0
邻苯二甲酸盐	5.6	取邻苯二甲酸氢钾10g,加水900mL,搅拌使溶解,用氢氧化钠试液(必要时用稀盐酸)调节pH值至5.6,加水稀释至1000mL,混匀
磷酸盐	5.8	取磷酸二氢钾8.34g与磷酸氢二钾0.87g,加水使溶解成1000mL

缓冲溶液组成	pH	缓冲溶液配制方法
醋酸-醋酸铵	6.0	取醋酸铵100g，加水300mL使溶解，加冰醋酸7mL，摇匀
醋酸-醋酸钠	6.0	取醋酸钠54.6g，加1mol/L醋酸溶液20mL溶解后，加水稀释至500mL
枸橼酸盐	6.2	取枸橼酸4.2g，加1mol/L的20%乙醇制氢氧化钠溶液40mL使溶解，再用20%乙醇稀释至100mL 枸橼酸盐缓冲液(pH=6.2)：取2.1%枸橼酸水溶液，用50%氢氧化钠溶液调节pH值至6.2
磷酸盐	6.5	取磷酸二氢钾0.68g，加0.1mol/L氢氧化钠溶液15.2mL，用水稀释至100mL
磷酸盐	6.6	取磷酸二氢钾1.74g、磷酸氢二钠2.7g与氯化钠1.7g，加水使溶解成400mL
磷酸盐缓冲液（含胰酶）	6.8	取磷酸二氢钾6.8g，加水500mL使溶解，用0.1mol/L氢氧化钠溶液调节pH值至6.8；另取胰酶10g，加水适量使溶解，将两液混合后，加水稀释至1000mL
磷酸盐	6.8	取0.2mol/L磷酸二氢钾溶液250mL，加0.2mol/L氢氧化钠溶液118mL，用水稀释至1000mL，摇匀
磷酸盐	7.0	取磷酸二氢钾0.68g，加0.1mol/L氢氧化钠溶液29.1mL，用水稀释至100mL
磷酸盐	7.2	取0.2mol/L磷酸二氢钾溶液50mL与0.2mol/L氢氧化钠溶液35mL，加新沸过的冷水稀释至200mL，摇匀
磷酸盐	7.3	取磷酸氢二钠1.9734g与磷酸二氢钾0.2245g，加水至1000mL，调节pH值至7.3
巴比妥	7.4	取巴比妥钠4.42g，加水使溶解并稀释至400mL，用2mol/L盐酸溶液调节pH值至7.4，滤过
磷酸盐	7.4	取磷酸二氢钾1.36g，加0.1mol/L氢氧化钠溶液79mL，用水稀释至200mL
磷酸盐	7.6	取磷酸二氢钾27.22g，加水使溶解成1000mL，取50mL，加0.2mol/L氢氧化钠溶液42.4mL，再加水稀释至200mL
巴比妥-氯化钠	7.8	取巴比妥钠5.05g，加氯化钠3.7g及水适量使溶解，另取明胶0.5g加水适量，加热溶解后并入上述溶液中。然后用0.2mol/L盐酸溶液调节pH值至7.8，再用水稀释至500mL
磷酸盐	7.8	甲液：取磷酸氢二钠35.9g，加水溶解，并稀释至500mL。乙液：取磷酸二氢钠2.76g，加水溶解，并稀释至100mL。取上述甲液91.5mL与乙液8.5mL混合
磷酸盐	7.8~8.0	取磷酸氢二钠5.59g与磷酸二氢钾0.41g，加水使溶解成1000mL
三羟甲基氨基甲烷	8.0	取三羟甲基氨基甲烷12.14g，加水800mL，搅拌溶解，并稀释至1000mL，用6mol/L盐酸溶液调节pH值至8.0
硼砂-氯化钙	8.0	取硼砂0.572g与氯化钙2.94g，加水约800mL溶解后，用1mol/L盐酸溶液约2.5mL调节pH值至8.0，加水稀释至1000mL
氨-氯化铵	8.0	取氯化铵1.07g，加水使溶解成100mL，再加稀氨溶液(1→30)调节pH值至8.0
三羟甲基氨基甲烷	8.1	取氯化钙0.294g，加0.2mol/L三羟甲基氨基甲烷溶液40mL使溶解，用1mol/L盐酸溶液调节pH值至8.1，加水稀释至100mL
巴比妥	8.6	取巴比妥5.52g与巴比妥钠30.9g，加水使溶解至2000mL
三羟甲基氨基甲烷	9.0	取三羟甲基氨基甲烷6.06g，加盐酸赖氨酸3.65g、氯化钠5.8g、乙二胺四乙酸二钠0.37g，再加水溶解使成1000mL，调节pH值至9.0
硼酸-氯化钾	9.0	取硼酸3.09g，加0.1mol/L氯化钾溶液500mL使溶解，再加0.1mol/L氢氧化钠溶液210mL
氨-氯化铵	10.0	取氯化铵5.4g，加水20mL溶解后，加浓氨水35mL，再加水稀释至100mL
硼砂-碳酸钠	10.8~11.2	取无水碳酸钠5.30g，加水使溶解成1000mL；另取硼砂1.91g，加水使溶解成100mL。临用前取碳酸钠溶液973mL与硼砂溶液27mL，混匀

附录4 常用酸碱指示剂及配制方法

名称	变色范围(pH)	颜色变化	配制方法
甲酚红（第一次变色范围）	0.2～1.8	红—黄	0.1%乙醇溶液
甲酚红（第二次变色范围）	7.2～8.8	黄—紫红	0.1%乙醇溶液
百里酚蓝（第一次变色范围）	1.2～2.8	红—黄	0.1%乙醇溶液,加入0.05mol/L NaOH 4.3mL
百里酚蓝（第二次变色范围）	8.0～9.6	黄—蓝	0.1%乙醇溶液,加入0.05mol/L NaOH 4.3mL
甲基橙	3.0～4.4	红—橙黄	0.1%水溶液
溴酚蓝	3.0～4.6	黄—蓝	0.1%乙醇溶液,加入0.05mol/L NaOH 4.3mL
刚果红	3.0～5.2	蓝紫—红	0.1%水溶液
茜素红S（第一次变色范围）	3.7～5.2	黄—紫	0.1%水溶液
茜素红S（第二次变色范围）	10.0～12.0	紫—淡黄	0.1%水溶液
甲基红	4.4～6.2	红—黄	0.1%乙醇溶液
石蕊	5.0～8.0	红—蓝	0.1%乙醇溶液
溴百里酚蓝	7.2～8.8	黄—紫红	0.1%乙醇溶液
酚酞	8.2～10.0	无色—紫红	0.1%乙醇溶液
鞑靼黄	12.0～13.0	黄—红	0.1%水溶液

附录5 弱酸、弱碱的解离常数

（1）无机酸在水溶液中的解离常数（25℃）

序号	名称	化学式	K_a	pK_a
1	偏铝酸	$HAlO_2$	6.3×10^{-13}	12.2
2	亚砷酸	H_3AsO_3	6.0×10^{-10}	9.22
3	砷酸	H_3AsO_4	$6.3\times10^{-3}(K_1)$ $1.05\times10^{-7}(K_2)$ $3.2\times10^{-12}(K_3)$	2.2 6.98 11.5
4	硼酸	H_3BO_3	$5.8\times10^{-10}(K_1)$ $1.8\times10^{-13}(K_2)$ $1.6\times10^{-14}(K_3)$	9.24 12.74 13.8
5	次溴酸	$HBrO$	2.4×10^{-9}	8.62
6	氢氰酸	HCN	6.2×10^{-10}	9.21
7	碳酸	H_2CO_3	$4.2\times10^{-7}(K_1)$ $5.6\times10^{-11}(K_2)$	6.38 10.25
8	次氯酸	$HClO$	3.2×10^{-8}	7.5
9	氢氟酸	HF	6.61×10^{-4}	3.18
10	锗酸	H_2GeO_3	$1.7\times10^{-9}(K_1)$ $1.9\times10^{-13}(K_2)$	8.78 12.72
11	高碘酸	HIO_4	2.8×10^{-2}	1.56
12	亚硝酸	HNO_2	5.1×10^{-4}	3.29

续表

序号	名称	化学式	K_a	pK_a
13	次磷酸	H_3PO_2	5.9×10^{-2}	1.23
14	亚磷酸	H_3PO_3	$5.0\times10^{-2}(K_1)$ $2.5\times10^{-7}(K_2)$	1.3 6.6
15	磷酸	H_3PO_4	$7.52\times10^{-3}(K_1)$ $6.31\times10^{-8}(K_2)$ $4.4\times10^{-13}(K_3)$	2.12 7.2 12.36
16	焦磷酸	$H_4P_2O_7$	$3.0\times10^{-2}(K_1)$ $4.4\times10^{-3}(K_2)$ $2.5\times10^{-7}(K_3)$ $5.6\times10^{-10}(K_4)$	1.52 2.36 6.6 9.25
17	氢硫酸	H_2S	$1.3\times10^{-7}(K_1)$ $7.1\times10^{-15}(K_2)$	6.88 14.15
18	亚硫酸	H_2SO_3	$1.23\times10^{-2}(K_1)$ $6.6\times10^{-8}(K_2)$	1.91 7.18
19	硫酸	H_2SO_4	$1.0\times10^{3}(K_1)$ $1.02\times10^{-2}(K_2)$	-3 1.99
20	硫代硫酸	$H_2S_2O_3$	$2.52\times10^{-1}(K_1)$ $1.9\times10^{-2}(K_2)$	0.6 1.72
21	氢硒酸	H_2Se	$1.3\times10^{-4}(K_1)$ $1.0\times10^{-11}(K_2)$	3.89 11
22	亚硒酸	H_2SeO_3	$2.7\times10^{-3}(K_1)$ $2.5\times10^{-7}(K_2)$	2.57 6.6
23	硒酸	H_2SeO_4	$1\times10^{3}(K_1)$ $1.2\times10^{-2}(K_2)$	-3 1.92
24	硅酸	H_2SiO_3	$1.7\times10^{-10}(K_1)$ $1.6\times10^{-12}(K_2)$	9.77 11.8
25	亚碲酸	H_2TeO_3	$2.7\times10^{-3}(K_1)$ $1.8\times10^{-8}(K_2)$	2.57 7.74

（2）有机酸在水溶液中的解离常数（25℃）

序号	名称	化学式	K_a	pK_a
1	甲酸	$HCOOH$	1.8×10^{-4}	3.75
2	乙酸	CH_3COOH	1.74×10^{-5}	4.76
3	乙醇酸	$CH_2(OH)COOH$	1.48×10^{-4}	3.83
4	草酸	$(COOH)_2$	$5.4\times10^{-2}(K_1)$ $5.4\times10^{-5}(K_2)$	1.27 4.27
5	甘氨酸	$CH_2(NH_2)COOH$	1.7×10^{-10}	9.78
6	一氯乙酸	$CH_2ClCOOH$	1.4×10^{-3}	2.86
7	二氯乙酸	$CHCl_2COOH$	5.0×10^{-2}	1.3
8	三氯乙酸	CCl_3COOH	2.0×10^{-1}	0.7
9	丙酸	CH_3CH_2COOH	1.35×10^{-5}	4.87
10	丙烯酸	$CH_2=CHCOOH$	5.5×10^{-5}	4.26
11	乳酸	$CH_3CHOHCOOH$	1.4×10^{-4}	3.86
12	丙二酸	$HOCOCH_2COOH$	$1.4\times10^{-3}(K_1)$ $2.2\times10^{-6}(K_2)$	2.85 5.66
13	2-丙炔酸	$HC\equiv CCOOH$	1.29×10^{-2}	1.89
14	甘油酸	$HOCH_2CHOHCOOH$	2.29×10^{-4}	3.64
15	丙酮酸	$CH_3COCOOH$	3.2×10^{-3}	2.49

续表

序号	名称	化学式	K_a	pK_a
16	α-丙胺酸	CH_3CHNH_2COOH	1.35×10^{-10}	9.87
17	β-丙胺酸	$CH_2NH_2CH_2COOH$	4.4×10^{-11}	10.36
18	正丁酸	$CH_3(CH_2)_2COOH$	1.52×10^{-5}	4.82
19	异丁酸	$(CH_3)_2CHCOOH$	1.41×10^{-5}	4.85
20	3-丁烯酸	$CH_2=CHCH_2COOH$	2.1×10^{-5}	4.68
21	异丁烯酸	$CH_2=C(CH_2)COOH$	2.2×10^{-5}	4.66
22	反丁烯二酸(富马酸)	$HOCOCH=CHCOOH$	$9.3\times 10^{-4}(K_1)$ $3.6\times 10^{-5}(K_2)$	3.03 4.44
23	顺丁烯二酸(马来酸)	$HOCOCH=CHCOOH$	$1.2\times 10^{-2}(K_1)$ $5.9\times 10^{-7}(K_2)$	1.92 6.23
24	酒石酸	$HOCOCH(OH)CH(OH)COOH$	$1.04\times 10^{-3}(K_1)$ $4.55\times 10^{-5}(K_2)$	2.98 4.34
25	正戊酸	$CH_3(CH_2)_3COOH$	1.4×10^{-5}	4.86
26	异戊酸	$(CH_3)_2CHCH_2COOH$	1.67×10^{-5}	4.78
27	2-戊烯酸	$CH_3CH_2CH=CHCOOH$	2.0×10^{-5}	4.7
28	3-戊烯酸	$CH_3CH=CHCH_2COOH$	3.0×10^{-5}	4.52
29	4-戊烯酸	$CH_2=CHCH_2CH_2COOH$	2.10×10^{-5}	4.677
30	戊二酸	$HOCO(CH_2)_3COOH$	$1.7\times 10^{-4}(K_1)$ $8.3\times 10^{-7}(K_2)$	3.77 6.08
31	谷氨酸	$HOCOCH_2CH_2CH(NH_2)COOH$	$7.4\times 10^{-3}(K_1)$ $4.9\times 10^{-5}(K_2)$ $4.4\times 10^{-10}(K_3)$	2.13 4.31 9.358
32	正己酸	$CH_3(CH_2)_4COOH$	1.39×10^{-5}	4.86
33	异己酸	$(CH_3)_2CH(CH_2)_3COOH$	1.43×10^{-5}	4.85
34	(E)-2-己烯酸	$H(CH_2)_3CH=CHCOOH$	1.8×10^{-5}	4.74
35	(E)-3-己烯酸	$CH_3CH_2CH=CHCH_2COOH$	1.9×10^{-5}	4.72
36	己二酸	$HOCOCH_2CH_2CH_2CH_2COOH$	$3.8\times 10^{-5}(K_1)$ $3.9\times 10^{-6}(K_2)$	4.42 5.41
37	柠檬酸	$HOCOCH_2C(OH)(COOH)$ CH_2COOH	$7.4\times 10^{-4}(K_1)$ $1.7\times 10^{-5}(K_2)$ $4.0\times 10^{-7}(K_3)$	3.13 4.76 6.4
38	苯酚	C_6H_5OH	1.1×10^{-10}	9.96
39	邻苯二酚	$o\text{-}C_6H_4(OH)_2$	$3.6\times 10^{-10}(K_1)$ $1.6\times 10^{-13}(K_2)$	9.45 12.8
40	间苯二酚	$m\text{-}C_6H_4(OH)_2$	$3.6\times 10^{-10}(K_1)$ $8.71\times 10^{-12}(K_2)$	9.3 11.06
41	对苯二酚	$p\text{-}C_6H_4(OH)_2$	1.1×10^{-10}	9.96
42	2,4,6-三硝基苯酚	$2,4,6\text{-}(NO_2)_3C_6H_2OH$	5.1×10^{-1}	0.29
43	葡萄糖酸	$CH_2OH(CHOH)_4COOH$	1.4×10^{-4}	3.86
44	苯甲酸	C_6H_5COOH	6.3×10^{-5}	4.2
45	水杨酸	$C_6H_4(OH)COOH$	$1.05\times 10^{-3}(K_1)$ $4.17\times 10^{-13}(K_2)$	2.98 12.38
46	邻硝基苯甲酸	$o\text{-}NO_2C_6H_4COOH$	6.6×10^{-3}	2.18
47	间硝基苯甲酸	$m\text{-}NO_2C_6H_4COOH$	3.5×10^{-4}	3.46
48	对硝基苯甲酸	$p\text{-}NO_2C_6H_4COOH$	3.6×10^{-4}	3.44
49	邻苯二甲酸	$o\text{-}C_6H_4(COOH)_2$	$1.1\times 10^{-3}(K_1)$ $4.0\times 10^{-6}(K_2)$	2.96 5.4
50	间苯二甲酸	$m\text{-}C_6H_4(COOH)_2$	$2.4\times 10^{-4}(K_1)$ $2.5\times 10^{-5}(K_2)$	3.62 4.6

续表

序号	名称	化学式	K_a	pK_a
51	对苯二甲酸	p-$C_6H_4(COOH)_2$	$2.9\times10^{-4}(K_1)$ $3.5\times10^{-5}(K_2)$	3.54 4.46
52	1,3,5-苯三甲酸	$C_6H_3(COOH)_3$	$7.6\times10^{-3}(K_1)$ $7.9\times10^{-5}(K_2)$ $6.6\times10^{-6}(K_3)$	2.12 4.1 5.18
53	苯基六羧酸	$C_6(COOH)_6$	$2.1\times10^{-1}(K_1)$ $6.2\times10^{-3}(K_2)$ $3.0\times10^{-4}(K_3)$ $8.1\times10^{-6}(K_4)$ $4.8\times10^{-7}(K_5)$ $3.2\times10^{-8}(K_6)$	0.68 2.21 3.52 5.09 6.32 7.49
54	癸二酸	$HOOC(CH_2)_8COOH$	$2.6\times10^{-5}(K_1)$ $2.6\times10^{-6}(K_2)$	4.59 5.59
55	乙二胺四乙酸(EDTA)	$CH_2-N(CH_2COOH)_2$ $CH_2-N(CH_2COOH)_2$	$1.0\times10^{-2}(K_1)$ $2.14\times10^{-3}(K_2)$ $6.92\times10^{-7}(K_3)$ $5.5\times10^{-11}(K_4)$	2 2.67 6.16 10.26

(3) 无机碱在水溶液中的解离常数（25℃）

序号	名称	化学式	K_b	pK_b
1	氢氧化铝	$Al(OH)_3$	$1.38\times10^{-9}(K_3)$	8.86
2	氢氧化银	$AgOH$	1.10×10^{-4}	3.96
3	氢氧化钙	$Ca(OH)_2$	$3.72\times10^{-3}(K_1)$ $3.98\times10^{-2}(K_2)$	2.43 1.4
4	氨水	NH_3+H_2O	1.78×10^{-5}	4.75
5	肼(联氨)	$N_2H_4+H_2O$	$9.55\times10^{-7}(K_1)$ $1.26\times10^{-15}(K_2)$	6.02 14.9
6	羟氨	NH_2OH+H_2O	9.12×10^{-9}	8.04
7	氢氧化铅	$Pb(OH)_2$	$9.55\times10^{-4}(K_1)$ $3.0\times10^{-8}(K_2)$	3.02 7.52
8	氢氧化锌	$Zn(OH)_2$	9.55×10^{-4}	3.02

(4) 有机碱在水溶液中的解离常数（25℃）

序号	名称	化学式	K_b	pK_b
1	甲胺	CH_3NH_2	4.17×10^{-4}	3.38
2	尿素(脲)	$CO(NH_2)_2$	1.5×10^{-14}	13.82
3	乙胺	$CH_3CH_2NH_2$	4.27×10^{-4}	3.37
4	乙醇胺	$H_2N(CH_2)_2OH$	3.16×10^{-5}	4.5
5	乙二胺	$H_2N(CH_2)_2NH_2$	$8.51\times10^{-5}(K_1)$ $7.08\times10^{-8}(K_2)$	4.07 7.15
6	二甲胺	$(CH_3)_2NH$	5.89×10^{-4}	3.23
7	三甲胺	$(CH_3)_3N$	6.31×10^{-5}	4.2
8	三乙胺	$(C_2H_5)_3N$	5.25×10^{-4}	3.28
9	丙胺	$C_3H_7NH_2$	3.70×10^{-4}	3.432
10	异丙胺	i-$C_3H_7NH_2$	4.37×10^{-4}	3.36
11	1,3-丙二胺	$NH_2(CH_2)_3NH_2$	$2.95\times10^{-4}(K_1)$ $3.09\times10^{-6}(K_2)$	3.53 5.51
12	1,2-丙二胺	$CH_3CH(NH_2)CH_2NH_2$	$5.25\times10^{-5}(K_1)$ $4.05\times10^{-8}(K_2)$	4.28 7.393

续表

序号	名称	化学式	K_b	pK_b
13	三丙胺	$(CH_3CH_2CH_2)_3N$	4.57×10^{-4}	3.34
14	三乙醇胺	$(HOCH_2CH_2)_3N$	5.75×10^{-7}	6.24
15	丁胺	$C_4H_9NH_2$	4.37×10^{-4}	3.36
16	异丁胺	$C_4H_9NH_2$	2.57×10^{-4}	3.59
17	叔丁胺	$C_4H_9NH_2$	4.84×10^{-4}	3.315
18	己胺	$H(CH_2)_6NH_2$	4.37×10^{-4}	3.36
19	辛胺	$H(CH_2)_8NH_2$	4.47×10^{-4}	3.35
20	苯胺	$C_6H_5NH_2$	3.98×10^{-10}	9.4
21	苄胺	C_7H_9N	2.24×10^{-5}	4.65
22	环己胺	$C_6H_{11}NH_2$	4.37×10^{-4}	3.36
23	吡啶	C_5H_5N	1.48×10^{-9}	8.83
24	六亚甲基四胺	$(CH_2)_6N_4$	1.35×10^{-9}	8.87
25	2-氯酚	C_6H_5ClO	3.55×10^{-6}	5.45
26	3-氯酚	C_6H_5ClO	1.26×10^{-5}	4.9
27	4-氯酚	C_6H_5ClO	2.69×10^{-5}	4.57
28	邻氨基苯酚	$o\text{-}H_2NC_6H_4OH$	$5.2\times10^{-5}(K_1)$ $1.9\times10^{-5}(K_2)$	4.28 4.72
29	间氨基苯酚	$m\text{-}H_2NC_6H_4OH$	$7.4\times10^{-5}(K_1)$ $6.8\times10^{-5}(K_2)$	4.13 4.17
30	对氨基苯酚	$p\text{-}H_2NC_6H_4OH$	$2.0\times10^{-4}(K_1)$ $3.2\times10^{-6}(K_2)$	3.7 5.5
31	邻甲苯胺	$o\text{-}CH_3C_6H_4NH_2$	2.82×10^{-10}	9.55
32	间甲苯胺	$m\text{-}CH_3C_6H_4NH_2$	5.13×10^{-10}	9.29
33	对甲苯胺	$p\text{-}CH_3C_6H_4NH_2$	1.20×10^{-9}	8.92
34	8-羟基喹啉(20℃)	$8\text{-}HOC_9H_6N$	6.5×10^{-5}	4.19
35	二苯胺	$(C_6H_5)_2NH$	7.94×10^{-14}	13.1
36	联苯胺	$H_2NC_6H_4C_6H_4NH_2$	$5.01\times10^{-10}(K_1)$ $4.27\times10^{-11}(K_2)$	9.3 10.37

附录6 金属离子与EDTA生成配位化合物的稳定常数（18~25℃，I= 0.1mol/L）

金属离子	$\lg K_{稳}$ EDTA	金属离子	$\lg K_{稳}$ EDTA	金属离子	$\lg K_{稳}$ EDTA
Ag^+	7.32	Fe^{3+}	25.1	Sc^{3+}	23.1
Al^{3+}	16.3	Ga^{3+}	20.3	Sn^{2+}	22.11
Ba^{2+}	7.86	Hg^{2+}	21.7	Sr^{2+}	8.73
Be^{2+}	9.2	In^{3+}	25.0	Th^{4+}	23.2
Bi^{3+}	27.94	Li^+	2.79	Ti^{2+}	17.3
Ca^{2+}	10.69	Mg^{2+}	8.7	Tl^{3+}	37.8
Cd^{2+}	16.46	Mn^{2+}	13.87	U^{4+}	25.8
Co^{2+}	16.31	Mo^{5+}	~28	V^{2+}	18.8
Co^{3+}	36	Na^+	1.66	Y^{3+}	18.09
Cr^{3+}	23.4	Ni^{2+}	18.62	Zn^{2+}	16.50
Cu^{2+}	18.80	Pb^{2+}	18.04	Zr^{4+}	29.5
Fe^{2+}	14.32	Pd^{2+}	18.5	稀土元素	16~20

附录7 原子表（1995，IUPAC）

元素	符号	原子量	元素	符号	原子量	元素	符号	原子量
银	Ag	107.8682	铪	Hf	178.49	铷	Rb	85.4678
铝	Al	26.98154	汞	Hg	200.59	铼	Re	186.207
氩	Ar	39.948	钬	Ho	164.9304	铑	Rh	102.9055
砷	As	74.9216	碘	I	126.9045	钌	Ru	101.07
金	Au	196.9655	铟	In	114.82	硫	S	32.06
硼	B	10.81	铱	Ir	192.22	锑	Sb	121.75
钡	Ba	137.33	钾	K	39.083	钪	Sc	44.9559
铍	Be	9.01218	氪	Kr	83.80	硒	Se	78.96
铋	Bi	208.9804	镧	La	138.9055	硅	Si	28.0855
溴	Br	79.904	锂	Li	6.941	钐	Sm	150.36
碳	C	12.011	镥	Lu	174.967	锡	Sn	118.69
钙	Ca	40.08	镁	Mg	24.305	锶	Sr	87.62
镉	Cd	112.41	锰	Mn	54.9380	钽	Ta	180.9479
铈	Ce	140.12	钼	Mo	95.94	铽	Tb	158.9254
氯	Cl	35.453	氮	N	14.0067	碲	Te	127.60
钴	Co	58.9332	钠	Na	22.98977	钍	Th	232.0381
铯	Cs	132.9054	钕	Nd	144.24	铊	Tl	204.383
铜	Cu	63.543	氖	Ne	20.179	铥	Tm	168.9342
镝	Dy	162.50	镍	Ni	58.69	铀	U	238.0289
铒	Er	167.26	镎	Np	237.0482	钒	V	50.9415
铕	Eu	151.96	氧	O	15.9994	钨	W	183.85
氟	F	18.998403	锇	Os	190.2	氙	Xe	131.29
铁	Fe	55.847	磷	P	30.97376	钇	Y	88.9059
镓	Ga	69.72	铅	Pb	207.2	镱	Yb	173.04
钆	Gd	157.25	钯	Pd	106.42	锌	Zn	65.38
锗	Ge	72.59	镨	Pr	140.9077	锆	Zr	91.22
氢	H	1.00794	铂	Pt	195.08			
氦	He	4.00260	镭	Ra	226.0254			

参考文献

[1] 王卫平,吴靓,刘卫东,等.分析化学实验(英汉双语版)[M].北京:科学出版社,2019.
[2] 王新宏.分析化学实验(双语版)[M].北京:科学出版社,2009.
[3] 李秀玲,梁冰,谭光群,等.分析化学实验[M].4版.北京:高等教育出版社,2015.
[4] 孙玉凤,刘春玲,厉安昕,等.分析化学实验[M].北京:清华大学出版社,2019.
[5] 胡广林,张雪梅,徐宝荣.分析化学实验[M].北京:化学工业出版社,2010.
[6] 武汉大学.分析化学实验[M].5版.北京:高等教育出版社,2011.
[7] 黄臻臻,贾琼,郑海娇,等.中国古代诗词作品在分析化学酸碱滴定法教学中的应用[J].大学化学,2023,38(02):239-242.
[8] 赵宇明,董广彬,时华捷,吕亮.同时测定酱油中苯甲酸和山梨酸含量的两种检测方法的比较研究[J].江苏调味副食品,2021(03):41-44.
[9] 王淑美.分析化学实验[M].9版.北京:中国中医药出版社,2013.
[10] 马强.基础化学实验化学分析实验分册[M].3版.北京:高等教育出版社,2019.
[11] 漆寒梅,周言凤,陈利娟,文静.硫酸钡重量法测定硫条件实验探讨[J].中国无机分析化学,2021,11(04):31-34.
[12] 王婕,展宗波.硫酸钡重量法测定银精矿中硫量[J].甘肃冶金,2019,41(03):127-129.
[13] 李艳,高美娟.国标测定白酒中甲醇含量的方法改进研究[J].粘接,2019,40(05):105-107.
[14] 练有南,化经纬,杜建中.分光光度法测定市售白酒中甲醇含量[J].湛江师范学院学报,2014,35(03):67-71.
[15] 龙德清,周新,王娟,邹强.分光光度法测定白酒中微量甲醇的研究[J].郧阳师范高等专科学校学报,2002(03):63-65.
[16] 万阳浴,刘银坤,许小红.高效液相色谱法测定丙戊酸钠中醋酸的含量[J].成都医学院学报,2021,16(01):17-19.
[17] 康平利.分析化学实验[M].2版.沈阳:辽宁大学出版社,2020.
[18] 贾闯,陈葵阳,沈锦芳.离子色谱法测定水环境中高浓度Na^+存在下的痕量NH_4^+[J].化学分析计量,2010,19(01):49-51.
[19] 张歆皓,邓阿利,王会东,桑雅丽,刘艳华,马晓光.儿童补钙剂中钙含量的测定研究[J].赤峰学院学报(自然科学版),2016,32(08):7-10.
[20] 加建斌.高锰酸钾滴定法测定补钙制品中的钙含量[J].安徽农业科学,2007(23):7076-7077.
[21] 贾文平.基础实验Ⅲ:分析化学实验[M].杭州:浙江大学出版社,2011.
[22] 席桂同,陈述.复方黄连素片中盐酸小檗碱含量测定[J].世界中医药,2015,10(02):252-257.
[23] 吕长淮.盐酸小檗碱含量测定方法概述[J].中国医药导报,2010,7(33):141-142.
[24] 林瑞群,邹正轩.紫外分光光度法测定小儿氨酚烷胺颗粒中对乙酰氨基酚的含量[J].今日药学,2008(04):28-29.
[25] 孙国良,王小东,詹丹丹,李铁胜.UV法测定氨咖黄敏胶囊中对乙酰氨基酚的含量[J].科技创新导报,2009(02):3.
[26] WANG K C, YAN Z Y, FU L L. Gemini ionic liquid modified nacre-like reduced graphene oxide click membranes for ReO_4^-/TcO_4^- removal[J]. Separation and Purification Technology,2022,302:122073.
[27] 鲁立强,王晗,曹怀元.基于氢键作用的金纳米粒子比色法检测尿酸[J].大学化学,2020,35(4):173-177.
[28] 吴松,张朝,龙明昊.电子废弃物中金元素的绿色溶解与提取[J].大学化学,2020,35(4):32-36.
[29] 唐湘毅.甜菜红素的稳态体系构建及其作用机理研究[D].广州:华南理工大学,2021.
[30] 刘军.中国国家历史(4)[M].北京:东方出版社,2016:96-105.
[31] 管彦波.文化与艺术——中国少数民族头饰文化研究[M].北京:中国经济出版社,2005:70-72.
[32] 武汉大学.分析化学:上[M].6版.北京:高等教育出版社,2016:66-69.
[33] Yu S L, Liu X, Zhang Z Y, Yu H J, Zhao G R. Advances of betalains biosynthesis and metabolic regulation[J]. *China Biotechnology*,2018,38(8):84-91.